Cognition and Addiction

Cognition and Addiction

Edited by

Marcus R. Munafò

Department of Experimental Psychology,
University of Bristol, UK

and

Ian P. Albery

Department of Psychology,
London South Bank University, UK

OXFORD

UNIVERSITY PRESS

*This book has been printed digitally and produced in a standard specification
in order to ensure its continuing availability*

OXFORD
UNIVERSITY PRESS

Great Clarendon Street, Oxford OX2 6DP

Oxford University Press is a department of the University of Oxford.
It furthers the University's objective of excellence in research, scholarship,
and education by publishing worldwide in

Oxford New York

Auckland   Cape Town Dar es Salaam   Hong Kong   Karachi
Kuala Lumpur Madrid Melbourne Mexico City   Nairobi
New Delhi Shanghai Taipei Toronto
With offices in
Argentina Austria Brazil Chile Czech Republic France Greece
Guatemala Hungary Italy Japan South Korea Poland Portugal
Singapore Switzerland Thailand Turkey Ukraine Vietnam

Oxford is a registered trade mark of Oxford University Press
in the UK and in certain other countries

Published in the United States
by Oxford University Press Inc., New York

ISBN 978-0-19-856930-5

Contents

Foreword

Addiction research has a long history, but it is only recently that the cognitive aspects of addictive behaviours have begun to be investigated by experimental psychologists and neuroscientists. This has revealed a complex inter-play of cognitive mechanisms that subserve subjective experiences associated with addiction, such as drug craving. This has led to a marked increase in interest in the potential of such research to elucidate, for example, the processes that may lead to relapse following abstinence. As well as being of theoretical interest, such research clearly has profound clinical implications, and carries the potential to offer improved or novel treatment for addictive behaviours.

Although research into the relationship between cognitive processes and addictive behaviours is currently an area of substantial growth and interest, there are few volumes that bring together the state-of-the-art in this research. As the field matures, such a monograph is timely and will serve to capture the current state of knowledge, as well as identifying directions for future research. We have attempted to do this by inviting leading authorities in the field to contribute specialist chapters which synthesize the current state of their research, in the context of dominant theoretical models.

While the chapters are complementary, they are intended to stand alone as a snapshot of the current state-of-the-art of the research which they describe. Current research and theoretical models have been synthesized by leading authors in the field of cognition and addiction, with particular emphasis on widely investigated substances of abuse such as alcohol, nicotine, cocaine and opiates. Individual authors, all of whom are high profile researchers of international standing, have provided a series of chapters that cover mechanisms that subserve cognitive processes in addiction and their application to specific addictive behaviours.

In inviting authors to contribute to this monograph, we envisaged that the eventual readership would comprise postgraduate students, research workers and other academic staff, as well as current workers in the field of addiction. The first chapter describes the theoretical perspectives and approaches adopted in the subsequent chapters, which describe evidence for the role of specific cognitive and neurobiological mechanisms in the context of various addictive behaviours. The final chapters discuss the potential clinical implications offered by a greater understanding of the role of cognitive processes in addiction, and the likely novel treatments which may emerge from this field in the near future.

Marcus R. Munafò
Ian P. Albery

March 2006

Contributors

Ian P. Albery,
Department of Psychology,
London South Bank University, UK

Karen O. Brandon,
Department of Psychology,
University of South Florida, USA

W. Miles Cox,
School of Psychology,
University of Wales, Bangor, UK

Jack Darkes,
Department of Psychology,
University of South Florida, USA

Javad Salehi Fadardi,
School of Psychology,
University of Wales, Bangor,
UK and Department of
Psychology, Ferdowsi
University of Mashad, Iran

Matt Field,
School of Psychology,
University of Liverpool,
Liverpool, UK

Mark S. Goldman,
Department of Psychology,
University of South Florida, USA

Brian Hitsman,
Department of Psychiatry and Human
Behavior, Brown University, USA

Barry T. Jones,
Department of Psychology,
University of Glasgow, UK

Eric Klinger,
Division of Social Sciences,
University of Minnesota,
Morris, USA

Adam Leventhal,
Department of Behavioural Science,
University of Texas M.D. Anderson
Cancer Center, Texas, USA

Chris McCusker,
School of Psychology,
The Queen's University Belfast, UK

Antony C. Moss,
Department of Psychology,
London South Bank University, UK

Marcus R. Munafò,
Department of Experimental
Psychology, University of Bristol, UK

Asli Niazi,
Department of Psychology,
London South Bank University, UK

Anthony Nutting,
Queensland Department of Health,
Australia

Richard R. Reich,
Department of Psychology,
University of South Florida, USA

Frank Ryan,
School of Psychology, Birkbeck
College, University of London and
Central & North West London NHS
Mental Health Trust, UK

Alan W. Stacy,
Institute for Prevention Research,
Department of Preventive Medicine,
University of Southern California,
USA

Dinkar Sharma,
Department of Psychology,
University of Kent at Canterbury, UK

Ross McD. Young,
School of Psychology
and Counselling,
Queensland University of Technology,
Australia

Carey Walmsley,
School of Psychology
and Counselling,

Queensland University of Technology,
Australia

Andrew J. Waters,
Department of Behavioural Science,
University of Texas M.D. Anderson
Cancer Center, Texas USA

Reinout W. Wiers,
Experimental Psychology,
Universiteit Maastricht,
Behavioural Science Institute (BSI)
Radboud Universiteit Nijmegen,
IVO (Rotterdam Addiction Research
Institute),
The Netherlands

1

Theoretical perspectives and approaches

Ian P. Albery, Dinkar Sharma, Asli Niazi and Antony C. Moss

1.1 Introduction

Over the past few decades, there has been a move away from a behaviouristic perspective of human motivation and behaviour to one that identifies a cognitive or information-processing account of human behaviour. This has resulted in a revolution in the types of questions asked of research, theoretical proposals of human behaviour and also in the research techniques used to explore psychological domains such as the development, maintenance and cessation of addictive behaviours. In this chapter, we explore the role of cognition and cognitive biases in the understanding of concepts related to addiction, such as craving, from a number of theoretical stances. These include the dual-affect model (Baker *et al.* 1987), incentive sensitization theory (Robinson and Berridge 2001, 2003), social learning and expectancy approaches (e.g. Goldman *et al.* 1999; Rather and Goldman 1994; Fillmore and Vogel-Sprott 1995*a*), and finally the cognitive model of drug urges and drug-use behaviour (Tiffany 1990; Tiffany and Conklin 2000).

The second part of the chapter explores methodologies and research methods that have been used to test these various cognitive accounts of addictive behaviour. Recently researchers have been applying techniques and paradigms from mainstream experimental psychology to examine cognitive processes in addiction such as memory and priming-based paradigms (e.g. lexical decision tasks), attentional bias procedures (the modified Stroop approach, dot-probe and flicker-induced change paradigm), and association tasks (e.g. the implicit association test, the go–no-go task and the extrinsic affective Simon task). It is fair to say that the last decade or so has seen a shift in methodology from paper and pencil procedures to those that expect the participant to respond to environmentally generated stimuli in a controlled experimental setting. While the former can be seen as tapping something cognitive in addictive behaviour, the latter explicitly identifies core psychological processes including cognitive biases (e.g. attentional bias for addiction-related stimuli) that may be of significance for understanding why people develop and maintain addictive or habitual behaviours. The latter has been particularly useful in identifying a 'levels of processing' account of addiction. In essence, researchers who are interested in the cognition of addictive behaviour using experimental procedures are not interested in 'what people "say" about what they think, but rather make inferences about cognitive processes and structures based on behavioural responses' (McCusker 2001, pp. 49–50). These procedures have enabled addiction theorists to make the distinction between explicit thinking processes, that the individual has knowledge of, can access and control, and those cognitive processes that are outside of an individual's immediate awareness and control also known as implicit cognitions (Wiers *et al.* 2002). It will be argued that this development in the methods used to study cognition in addiction has reaped significant reward in terms of extending and enhancing our understanding of addictive behaviour through theory and the implications such advancement necessarily has had for the development of addiction-related interventions.

The final part of the chapter provides some recent evidence from our group in assessing cognitive processing biases in addiction. The evidence included provides details of work designed to examine the levels of frequency of exposure to addiction-related stimuli for attentional bias and also a conceptualization of attentional biases in terms of slow and fast processing components.

Initially we need to derive a definition of addiction that not only is relevant for the purposes of the present chapters but that is applicable to the detail contained in other chapters in this book.

1.2 Defining addiction

Addiction is a term used to describe a person's physical and psychological dependency on a behaviour. It may or may not involve the ingestion of a mood-altering psychoactive drug such as alcohol, nicotine, cannabis, opiates or amphetamines. Addiction is characterized by a number of key features. These include a strong desire to participate in a particular behaviour, an impaired capacity to control the behaviour, some discomfort or distress when the behaviour is prevented temporarily or permanently from occurring, and the reoccurrence of the behaviour despite evidence that the behaviour is causing physical and psychological harm to the individual (West 2006). Addiction comprises psychological and biological components, including some genetic influence (e.g. Lusher *et al.* 2001; Munafò and Hitsman Chapter 8 of this volume). The psychological processes involved in the addictive behaviour include learning-based factors and motivational processes based on thinking or cognitive processes (e.g. Klinger and Cox 2004; Cox *et al.* 2006*a* and Chapter 4 of this volume; West 2006). The cognitive psychologist views behaviour as occurring directly due to distinct and predictable ways of processing and experiencing environmentally based information sources. This may involve the manner in which people who undertake addictive behaviour selectively represent information cognitively in memory, how these representations may or may not be accessed and retrieved, and whether there are distinct biases in important cognitive factors involved in the processing of perceived information (e.g. biased attentional processes) (Stacy and Wiers Chapter 2; Field Chapter 3; McCusker Chapter 5; Goldman *et al.* Chapter 6; Young *et al.* Chapter 7 of this volume). Information gleaned from the conceptualization of addiction as a motivational problem grounded in biased information processing has implications for the design and evaluation of interventions to prevent individuals from undertaking the addictive behaviour initially and also in the treatment of the addictive behaviour (Musher-Eizenman and Kulick 2003; Wiers *et al.* 2004; Waters and Leventhal Chapter 9; Ryan Chapter 10 of this volume).

1.3 **Cognitive approaches**

Learning and association theorists argue that behaviour develops and is maintained through the dynamic experience of one's environment and reinforcement schedules of the behaviour emitted. These approaches have been developed by cognitive psychologists interested in the precise thinking-based mechanisms and processes that mediate the experience of reinforcement and predict actual behaviour. In this section, the models and theories generated from a number of approaches will be presented. These include the dual-affect model, cognitive labelling model, the cognitive model of drug use and drug-use behaviour (Tiffany 1990), and the incentive sensitization model (Robinson and Berridge 2003).

1.3.1 **Dual-affect model**

According to Baker *et al.* (1987), the experience of addiction-related craving is governed by dynamic emotion-processing systems. These processing systems influence physiological responses, as well as self-reports of craving and drug-use behaviour (Tiffany 1999). It is argued that craving is expressed as the interaction between negative and positive affect systems which are characterized as being mutually inhibitory, i.e. the activation of one system will inhibit activation of the other system. In this sense, an individual cannot experience both positive and negative affective feedback at one and the same time. The model extends this reasoning to comment on the representational organization of craving. Specifically, craving systems are organized cognitively as networks that store emotion-based information generated from physiological and semantic information. A network is said to become activated (and the other network inhibited) when environmentally based information is processed as matching information stored in that network (Zinser *et al.* 1992). The negative affect system is activated by aversive cues (e.g. withdrawal feedback, drug unavailability, negative emotional states) that act as an incentive for drug use through a withdrawal-relief representation. The positive system is contingent upon positive emotional states, drug availability and during non-abstinent episodes (Baker *et al.* 1987).

This model identifies that two distinct types of craving experience exist and that the type experienced is dependent on the nature of current environmental stimuli (i.e. the context in which the user finds themselves.) Consistent with the theory, it has been shown that the experience of stress has greater impact on negative as opposed to positive affect craving in drug-deprived individuals compared with those who are non-deprived (Zinser *et al.* 1992). Other research, however, has not provided supportive evidence for this model (e.g. Maude-Griffin and Tiffany 1996). Tiffany (1995) details findings of self-report data that have identified the simultaneous activation of positive and negative reinforcement factors—not consistent with the dual account. While there appears to be an inherent inconsistency in this evidence, future research should employ contemporary cognitive approaches for assessing the cognitive structure within the parameters of the dual-affect model.

A similar conceptualization comes from the cognitive labelling model (Schacter and Singer 1962) in which an emotional experience results from the interaction of experienced physiological arousal and a cognitive interpretation of that arousal. This interpretation leads to a semantically based emotional label which in turn determines the emotional state experienced. The theory also argues for a linear relationship between the intensity of the experienced emotion and the extent of arousal. In terms of craving, the model proposes that cues can create conditioned physiological arousal states (e.g. increased heart rate) and activate related mental processes (e.g. memories of previous drug use behaviour) leading to a label or representation of a physiological state as craving. Whilst it has been argued that labelling *per se* may be important in the description of and identification with craving states (Tiffany 1995), other work has shown no association between the extent of cue-elicited physiological reactions and cue-elicited craving (see Tiffany 1990). The dynamic regulatory model also argues that craving can be best conceptualized as occurring due to the combination of conditioned responses to related cues with affect (Drummond 2001). The premise is that drug use reinforces further drug use through its pharmacological effects on affect (Niaura *et al.* 1987).

We now turn to two models that have been the primary focus of work exploring the cognitive mediation of the addictive experience and addictive behaviours, namely the cognitive model of drug use and drug-use behaviour (Tiffany 1990), and incentive sensitization theory (Robinson and Berridge 1993, 2000, 2003).

1.3.2 Cognitive model of drug use and drug-use behaviour

Tiffany's (1990) cognitive model of drug use and drug-use behaviour argues that responses to drug and drug-related cues involves various levels of cognitive processing. The model, also labelled the cognitive processing model, suggests that alcohol use is largely controlled by automatic processes, whereas craving operates at a predominantly non-automatic level. Based on Schneider and Shiffrin's (1977) work, Tiffany proposed several properties as key features of automatic processing. (1) With practice, automatized actions speed up and become less variable, i.e. they become fast acting. (2) The presentation of a stimulus may be sufficient to cause the enactment of an automatized action sequence, i.e. these processes feature autonomy or being stimulus bound. (3) Once initiated, it is difficult to prevent the completion of an automatized sequence and, as such, these processes lack explicit control. (4) Automatized actions are cognitively effortless and cognitively non-demanding. (5) Automatic processing occurs without conscious awareness (Tiffany 1990; Tiffany and Conklin 2000). Non-automatic processes are defined as slow, dependent on attention and intention, effortful and operating under conscious control (Shiffrin and Schneider 1977). Non-automatic processes are those that take place during the learning of a new skill or the overcoming of a difficult problem. They are slow and decisive and require cognitive effort. For instance, when one learns any new skill, concentration is required to master it. Over time and repeated association between stimuli and behavioural response, these associations can transform from explicit or non-automatic to automatic processes (Shiffrin and Schneider 1977; Tiffany 1999).

Tiffany argues that this mechanism describes the development and maintenance of addictive behaviours. Many of the actions of a problem drinker or smoker may be viewed as examples of the kinds of behaviour exhibited by an automatized skill (Tiffany and Carter 1998) such that after repeated practice, the alcohol consumption of a problem drinker can be seen as stimulus bound, difficult to control, effortless and without awareness. Tiffany further argues that when an individual is somehow prevented from undertaking the addictive behaviour, non-automatic (explicit) cognitive processes are manifest and the feelings of craving are apparent. Non-automatic processing gives rise to urges/ cravings under two conditions: either when automaticity is blocked or obstructed by external factors (e.g. running out of cigarettes and there are no shops open to buy some); or when the individual is attempting to change drug use or maintain abstinence (e.g. in treatment). This craving may be experienced as physiological changes, such as increased heart rate or sweatiness (Tiffany 1990, 1999).

Several studies have shown the activation of non-automatic processing during an urge (Tiffany 1990; Baxter and Hinson 2001). For example, Baxter and Hinson (2001) demonstrated that, in a probe reaction time paradigm, smokers were slower to respond to trials where the automatized smoking action plans were interrupted compared with trials where they were not disturbed. In addition, experimental manipulations of drinking urges have produced support for the argument that urges have adverse effects on cognitive processing. Several studies have investigated the effects of addiction-related stimuli (cue exposure) on addicts' performance on secondary cognitive tasks which demand non-automatic attentional processing. These studies have consistently shown that being exposed to these stimuli impairs performance on the task. Sayette and colleagues (1994) demonstrated that alcoholics showed significantly longer reaction times to auditory probes when exposed to an alcoholic beverage than to a non-alcoholic beverage. McCusker and Brown (1990) have also found that non-alcohol-dependent participants report increases in urges to drink when they were presented with alcohol cues.

A number of reports have supported the idea that cognitive biases may operate at an automatic level for addictive behaviours (e.g. Johnsen et al. 1994; Sayette et al. 1994; Tiffany 1995; McCusker and Gettings 1997; Stormark et al. 2000; McCusker 2001; Sharma et al. 2001; Erhman et al. 2002; Mogg et al. 2003; Munafò et al. 2003; Bruce and Jones 2004; Lusher et al. 2004; Field et al. 2005a,b, 2006; Noel et al. 2005). In general, these studies have revealed that problem drinkers, gamblers, smokers and opiate addicts display attentional bias for related stimuli and that this processing is likely to result from the operation of implicit or automated cognition.

In addition, evidence is beginning to accumulate for the operation of these biases in non-problem individuals. Differences have been reported between heavy and light social drinkers (Cox et al. 1999; Sharma et al. 2001;

Field *et al.* 2005*b*). For instance, Cox and his colleagues found that alcohol-related stimuli significantly affected performance among heavy social drinkers (Cox *et al.* 1999). Similarly, in a modified Stroop paradigm, Sharma *et al.* (2001) demonstrated attentional bias for problem drinkers and also non-problem drinkers, while in a visual probe task Field *et al.* (2005*b*) showed attentional bias for social drinkers with increased levels of craving. Taken together, these studies suggest the operation of an attentional bias for different groups of users and that as dependence develops, automatized drug-related cognitions become increasingly integrated in memory and increasingly activated (Tiffany 1990; Sharma *et al.* 2001; Brandon *et al.* 2004).

1.3.3 Incentive sensitization model

It has long been debated in the field of addiction whether drugs are taken to achieve pleasant outcomes or to escape aversive conditions (Robinson and Berridge 2003). Some models argue that the most likely reason for drug taking involves positive reinforcement (Glickman and Schiff 1967), others argue that initial drug taking occurs as a result of negative reinforcement (Khantzian 1985) whereas other theories posit that the initial experience with drugs could be both positive and negative (Solomon 1980).

Positive reinforcement models argue that drugs act as primary reinforcers which in turn motivate continued drug use. Psychomotor stimulant theory (Glickman and Schiff 1967; Wise and Bozarth 1987) is one of the positive reinforcement models and suggests that drugs activate neural systems which induce appetite and produce positive affective conditions. Several studies have shown that addicts report 'euphoria' as a primary reason for drug taking (e.g. Johanson and de Wit 1989 for amphetamines; de Wit *et al.* 1989 for alcohol).

Negative reinforcement models propose that the experience of withdrawal symptoms is unpleasant, and reduction in these effects acts as reinforcers for the addict—in this case negative reinforcement. Negative reinforcement models, such as the self-medication model (Khantzian 1985), suggest that drug use is controlled by the user's attempt to escape or avoid aversive conditions (e.g. anxiety) and that learning to escape or avoid these conditions will motivate continued drug use. Although Colder (2001) studied stress and negative emotionality and their interaction as predictors of drinking motives and found some support for the self-medication model of alcohol use, this model does not appear to explain the high rate of substance misuse among people with severe mental health problems.

Opponent process theory (Solomon 1980) attempts to explain why behaviours for which the initial reinforcers are both positive (e.g. euphoria) and negative (e.g. pain) can become addictive. The opponent process theory argues that following a positive or negative affective state, there is often an opposite affective state (emotional contrast). Solomon argues that the person left in the opposite state might become addicted to this state either because he or she will like this state (when in a positive state) and would like to maintain it, or he or she will dislike it (when in a negative state) and is motivated to attempt a return to a positive state again, which will again lead to the opposing negative state. Both of these scenarios would form addictions (Solomon 1980). Therefore, the initial pleasant drug use leads to unpleasant withdrawal symptoms and, in order to avoid these symptoms, drug taking is maintained (e.g. shown for cigarette addiction by Solomon and Corbit 1973). Incentive sensitization theory (Robinson and Berridge 1993), in contrast, maintains that neither the initiation nor maintenance of drug-use behaviour are motivated by the desire either to get pleasure or to ease withdrawal symptoms. This theory focuses on three aspects of addiction: why craving occurs; why addicts crave after long periods of cessation of drug use; and the relationship between liking (pleasurable effects) and wanting (sensitization of neural systems) of drugs (Robinson and Berridge 1993). The theory suggests that repeated administration of drugs and alcohol leads to a sensitization of the dopamine system which results in drugs and drug-related stimuli acquiring incentive salience. This incentive salience results in drug-associated stimuli becoming more attractive and grabbing attentional resources (i.e. become 'wanted') (Robinson and Berridge 2003). As such, the theory posits that addictive behaviours develop as a result of repeated neuroadaptations caused by repeated drug use and that this leads to permanent or semi-permanent changes in the addict's brain. They also suggest that such processes can occur unconsciously, activate the wanting system implicitly and do not depend on controlled processes (Fishman and Foltin 1992; Robinson and Berridge 1993).

The difference between 'liking' and 'wanting' is fundamental to incentive sensitization theory. According to Robinson and Berridge (1993, 2001, 2003), 'liking' are the pleasurable effects of drugs whereas 'wanting' is the sensitization of the dopamine-regulated neural systems. They also argue that sensitization leads to an increase in wanting but to no change or to some decrease in liking (i.e. the pleasurable effects of drugs) (Robinson and Berridge 1993). In contrast, Koob and LeMoal (1997) argue that sensitization causes an increase not only in the wanting system but also in liking. According to Robinson and Berridge, it is this increase in wanting experience that produces craving, which then leads to drug-seeking and drug-taking behaviour.

There is extensive experimental evidence supporting the incentive sensitization theory of addiction. First of all, it has been demonstrated that many different addictive drugs (e.g. amphetamine, cocaine, opiates, nicotine, alcohol and cannabis) activate the release of dopamine in the nucleus accumbens (Di Chiara 1995; White 1996). It has also been shown that repeated drug use leads to hypersensitivity to drugs and drug-related stimuli (Robinson and Berridge 1993). There is also evidence that repeated administration of psychostimulants results in long-lasting sensitizations (Robinson and Becker 1986; Kalivas and Stewart 1991). Finally, in a recent series of studies, Hobbs *et al.* (2005) have demonstrated the dissociation between wanting and liking using self-report measures and alcohol priming doses, thus supporting incentive sensitization predictions. Overall, incentive sensitization has very important implications in understanding and explaining certain contradictions in the experience of addictive behaviours, such as why smokers do not 'like' smoking cigarettes but still have strong cravings (wanting) to smoke. This theory also acknowledges the significance of environmental and psychological variables which might affect the development of sensitization to drugs (Robinson and Berridge 2000).

1.3.4 Expectancy theory

We now turn to a brief overview of expectancy theory. This has been one of the most researched cognitive approaches in the study of addictive behaviours especially in relation to understanding motivations to consume alcohol. One might usefully define expectancies as simple knowledge structures about certain behaviours and actions which allow us to surmise the possible outcomes of various courses of action. Bandura (1977) made the distinction in the context of social learning theory between outcome and self-efficacy expectancies—put simply, 'what will happen if I do X, and do I feel capable of doing X?' More recently, Kirsch (1985) drew attention to response expectancies, or 'anticipations of automatic reactions to particular situational cues' (Kirsch 1999, p. 3). Importantly, these expectancies, as we shall see shortly, do not operate only within the explicit ('if I do X, then Y will follow') level of cognitive functioning—the research outlined below lends strong support to both the explicit and implicit operation of these structures on thought and behaviour.

The early work of Fillmore and Vogel-Sprott (1995a) indicated that performance on a simple motor task (the pursuit rotor) was mediated by the expected effects of alcohol consumption (i.e. expecting greater impairment led to greater impairment), and importantly this expectancy effect was also seen in placebo groups, suggesting that such expectancy effects are not reliant, exclusively, upon any actual physiological impairment from alcohol. In other work (Fillmore Mulvihill and Vogel-Sprott 1994, 1996), studies were conducted utilizing expectancy challenges (information which contradicts one's actual belief about the effects alcohol will have on some ability) to see if behavioural impairment would be affected if participants were told to expect an intense amount of impairment on the pursuit rotor task. Fillmore *et al.* (1994) first tested this under placebo conditions and found that when intense impairment was expected, participants' performance would improve significantly, thereby suggesting that they were compensating for this expected impairment. Subsequently, Fillmore *et al.* (1996) tested this effect again after actual alcohol consumption, and found that this compensatory effect to an expectancy challenge was still present when alcohol had been consumed. Furthermore, their research also indicated that the size of this effect was mediated by drinking experience (e.g. Fillmore and Vogel-Sprott 1995*b*), suggesting that the more direct experience one has with drinking and the effects of alcohol, the less susceptible one will be to information contradicting one's expectancies. In another study (Fillmore and Vogel-Sprott 1995*c*), participants were told they would receive both alcohol and caffeine; in line with these other findings, the level of impairment expected predicted performance on a pursuit rotor task irrespective of whether the substances were administered or not. One final finding worth mentioning comes from a later study, again by Fillmore *et al.* (1998), which demonstrated that cognitive impairment (on the rapid information processing task) was elicited in placebo groups as a function of participants' expectancies about the impairing effects alcohol would have (though the impairment was less than in the alcohol group where, presumably, there were additive effects from alcohol and expectancies).

Although highlighting the important relationship between self-reported beliefs about the effects of alcohol on a variety of tasks, the work just described falls short of addressing the influence which expectancies have on an implicit or automatic level, so failing to answer the question of whether expectancies can influence behaviour outside of our conscious awareness. More recently, Friedman *et al.* (2005) conducted a piece of research looking at the automatic effects of activating the concept 'alcohol' on participants' ratings of sexual attractiveness, and found that subliminal priming of the general alcohol concept led to higher ratings of sexual attractiveness in participants who had previously been assessed as having 'increased sexual desire' as part of their expectancy of consuming alcohol. Furthermore, Kramer and Goldman (2003) were able to demonstrate implicit attentional biases towards arousing and sedating alcohol words using a modified Stroop task. Importantly, their findings discriminated between heavy and light drinkers who, in previous studies (e.g. Rather and Goldman 1994), have reported differential expectations about the effects of alcohol, such that heavier drinkers expect more physiological arousal than lighter drinkers. Taken together, this work begins building a strong case for the importance of both explicit and implicit operation of expectancies on behaviour.

This brief review raises two questions: is there evidence for a common mechanism which would explain both sets of findings; and to what extent may the conclusions from the alcohol literature be generalized to other substances, particularly at levels of misuse and abuse? The first question begins to find a response from the work of Rather and Goldman (1994) where two main dimensions of alcohol expectancies were revealed to exist in terms of 'cognitive space', namely bad and good outcomes of drinking, and sedating or arousing effects of alcohol consumption. According to this cognitive neuroscientific model, an individual's expectancy will lie somewhere between each of these orthogonal scales, and as such will demarcate their expectancy as, for example, 'positive and sedating' (i.e. becoming calm and relaxed when I drink). Depending upon where the concept 'alcohol' (or, presumably, any other substance or concept) lies, other concepts (e.g. lazy, horny, sociable, funny) will be activated more or less to an extent depending upon their closeness to the originating concept in cognitive space. This notion of spreading activation (cf. Collins and Loftus 1975) in expectancy networks leads to a suggestion that, as a result of various experiences, our minds will literally become 'wired' to generate a certain pattern of cognitive and affective responses, which, in the absence of conscious control efforts, will lead to associated patterns of behaviour. The notion that differences in cognitive architecture will lead to differential behavioural patterns is echoed by findings in the attentional bias literature, where it has been found that the distribution of attention varies amongst drug users, importantly, as a function of the type of user they are (i.e. non, light, heavy or problem users) (e.g. alcohol, Sharma et al. 2001; Townshend and Duka 2001; Cox et al. 2003; nicotine, Bradley et al. 2003; Waters et al. 2003b; cannabis, Jones et al. 2002), implying a link between an underlying attentional process (which presumably is reflected in 'cognitive space' somewhere) and behaviour. Explanations for these types of biases, and the processes by which they come to influence drug-use behaviours, tend to focus on the development of learned associations between drug use and desirable outcomes, such that the urge to use drugs in the presence of certain cues may become automatized and very difficult to control (Tiffany 1990).

The key commonality here is that both bodies of work seem to be aiming towards an explanation of either drug-use behaviour or drug-related behaviour in terms of underlying cognitions, which do not necessarily have to be within the conscious awareness of an individual. The effects of implicit cognitions on behaviour have been thoroughly documented in recent years (e.g. Bargh 1997; Bargh and Chartrand 1999), and it would seem that in expectancies, we have a subset of pervasive cognitions which have a great influence over behaviours related not only to drug use, but also to the kinds of health risk behaviours commonly associated with drug use (e.g. drink driving).

1.4 Levels of cognitive processing in addiction

Contemporary work in experimental cognitive psychology has recently started to distinguish between two types of cognitive processes that may or may not operate together in the processing of environmental stimuli. These processes have been labelled implicit processing and explicit processing. Explicit processes, or non-automatic processes, are characterized by more cognitive effort and draw upon more limited capacity resources. These processes are intentional, controllable, are ultimately modifiable and are relatively slow acting. Finally, these processes are available and subject to conscious awareness. In contrast, the key components of an implicit process, or automatic cognitive process, are that it requires little cognitive effort and as such is not as taxing on limited cognitive resources. Automatic processing is effortless, is carried out without intention or control and is relatively stable and difficult to change. Finally, and importantly, automatic cognitive processes operate with the conscious awareness of the perceiver (Shiffrin and Schneider 1977; Bargh 2005). In social psychology, there has been significant progress in the implicit and explicit processing of attitude- and belief-based environment stimuli (e.g. Fazio and Olson 2003), and consideration of how these processes guide social behaviour and socially relevant responding (e.g. Bargh *et al.* 1996; see Bargh 2005; Hassin *et al.* 2005). The same is true in the field of addiction and addiction-related behaviour (Stacy 1997; Sheeran *et al.* 2005). Evidence has been published that suggests that one of the key characteristics of any addictive behaviour is that it is guided by both automatic (implicit) and non-automatic (explicit) cognitive processes (e.g. Huijding *et al.* 2005) and to change an individual's addictive behaviour may actually require a change in these explicit and implicit cognitions (Cox *et al.* 2003; Waters *et al.* 2003a; Wiers *et al.* 2004).

The perennial issue for psychologists exploring the role of implicit and explicit cognition in social perception and social behaviour is whether a specific behaviour is best accounted for by one or other process, or whether dual processes occur under different environmental conditions (e.g. cognitive load, stressful conditions). In essence, can we distinguish reliably between the two processes and, if so, what role do they have in predicting the likelihood that a person will behave in a particular manner under specific environmental circumstances? For instance, some have argued that recent social psychological research has been mainly theoretical and focusing on methodological aspects of the distinction between implicit and explicit cognitions processes (Fazio and Olson 2003). Wiers *et al.* (2004) have questioned whether the implicit–explicit distinction, and their role in addictive behaviour, is actually about distinct processes or is best described solely in terms of the type of measurement used in related studies (Wiers *et al.* 2004).

Numerous models from experimental social psychology in areas such as persuasion (e.g. Petty and Cacioppo 1986; Chaiken *et al.* 1989) and the attitude–behaviour relationship (e.g. Fazio 1990) have proposed dual-process models encompassing the operation of both processing types under predictable environmental conditions. In his MODE model (Motivation and Opportunity as Determinants), Fazio (1990) proposed that people hold effectively a *single* attitude/belief about an object and that this attitude is accessed and assessed in different implicit ways. So for Fazio and colleagues, attitudes affect behaviour through relatively spontaneous or more deliberative cognitive processes, and the operation of these processes is determined by a motivation to behave in an individual and the opportunity to behave in that way. If both of these factors are present, behaviour will be predicted by the operation of deliberative appraisal. When either is low at the time an attitude is to be accessed from memory, people undertake more implicit processing— they are more likely to process the attitude in a spontaneous (automatic) manner (Fazio and Olson 2003). As such, social behaviour, including addictive behaviour, operates according to either implicit or explicit processing of a common attitude- or expectancy-based representation, and the type of processing is dependent on situation-specific factors including motivational sources (Stacy 1997; Rudman 2004; Hendricks and Brandon 2005; Huijding *et al.* 2005; Cox *et al.* 2006a).

1.5 Measuring cognitive processes in addictive behaviour

Most cognitive addiction research has relied almost entirely on the self-report information gleaned from asking participants about the beliefs and perceptions vis-à-vis their behaviour and about environmental factors deemed as important for the behaviour. This can be seen in the plethora of research that has examined drug and alcohol users' perceptions about the behaviours associated with their use (e.g. Sutton 1987; Albery and Guppy 1996; Albery *et al.* 2000; Morris and Albery 2000). These self-report procedures are known as explicit tasks such that participants are asked explicitly to recall and report upon their perceptions and beliefs. This assumes that participants 'know' their beliefs and perceptions. More recent work in addiction has extended this approach to address information processes that each individual operates but to which they have little or no access or of which they have little or no knowledge (Sayette 1999). These are known as implicit cognitive processes (Wiers *et al.* 2002). Measures of implicit cognition (e.g. implicit memory tasks) are designed to examine performance of tightly controlled experimental tasks such that researchers can glean some insight into human internal information processing. Sayette (1999) has argued that measures based on individual introspection, such as those based on self-report methodology, focus on the product of the outcome of a number of cognitive processes, or in his terms 'cognitive products' (p. 248). There are many examples of the explicit measurement of, for example, drinking habits, alcohol misuse and abuse as well as levels of dependence for other drugs. In general, participants are required to make a response by marking a scale on a questionnaire or they are asked directly by a trained interviewer. These include the severity of dependence scale (Gossop *et al.* 1995), the Alcohol Use Disorders Identification Test (AUDIT) (Saunders *et al.* 1993), the Alcohol Problems Questionnaire (Drummond 1990), the Alcohol Expectancy Questionnaire (Brown *et al.* 1987), the Negative Alcohol Expectancy Questionnaire (McMahon and Jones 1993), the Fagerstrom Test for Nicotine Dependence (Heatherton *et al.* 1991), the Smoking Consequences Questionnaire (Brandon and Baker 1991) and the Questionnaire on Smoking Urges (Tiffany and Drobes 1991). In addition, there are examples of many studies in the psychological literature that have used self-report questionnaires to assess belief, attitudes, behavioural intentions and behaviour based on decision-making models derived from social cognition. These include, among others, those based on the Theory of Reasoned Action, the Theory of Planned Behaviour and protection motivation theory (e.g. Sutton 1987; Petraitis *et al.* 1995). The main issues with such methods are those that open up threats to the validity of the study such as response bias and impression formation processes (McCusker 2001).

In contrast to these issues, implicit methods based on cognitive performance provide evidence for the operation and organization of differential cognitive structures, such as how alcohol expectancies are organized in long-term memory and attentional biases for addiction-related stimuli (Sharma *et al.* 2001; Kramer and Goldman 2003; Munafò *et al.* 2003; Reich *et al.* 2004; Hendricks and Brandon 2005). It is to this latter issue that we now turn.

1.6 **Attentional bias for addiction-related stimuli**

As we have already seen, the role of automatic processes for the cognition of alcohol-related cues has been the subject of theoretical debate (Tiffany 1990; Stacy 1995; McCusker 2001). It is argued that problem drinkers and people who may be experiencing problems with other addictive behaviours (e.g. gambling, McCusker and Gettings 1997; opiate use, Mogg *et al.* 2003) have a memory structure for alcohol-related concepts that is generated at an implicit level (Weingardt *et al.* 1996; Stacy 1997). In other words, alcohol users, and other substance abusers, do not have control over attention to relevant stimuli and activation of appropriate memory structures that, in turn, may guide behavioural responses to such cues (McCusker and Gettings 1997; Stacy 1997; Leung and McCusker 1999; Ingjaldsson *et al.* 2003*a,b*; Munafò *et al.* 2003). If this is the case, then alcohol users should show greater pre-occupation with alcohol-related stimuli compared with non-alcohol-related stimuli. This effect has been shown to be consistent across studies using free association memory activation paradigms among alcohol users and other substance users (e.g. Stacy 1997; Leung and McCusker 1999), psychobiological measures (e.g. Ingjaldsson *et al.* 2003*a,b*) and other implicit correlates of alcohol-related problems (e.g. Johnsen *et al.* 1994; Stetter *et al.* 1995; Sharma *et al.* 2001; Townsend and Duka 2001; Pothos and Cox 2002; Cox *et al.* 2003; Jones *et al.* 2003; Bruce and Jones 2004).

Among other measures (e.g. the dot-probe task, Mogg and Bradley 2002; Field *et al.* 2005*a*; flicker-induced change paradigm, Jones *et al.* 2002, 2003; the implicit association test, Huijding *et al.* 2005; the go–no-go association task, Dabbs *et al.* 2003), researchers have utilized a modified Stroop task (Stroop 1935) in which participants are asked to ignore a presented word and respond to the colour in which the word is presented. Typically it is found that alcohol-related words show increased response latencies in comparison with neutral words among problem drinkers (e.g. Bauer and Cox 1998; Sharma *et al.* 2001; Cox *et al.* 2006*b*). Theoretically this has been explained in terms of the automatic activation of a semantic network related to alcohol (e.g. Stetter *et al.* 1995; Bauer and Cox 1998; Sharma *et al.* 2001). If this explanation were feasible, it would predict that such an effect would also be apparent among a subgroup of non-problem drinkers. Two studies have addressed this issue by comparing high and low consuming non-problem drinkers. Cox *et al.* (1999) reported no interference from alcohol-related words in either group. However, Sharma *et al.* (2001) demonstrated that within a high consuming group of non-problem drinkers there was significant interference from alcohol-related words.

We have recently found further evidence for interference from alcohol-related words in a high consuming group of non-problem drinkers (Albery *et al.* 2006). Although the preferred explanation for interference is one of an automatic activation of a semantic network related to alcohol, other not incompatible explanations are possible. One relates to the frequency of exposure to alcohol-related stimuli. This frequency of exposure explanation suggests that greater pre-exposure to alcohol-related stimuli acts to prime the related semantic network. Although a precise definition of frequency of exposure can take a number of forms, in this study we aimed to distinguish between two forms: frequency that relates predominantly to exposure to alcohol-related cues in the environment, and frequency that relates predominantly to the consumption of alcohol. Participants were divided into low exposure ($n = 22$) and high exposure ($n = 21$) groups based on whether they currently worked in a bar or pub. The high exposure group (mean = 18.14 h per week, SE = 0.90, range 11–26 h per week) reported a significantly greater number of hours spent in bars/nightclubs/pubs (including work time) than the low exposure group (mean = 7.77 h per week, SE = 0.61, range 1–10 h per week), $t(41) = 9.62$, $P < 0.001$. With regards to the consumption of alcohol, a quantity–frequency measure of alcohol consumption was derived from multiplying two questions of the AUDIT questionnaire [i.e. 'How often do you have a drink containing alcohol?' (scored 0–4) and 'How many drinks containing alcohol do you have on a typical day when you are drinking?' (each scored 0–4)]. The possible range of scores for this measure was 0–16. Participants were divided into either high consumption ($n = 21$, mean = 7.38, SE = 0.55, range 4–12) or low consumption ($n = 22$, mean = 1.22, SE = 0.25, range 0–3) groups accordingly.

One prediction is that both these measures of frequency will increase interference from alcohol-related cues. However, with regards to frequency of exposure, there is some evidence that repetition of alcohol stimuli may have no effect or can lead to habituation of the interference (Sharma *et al.* 2001). This is consistent with evidence from studies of other psychopathologies that show a reduction in the emotional Stroop effect with repetition (Williams *et al.* 1996). For example, a decrease in interference for phobic-related words has been shown after desensitization treatment (Watts *et al.* 1986; Levy *et al.* 1994). This issue has also been investigated by comparing spouses of patients with a control group. Spouses of patients are assumed to have been exposed more frequently to concern-related cues than a control group. One study showed that there was no greater interference for gambling-related stimuli in a group of spouses of gamblers and a control group (McCusker and Gettings 1997).

Our results showed that interference from alcohol words was moderated by both frequency of exposure and frequency of consumption, $F(1,39) = 4.62$, $P < 0.040$ (see Fig. 1.1 and Table 1.1).

Further analysis showed that interference in the low consumption/low exposure group was lower than in any of the other three groups (which did not differ from each other).

The pattern of results indicates that for the high consumption group, there is significant attentional bias and that exposure to alcohol-related environmental cues through working in a bar did not change the magnitude of interference. However, the pattern is different for low consumers who are exposed to environmental cues through working in a bar. They showed significantly greater interference than equivalent drinkers who had not been exposed to such cues. It seems, therefore, that the role of environmental cue exposure for attentional disruption is especially important for those individuals who do not consume much alcohol.

There are a number of ways to interpret this result. One concerns the nature of the cues to which a person is exposed. Drinking behaviour and working in a bar differ in as much as the former can be seen as involving a person actively manipulating alcohol-related cues, while the latter can be seen as a person being the passive recipient of alcohol-related cues. If this is the case, the argument is that low consumers appear to be more influenced by cues that they receive passively, whereas high consumers are not as influenced by these

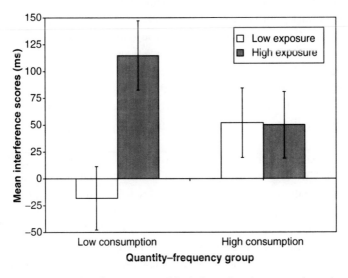

Fig. 1.1 Mean correct interference scores (alcohol reaction time–neutral reaction time) for consumption groups and exposure groups. Error bars are ±1 SEM.

Table 1.1 Mean correct reaction times in ms (SE in parentheses) to respond to the colour of alcohol and neutral words for combinations of exposure and consumption groups

Exposure–consumption group	Word type		Interference score (alcohol–neutral)
	Neutral	Alcohol	
Low consumption–low exposure ($n = 12$)	1027.01 (44.73)	1008.78 (52.49)	−18.23 (29.62)
Low consumption–high exposure ($n = 10$)	975.90 (48.95)	1090.62 (57.68)	114.72 (32.43)
High consumption–low exposure ($n = 10$)	982.06 (49.00)	1033.89 (57.50)	51.83 (32.44)
High consumption–high exposure ($n = 11$)	939.61 (46.72)	989.49 (54.82)	49.88 (30.93)

passive cues. This suggests that some low consumers might have a semantic network related to alcohol that is characterized more by representations associated with passively received environmental cues, and it is these cues that create disruption in the attentional system. For high consumers, the effect of these environmental cues is not apparent. Their alcohol-related semantic network seems to be characterized by more actively based cues encountered as a result of actual drinking behaviour.

A second explanation concerns the saliency of the cue and its relationship to the attentional priority given to drink-related stimuli. Our results seem to suggest that for the low consumption group, the priority given to alcohol-related stimuli is in general quite low. However, when this group is exposed to a drink-related environment the priority given to drink-related cues increases and therefore activates the drink-related semantic network more automatically. However, the high consumption group already gives priority to alcohol-related stimuli and therefore working in a drinking environment does not change this priority.

An extension of this idea is concerned with the threatening nature of the cues. The low consuming individual who is also heavily exposed to environmental cues from working in a bar is confronted with an inconsistency in their behaviours. This idea rests on the assumption that people believe that the behaviour people normally do in a bar is drink alcohol. If they do not drink but cognitively represent the 'normal' behaviour in the context to drink alcohol, any cues that encourage drinking *per se* will feel threatening. In other words, for low consumers, environmental cues associated with drinking alcohol appear threatening because they do not normally conform to that behaviour, and it is this threat that interferes with responses to alcohol and neutral words. It might also be the case that interference is only produced when the semantic network relating to alcohol is in addition associated with negative emotional consequences. For low consumers of alcohol this would occur after being exposed to the consequences of other peoples drinking (as a result of working in a bar) whereas for high consumers this would occur as a result of the consequences of their drinking. This might suggest a relationship with negative expectancies. Indeed literature related to negative and positive alcohol expectancies has demonstrated that whereas positive expectancies may be influential for the establishment of drinking behaviour, negative expectancies provide the motivation for refraining from drinking (Jones *et al.* 2001).

Any theory suggesting the activation of an alcohol-related semantic network should consider to what extent an emotional network is also activated. One possibility is that when an alcohol-related network is activated, this automatically activates a negative emotional network. This raises the possibility that the activation of these two networks may have different temporal dynamics. To investigate this issue, we have recently taken advantage of a methodology used by McKenna and Sharma (2004) to trace the time course of interference in a modified Stroop task. They distinguished between a fast component (one in which the salient stimulus interferes during the trial in which it is presented) and a slow component (one in which a salient stimulus interferes in the immediately following trial). Sharma *et al.* (2006) tested 45 drinkers (24 receiving treatment for problem drinking). We hypothesized that if the interference from alcohol words is due to activating a semantic network then it could produce its effects during the fast component as well as the slow component. More importantly, to see if the network that is activated during the fast or slow components can be characterized as related to alcohol or negative emotion, we obtained ratings of semantic relatedness from the participants at the end of the modified Stroop task. Alcohol relatedness was measured on a 5-point scale (0, not related to alcohol, to 4, highly related to alcohol) as to what extent the target words were related to alcohol. Emotional relatedness was measured on a 9-point scale ranging from -4 (highly related to negative emotion) to $+4$ (highly related to positive emotion). In a linear regression analysis, these ratings were used as predictors to see if the interference during the fast or slow components was predicted by alcohol or emotional ratings (Fig. 1.2 summarizes this analysis).

Figure 1.2 illustrates that for both alcohol and negative emotional words, the interference during the fast component is predicted by alcohol ratings,

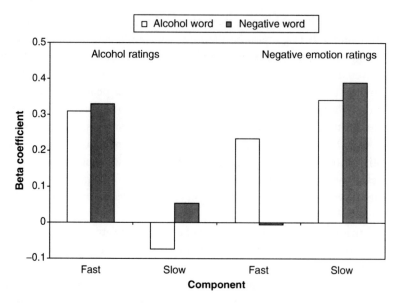

Fig. 1.2 Beta coefficients obtained from linear regression analyses in which the dependent variables are interference scores (from alcohol or negative emotional words during the fast or slow component) and the independent variables are ratings of alcohol relatedness and emotional relatedness (higher values represent higher negative values).

i.e. the higher the rating for alcohol relatedness the greater the interference (note that β coefficients >0.3 are significant). The alcohol ratings, however, do not predict the interference during the slow component. The negative emotional ratings, on the other hand, do predict interference during the slow component, i.e. the higher the negative rating, for negative or alcohol words, the greater the interference. These findings support the conclusion that interference in the fast component is largely driven by a semantic network that is related to alcohol, whereas interference in the slow component is largely driven by a semantic network related to negative emotion.

1.7 **Conclusions**

Contemporary research of addictive behaviours and addictive processes has begun to utilize methodologies derived from experimental psychology to understand the structure, organization and nature of important cognitive processes and predominantly biased cognitive processes. This chapter has been concerned with an overview of the types of positions psychologists have been embracing to understand the underlying cognition involved in undertaking addictive behaviours. Based on a few key theoretical positions derived from incentive sensitization processes, cognitive approaches concerned with drug urges and drug-use behaviour, and cognitively based motivational systems, this work has yielded important information on how the addicted individual (or those behaving in a habitual format without any classifiable addiction or dependence) develops and maintains the behaviour over time. In essence, this body of work argues that both the external environment in which an individual operates as well as the internal environment of the individual is characterized by a rich mixture of cues relevant to the behaviour. Once encountered, the processing of these cues, for example through the activation of expectancy structures in long-term memory, serves to guide related behaviour (such as taking an alcoholic drink or lighting a cigarette). Processing of such stimuli can occur both at a level characteristic as explicit in which, among other things, the individual can reason and access their thinking, and at an implicit level which operates more automatically such that the person is not aware of the 'thinking' involved. Recent work on the cognition of addiction has emphasized the implicit component emphasizing that ongoing addictive behaviour is determined by these less effortful, stimulus-bound and fast acting processes. This reasoning seems to make intuitive sense in that many addicted people report just finding themselves undertaking the behaviour without having consciously made the decision to act in that way. If this is the case, a number of questions are raised. What specifically characterizes these predominantly implicit processing systems? Are attentional process and/or neurobiological systems central in this conceptualization? Can this system be conceptualized in terms of a biased motivational system? What is the role of expectancies as a central cognitive component of addiction-related information processing? How can we best utilize the information derived from the study of cognition in addictive behaviours to create appropriate and dynamic interventions. Specifically, can we develop procedures that will maximize the likelihood of preventing people from undertaking addictive behaviours initially and also assisting individuals in their attempts to change their addictive behaviours? The chapters to follow in this book aim to provide answers to these points.

Acknowledgements

Correspondence concerning this chapter should be addressed to Professor Ian Albery, Department of Psychology, London South Bank University, SE1 OAA.

References

Albery, I.P. and Guppy, A. (1996). Drivers' biased perceptions of the adverse consequences of drink-driving. *Drug and Alcohol Review*, 15, 39–45.

Albery, I.P., Strang, J., Gossop, M. and Griffiths, P. (2000). Illicit drugs and driving: accident involvement among a cohort of current excessive drug users. *Drug and Alcohol Dependence*, 58, 197–204.

Albery, I.P., Sharma, D. and Noyce, S. (2006). Testing the frequency of exposure hypothesis in attentional bias for alcohol-related stimuli among social drinkers. Unpublished manuscript.

Baker, T.B., Morse, E. and Sherman, J.E. (1987). The motivation to use drugs: a psychobiological analysis of drugs. In P. C. Rivers (ed.), *The Nebraska Symposium on Motivation: alcohol use and abuse* Lincoln, NE: University of Nebraska Press, pp. 257–323.

Bandura, A. (1977). *Social learning theory*. Englewood Cliffs, NJ, Prentice-Hall.

Bargh, J.A. (1997). The automaticity of everyday life. In R.S. Wyer, Jr (ed.), *The automaticity of everyday life: advances in social cognition*, Vol. 10. Mahwah, NJ: Erlbaum, pp. 1–61.

Bargh, J. (2005). Bypassing the will: toward demystifying the nonconscious control of social behaviour. In R.R. Hassin, J.S. Uleman and J.A. Bargh (ed.), *The new unconscious*. Oxford: Oxford University Press, pp. 37–60.

Bargh, J.A. and Chartrand, T.L. (1999). The unbearable automaticity of being. *American Psychologist*, 54, 462–479.

Bargh, J.A., Chen, M. and Burrows, L. (1996). Automaticity of social behaviour: direct effects of the trait construct and stereotype activation on action. *Journal of Personality and Social Psychology*, 71, 230–244.

Bauer, D. and Cox, W.M. (1998). Alcohol-related words are distracting to both alcohol abusers and non-abusers in the Stroop color-naming task. *Addiction*, 93, 1539–1542.

Baxter, B.W. and Hinson, R.E. (2001). Is smoking automatic? Demands of smoking behavior on attentional resources. *Journal of Abnormal Psychology*, 110, 59–66.

Bradley, B.P., Mogg, K., Wright, T. and Field, M. (2003). Attentional bias in drug dependence: vigilance for cigarette-related cues in smokers. *Psychology of Addictive Behaviours*, 17, 66–72.

Brandon, T.H. and Baker, T.B. (1991). The smoking consequences questionnaire: the subjective expected utility of smoking in college students. *Psychological Assessment*, 3, 484–491.

Brandon, T.H., Herzog, T.A., Irvin, J.E. and Gwaltney, C.J. (2004). Cognitive and social learning models of drug dependence: implications for the assessment of tobacco dependence in adolescents. *Addiction*, 99 (Suppl. 1), 51–77.

Brown, S.A., Chrsitiansen, B.A. and Goldman, M.S. (1987). The alcohol expectancy questionnaire: an instrument for the assessment of adolescent and adult alcohol expectancies. *Journal of Studies on Alcohol*, 48, 483–491.

Bruce, G. and Jones, B. (2004). A pictorial Stroop paradigm reveals an alcohol attentional bias in heavier as compared to lighter social drinkers. *Journal of Psychopharmacology*, 18, 527–533.

Chaiken, S., Liberman, A. and Eagly, A. (1989) Heuristic and systematic information processing within and beyond the persuasion context. In J.S. Uleman and J.S. Bargh (ed.), *Unintended thought*. New York, NY: Guilford Press, pp. 212–252.

Colder, C.R. (2001). Life stress, physiological and subjective indexes of negative emotionality and coping reasons for drinking: is there evidence for a self-medication model of alcohol use? *Psychology of Addictive Behaviors*, 15, 237–245.

Collins, A.M. and Loftus, E.F. (1975). A spreading activation theory of semantic processing. *Psychological Review*, 82, 407–428.

Cox, W.M., Yeates, G.N. and Regan, C.M. (1999). Effects of alcohol cues on cognitive processing in heavy and light drinkers. *Drug and Alcohol Dependence*, 55, 85–89.

Cox, W. M., Brown, M.A. and Rowlands, L.J. (2003). The effects of alcohol cue exposure on non-dependent drinkers' attentional bias for alcohol-related stimuli. *Alcohol and Alcoholism*, 38, 45–49.

Cox, W.M., Fadardi, J.S. and Klinger, E. (2006a). Motivational processes underlying implicit cognition in addiction. In R.W. Wiers and A.W. Stacy (ed.), *Handbook of implicit cognition and addiction*. Thousand Oaks, CA: Sage, pp. 253–266.

Cox, W.M., Fadardi, J.S. and Pothos, E.M. (2006b). The addiction-Stroop test: theoretical considerations and procedural recommendations. *Psychological Bulletin*, 132, 443–476.

Dabbs, J.M., Bassett, J.F., Brower, A.M., Cate, K.L., DeDantis, J.E. and Leander, N.P. (2003). A portable version of the Go–No-Go Association Task (GNAT). Poster presentation at the 15th Annual Conference of the American Psychological Association, Atlanta, GA, USA, 2003.

de Wit, H., Pierri, J. and Johanson, C.E. (1989). Reinforcing and subjective effects of diazepam in nondrug-abusing volunteers. *Pharmacology, Biochemistry and Behavior*, 33, 205–213.

DiChiara, G.T. (1995). The role of dopamine in drug abuse viewed from the perspective of its role in motivation. *Drug and Alcohol Dependence*, 38, 95–137.

Drummond, D.C. (1990). The relationship between alcohol dependence and alcohol related problems in a clinical population. *British Journal of Addiction*, 85, 357–366.

Drummond, D.C. (2001). Theories of drug craving: ancient and modern. *Addiction*, 96, 33–46.

Ehrman, R.N., Robbins, S.J., Bromwell, M.A., Lankford, M.E., Monterosso, J.R. and O'Brien, C.P. (2002). Comparing attentional bias to smoking cues in current smokers, former smokers, and non-smokers using a dot-probe task. *Drug and Alcohol Dependence*, 67, 185–191.

Fazio, R.H. (1990). Multiple processes by which attitudes guide behaviour: the MODE model as an integrative framework. In M.P. Zanna (ed.), *Advances in experimental social psychology*. San Diego: Academic Press, pp. 75–109.

Fazio, R.H. and Olson, M.A. (2003). Attitudes: foundations, functions and consequences. In M.A. Hogg and J. Cooper (ed.), *The Sage handbook of social psychology*. London: Sage, pp. 139–160.

Field, M., Mogg, K. and Bradley, B.P. (2005a). Alcohol increases cognitive biases for smoking cues in smokers. *Psychopharmacology*, 180, 63–72.

Field, M., Mogg, K. and Bradley, B.P. (2005b). Craving and cognitive biases for alcohol cues in social drinkers. *Alcohol and Alcoholism*, 40, 504–510.

Field, M., Mogg, K. and Bradley, B.P. (2006). Attention to drug-related cues in drug abuse and addiction: component processes. In R.W iers, and A. Stacy (ed.), *Handbook of implicit cognition and addiction*. Thousand Oaks, CA: Sage, pp. 151–164.

Fillmore, M.T. and Vogel-Sprott, M. (1995a). Behavioral effects of alcohol in novice and experienced drinkers: alcohol expectancies and impairment. *Psychopharmacology (Berlin)*, **122**, 175–181.

Fillmore, M.T. and Vogel-Sprott, M. (1995b). Expectancies about alcohol-induced motor impairment predict individual differences in responses to alcohol and placebo. *Journal of Studies on Alcohol*, **56**, 90–98.

Fillmore, M.T. and Vogel-Sprott, M. (1995c). Behavioural effects of combining alcohol and caffeine: contribution of drug-related expectancies. *Experimental and Clinical Psychopharmacology*, **3**, 33–38.

Fillmore, M.T. and Vogel-Sprott, M. (1996). Evidence that expectancies mediate impairment under alcohol. *Journal of Studies on Alcohol*, **57**, 598–603.

Fillmore, M.T., Mulvihill, L.E. and Vogel-Sprott, M. (1994). Psychomotor performance under alcohol and under caffeine: expectancy and pharmacological effects. *Experimental and Clinical Psychopharmacology*, **2**, 319–327.

Fillmore, M.T., Carscadden, J.L. and Vogel-Sprott, M. (1998). Alcohol, cognitive impairment and expectancies. *Journal of Studies on Alcohol*, **59**, 174–179.

Fishman, M.W. and Foltin, R.W. (1992). Self-administration of cocaine by humans: a laboratory perspective. *CIBA Foundation Symposium*, **166**, 165–173.

Friedman, R.S., McCarthy, D.M., Förster, J. and Denzler, M. (2005). Automatic effects of alcohol cues on sexual attraction. *Addiction*, **100**, 672–681.

Glickman, S.E. and Schiff, B.B. (1967). A biological theory of reinforcement. *Psychological Review*, **74**, 81–109.

Goldman, M.S., Del Boca, F.K. and Darkes, J. (1999). Alcohol expectancy theory: the application of cognitive neuroscience. In K.E. Leonard and H.T.B lane (ed.), *Psychological theories of drinking and alcoholism*. New York, NY: Guilford Press, pp. 203–246.

Gossop, M., Darke, S. and Griffiths, P. (1995). The Severity of Dependence Scale (SDS): psychometric properties of the SDS in English and Australian samples of heroin, cocaine and amphetamine users. *Addiction*, **90**, 607–614

Hassin, R.R., Uleman, J.S. and Bargh, J.A. (2005). *The new unconscious*. New York, NY: Oxford University Press.

Heatherton, T., Kowlowski, L., Frecker, R. and Fagerstrom, K. (1991). The Fagerstrom test for nicotine dependence: a revision of the Fagerstrom tolerance questionnaire. *British Journal of Addiction*, **86**, 1119–1127.

Hendricks, P.S. and Brandon, T.H. (2005). Smoking expectancy associates among college smokers. *Addictive Behaviors*, 30, 235–245.

Hobbs, M., Remington, B. and Glautier, S. (2005). Dissociation of wanting and liking for alcohol in humans: a test of incentive-sensitisation theory. *Psychopharmacology*, **178**, 493–499.

Huijding, J., de Jong, P.J., Wiers, R.W. and Verkooijeb, K. (2005). Implicit and explicit attitudes toward smoking in a smoking and a nonsmoking setting. *Addictive Behavior*, 30, 949–961.

Ingjaldsson, J.T., Thayer, J.F. and Laberg, J.C. (2003a). Craving for alcohol and pre-attentive processing of alcohol stimuli. *International Journal of Psychophysiology*, **49**, 29–39.

Ingjaldsson, J.T., Thayer, J.F. and Laberg, J.C. (2003b). Preattentive processing of alcohol stimuli. *Scandinavian Journal of Psychology*, **44**, 161–165.

Johanson, C. and de Wit, H. (1989). The use of choice procedures for assessing the reinforcing properties of drugs in humans. *NIDA Research Monograph No. 92*. Washington, DC: Government Printing Office, pp. 171–210.

Johnsen, B.H., Laberg, J.C., Cox, W.M., Vakskal, A. and Hugdahl, K. (1994). Alcohol abusers' attentional bias in the processing of alcohol-related words. *Psychology of Addictive Behaviors*, **8**, 111–115.

Jones, B.T., Corbin, W. and Fromme, K. (2001). A review of expectancy theory and alcohol consumption. *Addiction*, **96**, 57–72.

Jones, B.C., Jones, B.T., Blundell, L. and Bruce, G. (2002). Social users of alcohol and cannabis who detect substance-related changes in a change blindness paradigm report higher levels of use than those detecting substance-neutral changes. *Psychopharmacology (Berlin)*, **165**, 93–96.

Jones, B.T., Jones, B.C., Smith, H. and Copley, N. (2003). A flicker paradigm for inducing change blindness reveals alcohol and cannabis information processing bias in social users. *Addiction*, **98**, 235–244.

Kalivas, P.W. and Stewart, J. (1991). Dopamine transmission in the initiation and expression of drug- and stress-induced sensitization of motor activity. *Brain Research Reviews*, **16**, 223–244.

Khantzian, E.J. (1985). The self-medication hypothesis of addictive disorders: focus on heroin and cocaine dependence. *American Journal of Psychiatry*, **142**, 1259–1294.

Kirsch, I. (1985). Response expectancy as a determinant of experience and behaviour. *American Psychologist*, **40**, 1189–1202.

Kirsch, I. (1999). Response expectancy: an introduction. In I.K irsch (ed.), *How expectancies shape experience*. Washington, DC: American Psychological Association, pp. 3–13.

Klinger, E. and Cox, W.M. (2004). Motivation and the theory of current concerns. In W.M. Cox and E. Klinger (ed.), *Handbook of motivational counseling: concepts, approaches and assessment*. New York, NY: Plenum, pp. 3–20.

Koob, G. and LeMoal, M. (1997). Drug abuse: hedonic homeostatic dysregulation. *Science*, **278**, 52–58.

Kramer, D.A. and Goldman, M.S. (2003). Using a modified Stroop task to implicitly discern the cognitive organization of alcohol expectancies. *Journal of Abnormal Psychology*, **112**, 171–175.

Leung, K.S. and McCusker, C.G. (1999). Accessibility and availability of smoking-related associations in smokers. *Addiction Research*, **7**, 213–226.

Levy, E., Van Oppen, P. and Van Den Hout, M. (1994). Selective processing of emotional information in obsessive compulsive disorder. *Behaviour Research and Therapy*, **32**, 243–246

Lusher, J., Chandler, C. and Ball, D. (2001). Dopamine D4 receptor gene (DRD4) is associated with novelty seeking (NS) and substance abuse: the saga continues.... *Molecular Psychiatry*, **6**, 497–499.

Lusher, J., Chandler, C. and Ball, D. (2004). Alcohol dependence and the alcohol Stroop paradigm: evidence and issues. *Drug and Alcohol Dependence*, **75**, 225–231.

Maude-Griffin, P.M. and Tiffany, S.T. (1996). Prediction of smoking urges through imagery: the impact of affect and smoking abstinence. *Experimental and Clinical Psychopharmacology*, **4**, 198–208.

McCusker, C.G. (2001). Cognitive biases and addiction: an evolution in theory and method. *Addiction*, **96**, 47–56.

McCusker, C.G. and Brown, K. (1990) Alcohol-predictive cues enhances tolerance to precipitate 'craving' for alcohol in social drinkers. *Journal of Studies on Alcohol*, **51**, 494–499.

McCusker, C.G. and Gettings, B. (1997). Automaticity of cognitive biases in addictive behaviours: further evidence with gamblers. *British Journal of Clinical Psychology*, **36**, 543–554.

McKenna, F.P. and Sharma, D. (2004) Reversing the emotional Stroop effect reveals that it is not what it seems: the role of fast and slow components. *Journal of Experimental Psychology: Learning, Memory, and Cognition*, **30**, 382–392.

McMahon, J. and Jones, B. (1993). The negative alcohol expectancy questionnaire (NAEQ) as an instrument to measure motivation to inhibit drinking and evaluating methods of treating problem drinkers. In E. Tongue and E. Martin (ed.), *Proceedings of the Thirty-Sixth International Conference on Alcohol and Drug Dependence*. Lausanne, Switzerland: ICAA/CIPAT, pp. 228–241.

Mogg, K. and Bradley, B.P. (2002) Selective processing of smoking-related cues in smokers: manipulation of three measures of processing bias. *Journal of Psychopharmacology*, **16**, 385–392.

Mogg, K., Bradley, B.P., Field, M. and De Houwer, J. (2003). Eye movements to smoking-related pictures in smokers: relationship between attentional biases and implicit and explicit measures of stimulus valence. *Addiction*, **98**, 825–836.

Morris, A.B. and Albery, I.P. (2001). Alcohol consumption and HIV risk behaviour: psychological formulations. *Addiction Research*, **9**, 73–86.

Munafò, M.R., Mogg, K., Roberts, S., Bradley, B.P. and Murphy, M.F.G. (2003). Selective processing of smoking-related cues in current smokers, ex-smokers and never-smokers on the modified Stroop task. *Journal of Psychopharmacology*, **17**, 311–317.

Musher-Eizenman, D.R. and Kulick, A.D. (2003). An alcohol expectancy-challenge prevention program for at risk college women. *Psychology of Addictive Behaviors*, **17**, 163–166.

Niaura, R.S., Rohsenrow, D.J., Binkoff, J.A., Monti, P.M., Pedraza, M. and Abrams, D.M. (1987). Relevance of cue reactivity to understanding alcohol and smoking relapse. *Journal of Abnormal Psychology*, **97**, 133–152.

Noël, X., Van der Linden, M. and d'Acremont, M. (2005). Cognitive biases toward alcohol-related words and executive deficits in polysubstance abusers with alcoholism. *Addiction*, **100**, 1302–1309.

Petraitis, J., Flay, B.R. and Miller, T.Q. (1995). Reviewing theories of adolescent substance use: organizing pieces in the puzzle. *Psychological Bulletin*, **117**, 67–86.

Petty, R.E. and Cacioppo, J.T (1986) *Communication and persuasion: cabral and peripheral routes to attitude change*. New, York, NY: Springer-Verlag.

Pothos, E.M. and Cox, M. (2002). Cognitive bias for alcohol-related information in inferential processes. *Drug and Alcohol Dependence*, **66**, 235–241.

Rather, B.C. and Goldman, M.S. (1994). Drinking-related differences in the memory organisation of alcohol expectancies. *Experimental and Clinical Psychopharmacology*, **2**, 167–183.

Reich, R.R., Goldman, M.S. and Noll, J.A. (2004). Using the false memory paradigm to test two key elements of alcohol expectancy theory. *Experimental and Clinical Psychopharmacology*, **12**, 102–110.

Robinson, T.E. and Becker, J.B. (1986). Enduring changes in brain and behavior produced by chronic amphetamine administration: a review and evaluation of animal models of amphetamine psychosis. *Brain Research Reviews*, **11**, 157–198.

Robinson, T.E. and Berridge, K. (1993). The neural basis of drug craving: an incentive-sensitization theory of addiction. *Brain Research Reviews*, **18**, 247–291.

Robinson, T.E. and Berridge, K. (2000). The psychology and neurobiology of addiction: an incentive-sensitization view. *Addiction*, **95** (Suppl. 2), S91–S117.

Robinson, T.E. and Berridge, K. (2001). Mechanisms of action of addictive stimuli: incentive-sensitisation and addiction. *Addiction*, **96**, 103–114.

Robinson, T.E. and Berridge, K. (2003). Addiction. *Annual Review of Psychology*, **54**, 25–53.

Rudman, L.A. (2004). Sources of implicit attitudes. *Current Directions in Psychological Science*, 13, 79–82.

Saunders, J.B., Aasland, O.G., Babor, T.F., De La Feunte, J.R. and Grant, M. (1993). Development of the Alcohol Use Identification Test (AUDIT): WHO collaborative project on early detection of persons with harmful alcohol consumption: II. *Addiction*, **88**, 791–804.

Sayette, M.A. (1999). Cognitive theory and research. In K.E. Leonard and H.Y. Blane (ed.), *Psychological theories of drinking and alcoholism*, 2nd edn. New York: Guilford Press pp. 247–291.

Sayette, M.A., Monti, P.M., Rohensow, D.J., Bird-Gulliver, S., Colbt, S., Sorota, A., Niaura, R.S. and Abrams, D.B. (1994). The effects of cue exposure on attention in male alcoholics. *Journal of Studies on Alcohol*, 55, 629–634.

Schachter, S. and Singer, J.E. (1962). Cognitive, social and physiological determinants of emotional state. *Psychological Review*, **69**, 379–399.

Schneider, W. and Shriffrin, R.M. (1977). Controlled and automatic human information processing: I.D etection, search and attention. *Psychological Review*, **84**, 1–66.

Sharma, D., Albery, I.P. and Cook, C. (2001). Selective attentional bias to alcohol-related stimuli in problem drinkers and non-problem drinkers. *Addiction*, **96**, 285–295.

Sharma, D., Albery, I.P. and Fernandez, A. (2006). What drives attention to addiction-related stimuli? Dissociating the role of affective and cognitive components. Unpublished manuscript.

Sheeran, P., Aarts, H., Custers, R., Rivis, A., Webb, T.L. and Cooke, R. (2005). The goal-dependent automaticity of drinking habits. *British Journal of Social Psychology*, **44**, 47–63.

Shiffrin, R.M. and Schneider, W. (1977). Controlled and automatic human information processing: II. Perceptual learning, automatic attending, and a general theory. *Psychological Review*, **84**, 127–190.

Solomon, R. (1980). The opponent-process theory of acquired motivation: the costs of pleasure and the benefits of pain. *American Psychologist*, **35**, 691–712.

Solomon, R.L. and Corbit, J.D. (1973). An opponent-process theory of motivation: II. Cigarette addiction. *Journal of Abnormal Psychology*, **82**, 158–171.

Stacy, A.W. (1995). Memory association and ambiguous cues in models of alcohol and marijuana use. *Experimental and Clinical Psychopharmacology*, 3, 183–194.

Stacy, A.W. (1997). Memory activation and expectancy as prospective predictors of alcohol and marijuana use. *Journal of Abnormal Psychology*, 106, 61–73.

Stetter, F., Ackerman, K., Bizer, A., Straube, E.R. and Mann, K. (1995). Effects of disease-related cues in alcohol abuser inpatients: results of a controlled 'Alcohol Stroop' study. *Alcoholism: Clinical and Experimental Research*, 19, 593–599.

Stormack, K.M., Laberg, J.C., Nordby, H. and Hugdahl, K. (2000). Alcoholics' selective attention to alcohol stimuli: automated processing? *Journal of Studies on Alcohol*, 61I, 18–23.

Stroop, J.R. (1935). Studies of interference in serial verbal reactions. *Journal of Experimental Psychology*, 18, 643–662.

Sutton, S. (1987). Social–psychological approaches to understanding addictive behaviours: attitude-behaviour and decision-making models. *British Journal of Addiction*, 82, 355–370.

Tiffany, S.T. (1990). A cognitive model of drug urges and drug-use behavior: role of automatic and nonautomatic processes. *Psychological Review*, 97, 147–168.

Tiffany, S.T. (1995). The role of cognitive factors in reactivity to drug cues. In D.C. Drummond, S.T. Tiffany, S.P. Glautier and B.R emington (ed.), *Addictive behaviour: cue exposure, theory and practice*. Chichester: Wiley, pp. 41–47.

Tiffany, S.T. (1999) Cognitive concepts of craving. *Alcohol Research and Health*, 23, 215–224.

Tiffany, S.T. and Carter, B. (1998). Is craving the source of compulsive drug use? *Journal of Psychopharmacology*, 12, 23–30.

Tiffany, S.T. and Conklin, C.A. (2000). A cognitive processing model of alcohol craving and compulsive alcohol use. *Addiction*, 95 (Suppl. 2), S145–S153.

Tiffany, S.T and Drobes, D.J. (1991). The development and initial validation of a question-naire on smoking urges. *British Journal of Addiction*, 86, 1467–1476.

Townshend, J.M. and Duka, T. (2001). Attentional bias associated with alcohol cues: differences between heavy and occasional social drinkers. *Psychopharmacology (Berlin)*, 157, 67–74.

Waters, A.J., Shiffman, S., Sayette, M., Paty, J., Gwaltney, C. and Balabanis, M. (2003a). Attentional bias predicts outcome in smoking cessation. *Health Psychology*, 22, 378–387.

Waters, A.J., Shiffman, S., Bradley, B.P. and Mogg, K. (2003b). Attentional shifts to smoking cues in smokers. *Addiction*, 98, 1409–1417.

Watts, F.N., McKenna, F.P., Sharrock, R. and Trezise, L. (1986). Colour naming of phobia-related words. *British Journal of Psychology*, 77, 97–108.

Weingardt, K.R., Stacy, A.W. and Leigh, B.C. (1996) Automatic activation of alcohol concepts in response to positive outcomes of alcohol use. *Alcoholism: Clinical and Experimental Research*, 20, 25–30.

West, R. (2006). *Theory of addiction*. Oxford, Blackwell.

White, N.M. (1996). Addictive drugs as reinforcers: multiple partial actions on memory systems. *Addiction*, 91, 921–949.

Wiers, R.W., de Jong, P.J., Havermans, R. and Jelicic, M. (2004). How to change implicit drug-use related cognitions in prevention: a transdisciplinary integration of findings from experimental psychopathoghy, social cognition, memory and experimental learn-ing psychology. *Substance Use and Misuse*, 39, 1625–1684.

Wiers, R.W., van Woerden, N., Smulders, F.T. and de Jong, P.J. (2002). Implicit and explict alcohol-related cognitions in heavy and light drinkers. *Journal of Abnormal Psychology*, 111, 648–658.

Williams, J.M.G., Mathews, A. and Macleod, C. (1996). The emotional Stroop task and psychopathology. *Psychological Bulletin*, 120, 3–24.

Wise, R.A. and Bozarth, M.A. (1987). A psychomotor stimulant theory of addiction. *Psychological Review*, 94, 469–492.

Zinser, M.C., Baker, T.B., Sherman, J.E. and Cannon, D.S. (1992). Relation between self-reported affect and drug urges and carvings in continuing and withdrawing smokers. *Journal of Abnormal Psychology*, 101, 617–629.

2

An implicit cognition, associative memory framework for addiction

Alan W. Stacy and Reinout W. Wiers

2.1 Introduction

This chapter outlines a framework that applies basic research on implicit cognition and associative memory to addictive behaviours. The framework helps provide a basis for continued development of cognitive theories of addiction, and suggests how the approach can foster prevention and cessation efforts. Findings and theories from neural systems, memory, implicit processes and addiction research are considered in an attempt to derive basic principles for the framework. Measurement domains are briefly summarized. Concepts from this framework are compared with related ideas, from expectancy and cue-reactivity research areas. This framework calls for a greater focus on the specific principles derived from basic cognitive research in multiple disciplines and encourages more attempts at integration across these areas.

Implicit processes relevant to addiction can be understood at different levels of explanation. They can be understood in terms of neural processes and systems, as well as a wide range of cognitive responses produced in divergent research paradigms from different disciplines. The first general section of this chapter focuses on implicit cognition and its relevance to addiction, acknowledging this diversity of support. Many implicit processes can also be understood in terms of more global theories of associative memory, a focus of the second general section of this chapter. Such theories provide explanations for the development and activation of associations that probably form an integral part of many implicit cognitions. The time seems right for a more thorough application of basic theories of implicit cognition and associative memory to addiction, for the sake of improved understanding and possible intervention.

2.2 Implicit cognition

Implicit cognition, as defined in this chapter, is revealed on tests that do not require or encourage deliberate or conscious recollection or introspections about the causes of one's behaviour. This definition is meant to provide a broad umbrella and is essentially an amalgamation of definitions from basic research on memory (Roediger 1990), social cognition (Greenwald and Banaji 1995) and cognitive neuroscience (Schott *et al.* 2005). Such tests use *indirect* assessment strategies that do not refer directly to the target constructs (Fazio and Olson 2003). Thus, memories, attitudes or emotions are assessed without directly asking about them. Implicit cognitions are activated without the need for engagement of executive decisions, deliberations or other conscious processes, although products of implicit cognition (e.g. related thoughts) can sometimes reach awareness. In this approach, implicit cognitions can affect behaviour without the individual's awareness or explicit knowledge of the process, essentially bypassing conscious weighing of pros and cons. As shown below, implicit cognition processes help explain perplexing characteristics of addictive behaviours.

2.2.1 Why study implicit cognition?

During the past decade, theories of implicit or automatic processes and indirect measures of cognition have been increasingly applied to the study of addictions and other health behaviours. Theories and allied measurement strategies from this perspective have provided new ways of thinking about the cognitive processes that may underlie addiction. The innovations did not emerge from a vacuum, but followed the lead of more basic research on memory, social cognition, learning, neuroscience and other areas. For example, an automatic, implicit associative system is seen as the 'default' system in current major theories on decision making (Evans 2003; Kahneman 2003). New ways of thinking such as this open up exciting avenues not only for improved understanding, but also for novel interventions or intervention components that may substantially improve prevention and cessation efforts (for an overview, see Wiers and Stacy 2006).

Implicit cognition approaches help the theorist and practitioner deal with an age old problem: why do people continue to do things they know are bad for them? Rather than explain perplexing, counterproductive behaviour by assuming there is a problem in rational processing, knowledge, will power or the like, an implicit cognition approach assumes that all people are prone to develop automatically activated cognitions that propel them toward certain behaviours, given specific circumstances and experiences. This does not preclude individual differences in other risk factors that may modify the development, or manifestation, of these tendencies.

Another important feature of many approaches to implicit cognition is the strong link between basic research and more applied research focusing on addiction. Most of the best approaches to implicit cognition in addiction have been directly 'translated' from more basic research, whether in areas of learning, memory, cognitive neuroscience or social cognition. Some principles have been derived from both human and non-human animal basic research. Common principles that emerge from diverse paradigms and research areas in both humans and animals may constitute fundamental insights that should not be overlooked in either basic or applied research.

2.2.2 Implicit cognition and related concepts

Beyond the general definition of implicit cognition advanced earlier, there are a variety of different senses of this term and presumed functional characteristics, well articulated in previous work (De Houwer 2006). Operational definitions of the term need to be addressed in empirical tests and specific theoretical explanations, some of which are outlined later in this chapter. Implicit cognition also needs to be considered in terms of related concepts, such as automaticity and unconscious influences of memory.

In the present approach, implicit cognition is used as a broad category of phenomena that match the general definition. Different manifestations of implicit cognition, often studied in different research areas or disciplines (e.g. social cognition, cognitive neuroscience, memory), often involve additional definitions and functional characteristics, while still sharing common ground with the general definition. Consistent with this approach, automatically activated cognitions are a subset of a broader class of implicit cognition, rather than the other way around (cf. De Houwer 2006). Implicit memory is also a subset of this broader class. Placing both implicit memory and automatic cognition under the umbrella of implicit cognition has certain advantages over subsuming implicit memory or cognition under automatic processes. For example, many compelling findings revealing implicit or unconscious memory have not been accompanied by evidence of *efficiency*, often attributed to automaticity (for a recent review of automaticity, see Moors and De Houwer 2006). Responses attributed to implicit memory are not necessarily always efficient because, although the memory may be activated rapidly, responses to some tasks require some cognitive resources and time (e.g. more than a second). Thus, implicit cognition (as defined here) is proposed in this chapter as the most general category, providing a useful multidisciplinary umbrella. We also include attentional bias and pre-conscious processes under this category.

The definition of implicit cognition used here is general enough to cover a broad set of responses that, at least, do not require extensive deliberation, intentional retrieval or judgements based on self-perceptions to operate.

On the basis of arguments available for some time (Nisbett and Wilson 1977; Feldman and Lynch 1988), the latter processes plague and confound the types of cognitive responses that have been the priority in decades of research in addiction and other health behaviour research. A search for alternative processes and measures is further justified by various lines of basic and applied research, outlined below.

2.2.3 Examples of findings and measures in drug abuse

There are several recent reviews of the application of implicit cognition concepts and indirect cognitive measures to addictive behaviours, focusing on different behaviours or different research paradigms (McCusker 2001; Ames *et al.* 2006; Bruce and Jones 2006; Cox *et al.* 2006; Houben *et al.* 2006; Waters and Sayette 2006). It is beyond the scope of this chapter to reiterate all of these findings. The most general conclusion from this research is that a variety of indirect measures implicating implicit processes have been found to predict, or correlate with, drug use. In a few instances, prospective effects have been investigated. These findings are consistent with the view that implicit processes play an important role in addictive behaviours.

Several examples reveal relevant findings from diverse paradigms. Some assessment paradigms rely on reaction time assessments in various types of tasks that do not directly inquire about drug-use behaviour. These tasks can be grouped into two large categories: (1) tests that have comprehensive support from basic research and include within-subject comparisons essential for adequate inference; and (2) tests that may measure reaction time to some type of judgement but have not received widespread support or do not include very informative comparisons. This section is restricted to the former class of test. Reaction time tasks of this class found to be predictive of drug use can be further subdivided into measures that assess aspects of an attentional bias for a drug and measures that assess automatic associations with drugs. Examples of measures of drug-related attentional biases are the modified Stroop task including drug-related words (for reviews, see Bruce and Jones 2006; Cox *et al.* 2006), different versions of dot-probe tests (Field *et al.* 2006), which assess whether a drug-related picture is responded to faster than a control (non-drug) picture when probed on a computer screen, and the more recently developed change blindness paradigm (Jones *et al.* 2002), in which problem drug users detect masked changes in scenes faster for drug-related changes than for changes in neutral materials. In this line of research, reaction time tests have been supplemented with registration of eye movements, which provide a measure of attentional subprocesses involved in attentional biases for drugs. Subprocesses include automatic *engagement* of attention versus latency in *disengagement* of attention from drug-related materials (Mogg *et al.* 2003; Field *et al.* 2005).

One example of reaction time tests of automatic or implicit associations is the Implicit Association Test (IAT; Greenwald *et al.* 1998), which has now been adapted for the assessment of drug-related associations, including alcohol (Wiers *et al.* 2002*b*, 2005; Jajodia and Earleywine 2003; Palfai and Ostafin 2003; De Houwer *et al.* 2004), cigarette smoking (Swanson *et al.* 2001; Sherman *et al.* 2003; Huijding *et al.* 2005) and cannabis smoking (Field *et al.* 2004). The advantage of the IAT is that it generates large and reliable effects; however, the interpretation of IAT effects is a topic of debate (see De Houwer 2002; Fazio and Olson 2003; Rothermund and Wentura 2004). Two recent studies tested to what extent alternative explanations such as the Figure-Ground account of IAT effects could account for previous findings concerning the alcohol-IAT (robust negative and arousal associations), and this proved to be partly but not fully the case: after accounting for Figure-Ground effects, the alcohol IATs remained predictive of behaviour (Houben and Wiers 2006*a,b*). There are a number of alternative reaction time measures well supported in basic cognitive research that can be used to assess drug-related associations, such as semantic priming with lexical decision (Hill and Paynter 1992; Zack *et al.* 1999; O'Connor and Colder 2005) or naming (Weingardt *et al.* 1996; Zack *et al.* 2003); a primed Stroop (e.g. Stewart *et al.* 2002); the Extrinsic Affective Simon Task (EAST; De Houwer 2003); the Go–NoGo Association Task (Nosek and Banaji 2001); and affective priming (Fazio 2001; Ostafin *et al.* 2003). However, among the well-researched reaction time tests relevant to implicit associations, the IAT has been studied the most with respect to addictive behaviour and psychometric properties. In general, reaction time tests of implicit processes have been found to be relevant correlates of drug use (for a more complete review, see Houben *et al.* 2006).

A different type of measure does not rely on reaction time tests but uncovers associations using word or picture association methods. Here again, there are essentially two classes of measures that vary in their relevance to implicit processes. In one class, measures are identical, or virtually identical, to those found to be effective in basic research on associative memory and implicit memory, while maintaining an *indirect* assessment of cognition and often including within-subject comparisons. Measures in this class have included free association (e.g. Kelly *et al.* 2006), where participants are asked to write the first word that comes to mind in response to single word cues (e.g. fun:___), a variant of this method using continuous association

(Szalay *et al.* 1999) to repeated cues (e.g. fun:___; fun:___; fun:___) and a variant using controlled association (Cramer 1968), in which responses are restricted in some fashion (e.g. respondents are asked for a category or verb). If measures within this class do not mention the target behaviour, or do anything to encourage recollection of previous events or introspections about the causes of one's behaviour, they clearly can be classified as *indirect* tests. A recent review of basic research on word association concluded that responses to indirect measures of this type are often implicated in implicit processes, although inferences of implicit processes range from strong to weak (or suggestive) depending on the study (Stacy *et al.* 2006*a*). Responses to such measures have been found to be good correlates of alcohol and marijuana use in a number of studies, despite indirect assessment that does not mention the target behaviour or encourage self-perception attributions about that behaviour (Palfai and Wood 2001; Ames *et al.* 2006; Kelly *et al.* 2006). The second class of measure asks for a first response to some question, but departs from indirect assessment, either by mentioning the target behaviour (and thereby encouraging a host of alternative processes; Feldman and Lynch 1988), by encouraging recollection of events or by fostering self-perceptions of behaviour. Such measures also may not be compatible with inferences drawn in basic research using word association, and may be more reasonably classified under the general category of open-ended survey responses (Krosnick 1999). Although a wide range of open-ended survey responses may be useful in addiction research because they can take advantage of important concepts such as top-of-mind awareness, self-generation and *relative* cognition (Stacy *et al.* 2004*b*), an implicit cognition, associative memory framework said to be derived from basic cognitive research should retain a close alignment to definitions and findings from that research.

A third type of assessment that can lead to reasonably good inferences about implicit cognition in addiction research comes from various types of memory testing paradigms that engage implicit processes across study/test trials. One of the best known paradigms in this domain is the testing of false memory (Deese 1959; Roediger and McDermott 1995). In an implicit processing perspective (McEvoy *et al.* 1999), false memories in recall and recognition arise from implicitly activated, associated cognitions (for a different explanation, see Brainerd and Reyna 1998). The pattern of recall or recognition can then uncover how this implicit process operates, as well as individual differences in the associations underlying the effects. In a recent extension of this paradigm to alcohol use, Reich *et al.*'s (2004) findings were consistent with implicit activation effects on illusory memory. Although they concluded that these findings were most usefully described in terms of expectancy processes (see also Chapter 6), they also acknowledged that the results could be explained within the associative memory theory of Nelson, McEvoy and colleagues (e.g. Nelson *et al.* 1998; McEvoy *et al.* 1999), which focuses on implicitly activated associations. Other memory-testing paradigms that have been applied to addiction include process dissociation (Fillmore *et al.* 1999), the famous name paradigm (Krank and Swift 1994) and extralist-cued recall (Stacy 1994), each of which can yield inferences of implicit or unconscious effects of memory (Jacoby *et al.* 1989; Jacoby 1998; Nelson *et al.* 1998). A range of other paradigms from basic cognitive research are applicable to inferences of implicit processes. In a few instances, addiction researchers have developed paradigms not contemplated in basic cognitive research, leading to some findings that are nevertheless difficult to explain without acknowledging implicit processes (e.g. Roehrich and Goldman 1995).

Although not enough research has yet addressed longitudinal effects of implicit cognition on addictive behaviours, a few prospective studies have been done. Using indirect tests of word association, two studies have found that these measures prospectively predict substance use (alcohol or marijuana), adjusting for the predictive effects of covariates such as previous drug habits, outcome expectancies, background characteristics and personality (Stacy 1997; Kelly *et al.* 2006). Importantly, Kelly *et al.* (2006) found prospective effects across 1-year intervals even though they carefully minimized various sources of contamination and confounding, such as hypothesis guessing. Using the IAT, Wiers and his colleagues (Wiers *et al.* 2002*b*) found that implicit associations predicted unique variance in 1-month prospective drinking in students, above explicit measures. Similar results have been found using different versions of the IAT (Houben and Wiers 2006*a,b*). In a recent study relevant to severe drinking patterns, Thush and Wiers (2006) found that adolescents' implicit alcohol associations predicted binge-drinking 1 year later. Although there are remaining issues regarding the optimal assessment of implicit drug-related cognitive processes, these studies make it clear that the assessments provide unique information that goes well beyond the information provided by explicit measures in terms of prospective prediction of substance use; unique effects have also been found in terms of prediction of treatment outcome (Cox *et al.* 2002; Waters *et al.* 2003).

Cross-sectional and prospective results are quite promising, but they are not always consistent and predictive effects are not always strong (Waters and Sayette 2006). It is likely that much stronger effects could be seen when assessments are refined and optimized. Improved measurement will probably require more attention to the functional properties of tests (Moors and De Houwer 2006). Improved measurement may also require advances in cognitive theory about the best concepts to apply to this area. As outlined below, concepts from cognitive neuroscience and associative memory may provide fertile ground for more detailed understanding of likely processes and presumed functional properties, as well as advances in assessment and improved prediction.

2.2.4 Emerging frameworks from cognitive neuroscience and the neural basis of addiction

Basic research on neural processes has attempted to pinpoint the precise nature and substrates of implicit cognition phenomena. This section provides a brief sample of this research, acknowledging a range of possibilities while providing neuropsychological evidence that distinct processes are likely to exist and thus should be investigated in research on cognition and addiction.

Moscovitch (2000) summarized several useful frameworks describing how implicit and explicit memory may be related, in terms of both functional (memory response) and structural (neural) levels of analysis. These models differ, for example, in whether implicit memory is a degraded, similar or distinct form of memory when compared with explicit memory. In more recent event-related functional magnetic resonance imaging (fMRI) research, Schott and his colleagues (Schott *et al.* 2005) found support for distinct neuroanatomy involved in implicit and explicit memory, as well as evidence for a further distinction involving intentional control of retrieval. Importantly, these investigators dissociated the neural correlates of intentional retrieval from those accompanying conscious recollection while showing that conscious recollection can occur in the absence of intentional control—an example of involuntary explicit memory (Richardson-Klavehn and Gardiner 1996). Further, some of the same neural correlates (e.g. decreased activation of bilateral occipital regions) accompanied both implicit memory and explicit (consciously recollected) memory, independent from intentional retrieval. Additional regions were affected during explicit (consciously recollected) memory (e.g. anterior prefrontal regions). These findings are compatible with the view that involuntary (or unintentional) explicit memory involves an initially automatic component, supported by at least some of the same regions that are involved in implicit memory, followed by conscious recollection, supported by distinct brain regions (Schacter and Badgaiyan 2001). Schott's findings are also consistent with the position that some memories may remain implicit under some primary senses of the term, i.e. activated without intention and unconscious, at least with respect to identification of the source of the memory. Indeed, some forms of implicit cognition may be completely distinct in representation or process (e.g. Eldridge *et al.* 2002).

For many issues in addiction and cognition, the critical question may not involve conscious awareness of the source of cognition but whether deliberation or intention is necessary for retrieval. If behaviour-relevant cognitions are spontaneously activated, even without intentional effort to retrieve a particular event from memory or direct questioning about a particular behaviour, then they may be influential under a wider variety of conditions or situations than cognitions that require such effort. Such memories may (but do not have to, see Bargh 2005) be accompanied by at least one of two forms of awareness: (1) awareness of content, as in a thought that spontaneously comes to mind; and (2) awareness of the source of the memory, as in involuntary explicit memory. Avoiding the first form of awareness is not a necessary criterion in cognitive neuroscience approaches to implicit memory, which often use indirect tests that require content to pop into mind (e.g. Vaidya *et al.* 1999; Levy *et al.* 2004; Schott *et al.* 2005). The second form of awareness (i.e. of source) is trickier to deal with, because involuntary explicit memory involves both an automatic (in the sense of unintentional) process and conscious awareness of source. If conscious awareness emerges only after an initial automatic process not involving awareness of source, the initial process could be characterized as implicit. Note that the point is not to argue for abandoning conscious recollection of source as a defining characteristic of involuntary explicit memory but to suggest that some sequences of cognition may involve both implicit and explicit processes. If one form of cognition happens to accompany or follow the other, it does not necessarily mean that one form has 'contaminated' the assessment of the other. It does mean that inferences are more difficult, since complete independence in the absence of other evidence is thus not used as a necessary or sufficient criterion.

A number of additional details about different forms of implicit processes have recently emerged from cognitive neuroscience. As examples, implicit habit learning, involving cue–outcome associations, has been found to remain intact in Alzheimer's patients, who show impaired implicit conceptual and explicit memory (Eldridge *et al.* 2002). The basal ganglia, presumably spared in Alzheimer's disease, are implicated in habit learning, whereas the association cortices and medial temporal lobe are implicated in implicit conceptual memory and explicit memory, respectively (e.g. Monti *et al.* 1996). Alzheimer's patients show normal levels of implicit perceptual memory (e.g. Park *et al.* 1998), whereas amnesic patients often show normal levels of both conceptual and perceptual implicit memory (for a review, see Gabrieli 1998). A range of additional dissociations indicative of different implicit processes have been revealed in Parkinson's patients (Knowlton *et al.* 1996; Moody *et al.* 2004), patients with specific damage to modality-specific areas related to perceptual priming (Fleischman *et al.* 1995) and patients with amygdala damage related to emotional conditioning (Bechara *et al.* 1995). In addition, recent fMRI research on implicit versus explicit evaluations has indicated that emotional stimuli automatically activate subcortical structures (including the amygdala) and that explicit processes supported by the frontal cortex and anterior cingulate can later moderate the automatic initial emotional response (Cunningham *et al.* 2003, 2004*b*). The automatic emotional responses also can occur when stimuli do not reach conscious awareness (e.g. Dehaene *et al.* 2003; Cunningham *et al.* 2004*a*). Overall, these data are consistent with the view that explicit memory and at least several different forms of implicit cognition can be distinguished neurologically and behaviourally. However, research on implicit cognition assessments in drug use have not yet been comprehensively linked to findings from neuroscience revealing diverse implicit processes, and these processes usually are not differentiated in either addiction theory or assessment. If the neural basis of responses to different types of indirect cognitive assessments in addiction are better understood, it could lead to important advances in understanding the different implicit cognition processes involved in addiction. The neuropsychological research suggests that a range of processes, with possibly distinct intervention implications, should be studied.

Several forms of implicit or automatic learning and memory processes have been directly implicated as reflections of the likely neural basis of addiction. Although there are a range of major views from neuroscience that can be applied to this specific topic in addiction (e.g. White 1996; Ahmed and Koob 2005; Everitt and Robbins 2005; Cox *et al.* 2006), space limitations restrict this chapter to only one of these theories. In the 'Incentive Sensitization Theory of Addiction' (IST), proposed by Robinson and Berridge (1993, 2003), addictive drugs enduringly alter nucleus accumbens-related brain systems that mediate a basic incentive-motivational function, the attribution of incentive salience to stimuli that signal reward. As a consequence, these neural circuits become enduringly hypersensitive (or 'sensitized') to specific drug effects and to drug-associated stimuli. This drug-induced brain change is called neural sensitization. As a result, this system automatically draws attention to drug-related stimuli (attentional bias) and can activate drug-seeking behaviour in the absence of conscious awareness (this process is labelled 'wanting' in quotation marks, to distinguish it from conscious wanting). The neural circuit involved, the mesolimbic dopamine (DA) system, has been related both to natural appetitive behaviours (e.g. food and sex) and to addiction (Berridge 2001). The general idea is that this subcortical system has evolved to signal cues for important resources (food, sex), by attributing incentive salience to the representations of cues that were associated with the appetitive reward. These cues then become 'wanted' themselves (Berridge 2001). As a result, behaviour can become 'irrational': cues to which incentive salience has been attributed automatically attract attention and induce an approach tendency, even when a deliberate cognitive evaluation of the situation would suggest staying away (Stacy 1997; Berridge 2001; Wiers *et al.* 2002*b*). According to this view, the crucial difference between an addicted and a not-addicted individual lies in an *enhanced mesolimbic dopaminergic reaction* both to the substance itself and to cues that have been associated with the substance. The activation of this system can but does not have to enter conscious awareness; if it does, this will be experienced as wanting or craving.

The 'wanting' response in animal research has been theoretically linked to an attentional bias in humans (Franken 2003) and to implicit alcohol–arousal associations (Wiers *et al.* 2002*b*, 2005), but these hypotheses need further confirmation (cf. Mogg *et al.* 2005; Cox *et al.* 2006). In the present framework, associations established between cues and reward can be considered part of a larger associative memory process that includes different types of implicit associations supported by different neural circuits. Consistent with Robinson and Berridge, as well as the previously mentioned work in cognitive neuroscience, at least several different systems may operate autonomously though they may also sometimes appear to act either in concert or in opposition.

2.3 Applying general theories of associative memory and connectionist frameworks

General theories of associative memory have only rarely been applied in any detail to addictive behaviours, either in theory or in intervention. Yet, these theories may provide ways to integrate otherwise disparate findings, as well as to begin to understand the cognitive processes underlying addiction and its amelioration. As shown later, these theories, with minor extensions, may help model a variety of implicit associations that are involved in addiction as well as prevention and treatment. It is beyond the scope of this chapter to reconcile these general theories with all the relevant research on the neural basis of implicit cognition and addiction. Rather, the present framework has the more humble goal of suggesting directions for integration. These possibilities suggest some common cognitive processes across ostensibly discrete areas of onset, progression and intervention.

2.3.1 Why associative memory should be studied in more detail

Many theories of memory and approaches to implicit cognition rely on conceptions of association. Although association in memory and cognition is certainly an old concept, dating at least back to Aristotle and traced through Hobbes, Locke, Bain and James (Warren 1921), its usefulness continues to evolve in diverse research from neural, to cognitive, behavioural and social levels of analysis (Nelson *et al.* 1998; Queller and Smith 2002; Phillips *et al.* 2003; Rescorla 2003). In one of the more balanced views of the history of the 'cognitive revolution' in the last century (Hintzman 1993), association was eschewed only temporarily, in favour of other concepts such as organization. After a relatively brief period in the backwaters of research on human cognition, association concepts re-emerged as important explanations of a variety of cognitive phenomenon (e.g. cued recall, semantic priming, false memory, implicit processes, social cognition, and so on). They also are used in conjunction with neural explanations of cognition and addiction as outlined earlier and are well integrated into more than a few general theories of memory, examples of which are outlined below. As Hintzman pointed out, the cognitive revolution was fought against a caricature and subset of the more extreme positions from early twentieth century learning theory; some of the most useful concepts from that early work have re-emerged in contemporary cognitive research, though often under different names or in an evolved state. In animal research, association concepts were never in the backwaters, and have provided useful explanations in neuroscience, memory, addiction and other research on behaviour. Association or connection concepts may persist, re-emerge and advance because they keep providing explanatory value across species and across levels of analysis, despite the fashion of the times.

There are several more specific reasons why theories of associative memory are worthy of intensive application to addiction theory and practice, though other general theories of cognition (e.g. schema, fuzzy trace) and non-cognitive theories are also often applicable. Associations in memory, in one form or another, have relevance for how one act of cognition may trigger another, how memory may be automatically cued, how perception of an object may trigger craving for a substance or a particular act of behaviour, and how habits may be perpetuated. General theories that model associative memory are useful because they provide postulates explaining how associations are developed and triggered. Yet, such theories have only rarely been applied to addictive behaviours. Because these theories are relevant to at least some of the forms of association that are likely to play a role in these behaviours, a greater focus on the propositions from these theories may prove important for this area. The theories outlined here not only explain certain types of associations relevant to addiction but also suggest cognitive processes that may be fundamental for prevention and treatment. To be effective, most interventions with individuals must teach something that is retained in memory at some level. Yet, specific memory processes consistent with major theories and research on memory are rarely a focus and seldom acknowledged. Indeed, an associative memory approach suggests that many of the same memory processes that are relevant to addiction are relevant to prevention and cessation of this behaviour. Such an approach needs to consider a wide range of potential associations in memory (Stacy *et al.* 2004*a*).

The present framework applies several approaches at different levels of analysis, not focusing only on a single theory or single type of association. This section focuses primarily on general theories of associative memory applicable to addiction, rather than addiction-specific theories, in hopes that the two types of theories can become better integrated in future research. One addiction-specific theory was outlined in the preceding section, and many have been thoroughly addressed in the chapters in this book as well as other recent work (Wiers and Stacy 2006). Application of general theories of memory to this area has been less common. The general theories vary in the extent to which implicit or automatic processes are formally covered, but in all instances the theories provide some process through which implicit or automatic cognitive processes can affect habitual or addictive behaviour and intervention outcomes. Some also provide possibilities for integrating findings from neuroscience. Rather than outline theories in detail, or list all relevant theories, the authors summarize some of the key processes from several examples of general theories of associative memory readily applicable to addictive behaviour. The focus is on processes that develop or strengthen associations, on how these associations trigger subsequent cognitions, on a general characterization of the representation or form of memory in each theory, and implications for addiction.

2.3.2 Spreading activation

One of the first concepts widely applied to automatic cognition phenomena was spreading activation (e.g. Posner and Snyder 1975; Anderson 1983; Neely 1991). Cognitive theories applying this concept are also applicable to a range of processes often characterized as implicit. Several approaches to addictive behaviour (e.g. Bovasso *et al.* 1993; Sayette 1993; Sayette *et al.* 2001), as well as theories of affect and attitudes (e.g. Fazio *et al.* 1986), have successfully applied the concept, and affective/emotional responses can be viewed as a form of implicit memory (Tobia *et al.* 1992). In spreading activation theories, concepts (e.g. alcohol, fun, party) are represented by nodes, which are units in memory. Nodes are connected by links in a network structure. The links represent an associative or semantic relationship (Neely 1991). When a node is activated in memory, it essentially triggers the activation of related nodes. So, if party and alcohol are strongly associated in memory, when the concept *party* is activated, activation will automatically spread to alcohol. Automatic activation of a related concept (alcohol) implies that stimuli representing this concept will be processed faster, or more fluently, and that strong associates of this second concept may then become activated as well (e.g. fun), in a mediational chain. Concepts (e.g. alcohol) that are more strongly related in memory to an initial concept (e.g. party) will 'win out' over weakly associated concepts, in terms of activation and influences on further processing. Activated concepts may or may not reach consciousness, but in either case they can colour or bias further cognition, which in turn can influence behaviour. This is plausible if ongoing cognition channels and limits the range of alternatives, a position long favoured in work on social behaviour (Bargh *et al.* 1986; Fazio *et al.* 1986).

Some models that postulate spreading activation also address a second system or process, such as a production system (Anderson 1983), beyond the scope of this chapter, or an expectancy process (Neely 1991). In predominant semantic priming approaches, highly relevant to cognitive biases, expectancy involves post-access decisions, constituting one of the most common explanations of non-automatic semantic/associative priming in basic research, whereas spreading activation is the most common explanation of an automatic component in semantic priming (Neely 1991). Although there are alternative explanations of semantic priming and other findings frequently attributed to spreading activation, it remains a popular theory of associative memory that is applicable to addiction. Several competing theories are addressed below, in terms of relevance to addictive behaviours.

2.3.3 Implications for addiction-related cognition

Since basic concepts from this theoretical approach have been previously applied to theories of addiction in a few instances, its relevance is only briefly highlighted here primarily as a reminder of these earlier works. Perhaps the primary implication of spreading activation is the inevitable focus on the precursors of cognition, i.e. where does ongoing cognition come from?

In this approach, it comes from activation spreading from concepts (or emotions) recently activated to associated concepts. Another primary implication is the ability of spreading activation networks to predict cognitive responses and related behaviour by understanding the pattern of associations in memory, in terms of such network variables as mediation (Sayette *et al.* 2001), constraint and density (Bovasso *et al.* 1993). More generally, associative and semantic network models have proven to be useful in several different approaches to addictive behaviours (Baker *et al.* 1986; Rather *et al.* 1992; Szalay *et al.* 1999). Because of their previous application, we turn now to less frequently applied approaches but later re-visit some common principles. These approaches are at least as applicable as spreading activation and provide novel heuristics through which to consider addiction.

2.3.4 Hopfield network

The Hopfield network (Hopfield and Tank 1986; Masson 1995) postulates a different representational structure for memory and implements a different explanation for the transition from one cognition to another. In this network, the units of memory are not concept nodes but are simpler elements or features, which could be single neurons or neural populations. Concepts are represented by a distributed pattern of activation across all elements, consistent with most contemporary connectionist models of cognition. Queller and Smith (2002) have provided a nice metaphor for this structure, applicable to most connectionist networks. In this analogy, the elements of memory are akin to pixels on a television screen, with a fixed but very large number. Any single pixel (element) has no meaning by itself. However, the pattern of brightness and colours of the pixels is able to represent an enormous variety of different scenes. In a connectionist network, concepts are represented in essentially the same way, through unique configurations of the activation states of elements. Thus, holistic patterns are represented by many simple elements.

Connection weights, or associations, are established in this model in accord with a Hebbian learning rule such that elements that are activated together become connected; Masson (1995) provides additional details. Thus, many elements, if strongly interconnected, can become activated together to form a larger meaningful unit, whether termed a concept, schema, script or other abstraction. One pattern of activation much more readily transitions to a second pattern if the two patterns have similar patterns of activation, which arise from similar connection weights. For example, if two concepts (e.g. fun and beer) are often encountered in a similar context, their patterns of activation will be similar because of the connection weights established among elements encoding that context. Thus, experiencing one concept may 'prime' the other, making the second concept more likely to be activated as well, essentially biasing subsequent processing.

2.3.5 Implications for addiction cognitions

Individual differences in patterns of activation relevant to addiction are likely because of variation in the contexts and affective states in which objects and events are encountered. For example, a college student who goes to drinking parties on the weekend will have very different patterns of activation with respect to certain concepts (party, drinking, fun, pot) than a Seventh-Day-Adventist whose parties involve church activities that provide non-alcoholic punch or food ('pot-luck'). Similarly, the phrase or thought of 'Friday night' may prompt a dramatically different pattern of activation in the two individuals: one sharing elements with alcohol (or other drug) concepts, and the other sharing elements with Sabbath activities. A host of other examples could be provided. A focus on the pattern of activation, and the interconnections among elements, forces a consideration of the context of meaning and likely individual differences across an entire pattern of activation. The Hebbian learning rule provides a plausible process through which differences in experiences get translated into these different patterns.

Although more traditional network models such as spreading activation also provide for a consideration of context and nuances of meaning, context and divergent meaning are not as obligatory as in the Hopfield network and other distributed models. A distributed pattern focuses on *many* elements simultaneously registered in memory, and the entire pattern must be considered in research and application. A distributed pattern across many elements also opens up the possibility of considering many different types of connections among elements having divergent neural substrates, rather than just those among concepts. This difference in focus could have major implications for increased understanding and intervention effects.

On the basis of distributed memory models, interventions that teach new skills or knowledge must take into account that transitions in cognition depend on similarity across the entire pattern of activation in memory. A desirable (intervention-related) pattern of activation will not arise during a behavioural choice point unless a transition to this pattern occurs, and the transition depends on similarity with the preceding cognitive pattern. The implication is that whatever is taught in prevention or treatment interventions must share a certain number of elements or features with the pattern of cognition that precedes a risky decision point. In this way, the patterns may be similar enough to promote a transition to the intervention-related pattern of activation in memory. A variety of strategies may enhance memory for programme-related material or skills in ways that overlap with the situations and cognitive state that precede a risky behaviour, essentially fostering new associations between the elements of these situations and the programme material (Stacy and Ames 2000; Stacy *et al.* 2004*a*).

2.3.6 Hintzman's multiple-trace theory

This theory posits a dramatically different form of representation from the preceding theories; yet it has many implications for addiction because of its focus on the details of experiences encoded in memory. In this approach, the active representation of an experience is registered in primary memory (PM), akin to working memory. Each experience is then encoded as a trace in secondary (long-term) memory (SM). Memory traces lie dormant in SM until they are prompted by a 'probe' (cue). When a probe is active in PM, its information is broadcast in parallel to all traces in SM. A single 'reply', termed an echo, returns to PM. The content of the echo depends on the similarity between the probe and all traces in SM. Traces having the most similarity to the features of the probe are most strongly activated in the echo. In other words, the common properties of the strongly activated traces will predominate the echo. In addition to content, another important property of the echo is its intensity. Intensity depends on the total amount of activation in SM, which is determined by both similarity (between probe and traces) and the number of activated traces. Intensity provides a type of familiarity signal, which may affect recognition, frequency, familiarity and other judgements about content.

Abstract knowledge, such as schemas or categories, is activated only during retrieval, when relevant traces are activated in the echo. A schema, category or other abstract, meaningful unit is represented by a configuration of many memory traces having certain elements or features in common; this approach can thus solve the problem of lack of operational definition or circularity in some schema approaches. Associative effects arise when a probe's features (e.g. feature set A) are shared with traces that not only involve set A, but also involve additional features (set B). The probe activates the traces in complete units, which include the elements in common with the probe (set A) as well as features not in common (set B). Thus, features from both set A and B become activated in the echo. Associative effects have been directly applied in the context of paired associate learning (Hintzman 1984) as well as false memory (Arndt and Hirshman 1998), and we see no reason why they would not apply to associative and semantic priming and other phenomena involving relations or transitions. In addition to schema abstraction, categorization, paired associates and false memory, the theory has been extended recently to decision making and heuristics and biases (Dougherty *et al.* 1999; Dougherty 2001; Bearden and Wallsten 2004), as well as to memory and probability judgements in ageing (Spaniol and Bayen 2005), underscoring its generality. A similar approach, at least in its instance representation, has been applied to social cognition (Smith and Zarate 1992; Castelli *et al.* 2004). The extensive generality of application of multiple-trace and other instance theories suggests that details are well worth considering for addiction.

Although applications of Hintzman's multiple-trace theory have rarely mentioned concepts of automaticity or implicit processes, it is clear that memory in this approach is thought to affect behaviour in an automatic fashion not dependent on executive or explicit processes.

> Because information is abstracted from concrete experiences at the time of retrieval rather than during learning, no sophisticated executive routine is needed to decide when and how to tune, reorganize or abandon memory structures. Reminding is not confined to predetermined structures, and changes in behaviour follow automatically from the indiscriminate accumulation of new episodic traces in memory
>
> (Hintzman 1986, p. 423)

Although Hintzman is presumably equating 'behaviour' with memory performance, a number of approaches suggest that activated memories can then spontaneously influence a range of behaviours (e.g. Bargh 2005), including addictive ones (for a review, see Wiers and Stacy 2006). Logan (1988) offered an instance theory that explicitly addresses automatization but is otherwise quite similar to Hintzman's model, especially in its representation of memory.

It is important to note that multiple-trace theory posits a single memory system. This system operates primarily at the level of traces of individual experiences, which influence memory activation and performance. However, Hintzman suggests a more rudimentary level that represents each trace, which is a record (possibly imperfect) of the activation of primitive properties that were engaged during an experience. Primative properties may involve sensory, emotional and simple relational features. It is reasonable to consider a parallel between these primative properties and the units in distributed connectionist networks, though Hintzman has not drawn this parallel. Indeed, given the evidence from cognitive neuroscience summarized earlier, primative properties probably have diverse neural substrates. If so, then memories for features that are represented by certain substrates may sometimes be impaired depending on individual differences or the encoding situation.

This view can be seen, at least heuristically, as a single memory system at the level of the memory trace and echo, consistent with Hintzman, but as multiple neural modules at the level of features and underlying primitive elements. In a simple matrix view of this scheme, memory traces (vectors) are the rows, and features are the columns. If different classes of features are organized into functionally discrete sets, then the columns may correspond to different 'systems' in some views from cognitive neuroscience (e.g. Squire 2004). At a macro level, a single organizational scheme could unify and blend different activated elements, but at a more elemental level some elements may not be relevant to the memory episode or may be impaired due to neurological constraints. For example, source memory, impaired in amnesic patients, may be encoded as a feature corresponding to activity in the medial temporal lobe. When this feature is not encoded in the memory trace, due to impairment, a probe or memory cue focusing on that feature will be unsuccessful at activating the trace. Other probes that share substantial similarity to the trace may lead to evidence that the trace was nonetheless encoded but only in terms of other features revealed on an indirect test of memory. In this way, a simple 're-organization' of the features in Hintzman's theory could explain why memory for content encoded in the past can be revealed even without conscious awareness of the source of that content. This idea is consistent with an earlier notion that multiple memory modules, including emotional ones, can be represented in a variety of ways, including memory traces and other single representational architectures (Johnson and Multhaup 1992). Further, although Hintzman does not address conscious awareness of memory processes, Logan (1988) assumes that people are not aware of the process in which multiple instances (traces) lead to automatically activated memories, though they are often aware of the activated content.

2.3.7 Implications for addiction

Some neurobiological approaches to addiction may be accommodated into this model by assuming that certain primitive elements in neural structures unique to addiction are activated as features of the memory trace. Thus, it is possible to integrate associations for sensitized cues (Berridge 2001) or cues that signal reward (Di Chiara *et al.* 2004) into the model.

However, possibly the most general implication of this theory for addiction is that it forces the investigator or practitioner to focus on the specific characteristics (features) of drug-use episodes, in the user, or pro-drug episodes, in the at-risk non-user. When these characteristics (whether environmental, social, reward, arousal or mood related) are experienced in the future, memories consistent with them will be most likely to be activated. A common theme in this chapter is that activation of pro-drug-related memories will often bias cognition in favour of drug use. The most obvious way to counteract these

biases in this model is to encode new (anti-drug or drug alternative) information or skills in terms of the features that normally trigger pro-drug memories.
This is an implication of Hintzman's similarity postulate, as well as similarity
within the Hopfield network. New memories may not be activated unless
something cues or prompts them, and the memories thus must be encoded in
terms of these prompts. In other words, in the multiple-trace approach, at least
some of the cues preceding or accompanying drug use must become features of
the memory traces that encode new information or skills.

Another implication is that the strength of activation of relevant content or
features prompted by the probe depends on the number of memory traces
sharing that content. This allows for the prediction that content encoded in
many activated memory traces will have a major advantage over content
encoded in only a few activated traces. In an example relevant to prevention,
many exposures among adolescents to the positive features of alcohol in the
media would overwhelm only a few exposures to preventive education, with
respect to the activation level of content, even when similarity to a related
probe is equal. Regarding populations in need of treatment, the prediction
makes it easy to understand why an addict with numerous sensitizing or
rewarding experiences with a drug but relatively fewer negative ones continues
to use a drug in the face of some negative events. Of course, the nature of these
experiences is also likely to play an important role (Baker *et al.* 1986). From
this perspective, it is also easy to see why treatment is difficult, since the
number of memory traces capable of facilitating the behaviour may number
in the thousands while those involving treatment may be counted on a few
hands. Prevention and treatment interventions that do not take into account
the similarity postulate, at least tacitly, would suffer a further dramatic
reduction in the amount of possible activation of important content when
behavioural decisions are made outside of the intervention setting.

The relevance of this approach is not really very different from the Hopfield
network in terms of the general implications of association, activation and
similarity. However, in our opinion, multiple-trace approaches provide a
more concrete representation and clearly provide the most focus on encoded
instances of experience. Thus, it may have more heuristic value.

2.3.8 Brief examples of other theories relevant to associative memory and addiction

There are a number of additional associative memory and connectionist
models applicable to addictive behaviours (for relevant reviews, see Hintzman
1990; Bechtel and Abrahamsen 2002; Dawson 2004), but only two additional
examples are briefly mentioned here. Plaut and Booth's (2000) distributed

connectionist network differs in a number of ways from the Hopfield network, in its reliance on 'hidden' layers, learning through backpropagation, and modelling the distinction between associative and semantic (similarity) effects. Although it is an open question whether multilayer networks and the use of backpropagation (e.g. Rumelhart *et al.* 1986) add substantially to the explanation of addiction-related cognitions provided by the much simpler Hopfield network, the distinction between association and semantics seems quite relevant. Further, although both the Plaut/Booth and Hopfield models rely on a common conception of similarity effects (arising from similar patterns of activation across the network), Plaut and Booth's model also focuses on transition learning as an associative process distinct from similarity. Indeed, this is one of the few models that explicitly differentiate associative from semantic effects in cognition. In transition learning, the neural network learns that pattern B follows pattern A, no matter what their semantic relationship or similarity. Once learned, transitions will readily occur in the future. In similarity effects, pattern B will more rapidly emerge from pattern A if the two patterns are similar in activated characteristics or features.

A very different model of memory (Processing Implicit and Explicit Representations, PIER 2) is provided by Nelson, McEvoy and their colleagues (e.g. Nelson *et al.* 1998, 2003). Since they recently applied their model to addictive behaviour (McEvoy and Nelson 2006), this chapter only highlights a few characteristics of the model. PIER 2 proposes independent implicit and explicit processes underlying a variety of memory effects. Since the model relies primarily on patterns of associations among concepts, rather than rule-based effects, it would be classified as a connectionist model under some definitions (Hintzman 1990; Bechtel and Abrahamsen 2002; Dawson 2004), even though it does not propose distributed representations characteristic of the most popular connectionist approaches (e.g. Rumelhart *et al.* 1986). The model postulates that a pattern of asymmetric associations in memory governs implicit activation. Various parameters (such as set size, association strength, connectivity and resonance) describe this pattern in ways that predict activation. One of the intriguing implications is that the pattern of interconnections among concepts that are not even directly experienced in a given situation will influence implicit processing and memory, if one or more of these concepts are associated in memory with something in the situation. Individual differences in what gets spontaneously activated in memory are likely to influence a range of behaviour, including drug use or healthy alternatives. This theory provides another example of an approach that is capable of operationally defining the elements of a schema-like pattern without being underspecified. Predictions from PIER 2 were found to be superior to spreading activation in an analysis of over 29 experimental studies (Nelson *et al.* 2003).

It is important to emphasize that a number of other associative memory and connectionist models are readily applicable to addiction (e.g. Hintzman 1990; Smith and DeCoster 1998; Eiser *et al.* 2003). In general, propositions from this whole class of theory are underutilized in addiction research. Specific propositions, as well as commonalities across theories, are just as applicable to onset and prevention as they are to escalation, maintenance, relapse and treatment.

2.3.9 Commonalities across models relevant to addiction

One of the major common elements of these approaches involves associations or connections. Although the different general theories applied here vary in the proposed unit of memory and other structural details, they all specify an architecture through which the elements of memory are interconnected and associated. This structure provides an engine for spontaneous fluctuations in cognition, whether conceived of as transitions in cognitive states, cognitive biases, patterns of activation in memory or activated conceptual nodes. Such cognitive transitions, especially those operating unintentionally, can help explain how an individual in a single day may switch from a non-drug-using pattern to a drug-using pattern, whether the drug-use event is the first time a drug has been used or the thousandth time. Thus, transitions, as spontaneous fluctuations in the pattern of activation, are quite relevant to primary and secondary prevention, as well as cessation and harm reduction.

Associations or connections among cognitive elements govern another, related commonality: similarity effects. One cognitive state is more likely to evolve into a second if they share a similar pattern of association, either among concept nodes, distributed elemental units, or among the features of relevant memory traces. Similarity effects are an important example of a 'compatibility' effect, in that what gets activated at time B depends on compatibility with the previous cognitive state at time A. For example, if something taught in an intervention needs to be activated at time B, then it has less chance if it is incompatible with the preceding cognitive state at time A. The importance of compatibility is also consistent with the far-reaching principle of transfer-appropriate processing (Morris *et al.* 1977), which is applicable to all forms of memory (Roediger *et al.* 2002). Other manifestations of compatibility can be seen in context effects on implicit associations in social cognition (Mitchell *et al.* 2003) and addiction (Krank and Wall 2006), in work that posits automatic schema activation in addiction (Tiffany 1990), alcohol expectancy approaches (Stacy *et al.* 1990; Goldman 2002) as well as in the encoding specificity principle, which is restricted to explicit forms of memory (Tulving 1983). However, suggesting that compatibility is important is not enough. The associative memory theories provide specific propositions giving guidance on how compatibility can be studied and applied in sufficient detail.

Frequency or repetition effects are inherent in most theories of associative memory, although variations in learning rules (e.g. Hebbian learning versus error correction) illustrate one of the important differences across theories. Frequency effects are likely to go hand-in-hand with major types of associative effects (Nelson and McEvoy 2000), probably because frequency effects can involve repetitions of regular co-occurrences (influencing association strength) as well as a greater variety of associative experiences (influencing associative network parameters such as set size and resonance). It may make little difference in theories such as multiple-trace whether frequency effects arise from real, imagined or previously retrieved episodes of experience, since elements processed together in working memory unite as a memory trace, and traces having common features summate to influence activation. Such theories can model findings explained as concept activation or chronic accessibility (Bargh *et al.* 1986; Stacy *et al.* 2006*b*), as well as various effects of associations (e.g. Kelly *et al.* 2006). To our knowledge, cognitive theories of addiction that do not focus on associative memory cannot readily explain the full range of effects.

Ability to model cognition at the concept level is inherent in theories of associative memory. However, distributed and multiple-trace models also model cognition at more elementary levels. For example, in multiple-trace theory, an image or feeling can be represented as features of a memory trace, although a summation of traces may form a schema-like or concept-like generalization or 'blending' of the most common features. The ability to model memory for a diversity of features makes many forms of memory relevant, not just those involving outcomes of drug use, sensitization or reinforcement. In understanding prevention or treatment, an array of features of an intervention can thus be considered. Some of these features may represent non-verbal images, emotions, aspects of learned facts, skills, intentions or goals. Whatever is processed can be considered in terms of the same underlying associative memory process. Indeed, recent work on addiction-related implementation intentions (Prestwich *et al.* 2006) and goals (Palfai 2006) is entirely compatible with theories of associative memory, because the importance of triggering cues and implicit activation is emphasized. In the present framework, *something* from an intervention must be spontaneously activated in memory at a behaviour choice point, and many different types of memories from a programme (not only goals or intentions) may serve as an important mediator or trigger of further intervention-related processing (Stacy *et al.* 2004*a*). This view is also in accord with a larger cognitive bias perspective derived from social cognition, underscoring the importance of cognitive transitions that channel processing in healthy directions. Memory for almost anything compatible with intervention goals may be capable of channelling such processing, leading to a potential intervention effect.

Another common element across the theories outlined in this chapter is the lack of reliance on processes that require intention or deliberate retrieval, though these may sometimes operate before behaviour ensues. The memory processes of focus here operate primarily at an implicit level, in accord with the definition outlined earlier in the chapter. Cognitive transitions are essentially spontaneous where A activates B without ongoing planning, deliberation or engagement of control processes. This commonality, as well as each of the preceding ones, has implications for intervention strategies that take into account implicit associations (Stacy and Ames 2000; Wiers *et al.* 2004, 2006*b*; Wiers and Stacy 2006).

A final commonality, though not an asset, is that the theories say little, if anything, about different forms of implicit or automatic processes. The neuropsychological evidence outlined earlier is consistent with the view that different types of associations in memory are supported by different systems in the brain. If one takes this multiple-system view, then these theories would not necessarily model all types of associations. On the other hand, a simple extension of some of the theories is quite feasible. One such extension was already offered for Hintzman's multiple trace theory. For the neural network models, one might propose that different sets of elements are represented in different modules or systems (cf. Johnson and Multhaup 1992), and sometimes whole sets of elements may not be activated (i.e. all elements in a module are 'off' in the Hopfield net). Nonetheless, the underlying principles (e.g. learning rules, activation parameters) from at least some models may apply to the full range of different cognitive processes (Plaut and Booth 2000). More work needs to be done to integrate general theories of associative memory fully with the neuroscientific findings in cognition and addiction.

2.4 Other concepts and associative memory

2.4.1 Cue reactivity and associative memory

In the present framework, common principles govern memory for all forms of learned information relevant to addictive behaviour. Thus, the reliance on association here may be seen as somewhat more general than conditioning approaches focused on the behaviour and reactivity of addicts (Carter and Tiffany 1999). However, in the future, it might be possible to subsume some conditioning phenomena in addiction under one or more of the theories outlined here or to otherwise integrate the approaches in some fashion (Hermans and Van Gucht 2006). Some integrations of information processing, conditioning research on addiction and other theories have been successful in previous work (Tiffany 1990; Baker *et al.* 1986; Curtin *et al.* 2006).

2.4.2 Expectancy and associative memory

The concept of expectancy has a rich heritage (Tolman 1932; Bolles 1975) and has been usefully linked to memory and addiction concepts in a variety of approaches (e.g. Stacy *et al.* 1990; Goldman *et al.* 1999; Krank and Wall 2006), as addressed more thoroughly in Chapter 6. Expectancy has been regarded as a unifying concept that at a high level of abstraction may provide an umbrella subsuming a variety of cognitive phenomena relevant to addiction (Goldman *et al.* 2006). However, expectancy can also be seen more narrowly as a certain class of association (Yin and Knowlton 2006), as a belief analogous to social learning concepts of outcome expectancies (Rotter 1954; Bandura 1977) or as a post-access (i.e. after memory access) decision or control process in basic cognitive research (Neely 1991). There are different views regarding what to do about these differences in conceptualization and definition.

A view closely aligned with basic research on automatic memory effects (Neely 1991) finds it especially difficult to reconcile or bridge the definitions, because expectancy is reserved for non-implicit processes. Since the present framework claims to maintain a close allegiance to these basic cognitive approaches, it may be confusing to use the term expectancy for several different psychological processes, including both implicit and explicit processes. There may also be confusion with the cognitive neuroscience and addiction research outlined earlier, because of the reviewed evidence for different neural systems linked to different aspects of both cognition and addiction. In the addiction domain, recent research has demonstrated that conditioned incentive salience (underlying 'wanting') can be dissociated from expected outcomes. Several independent studies have shown that one system can be manipulated, without affecting the other. For example, blocking the mesolimbic dopamine system blocks the incentive salience attribution but leaves the expectations unchanged, while prefrontal and insular lesions affect the expectations but not incentive salience (for a review, see Berridge 2001). Finally, more general dual-process models from different theoretical backgrounds all distinguish between rapid associative memory processes that are not constrained by capacity and limited capacity reasoning abilities, spanning domains such as social cognition (Strack and Deutsch 2004), cognitive psychology (Evans 2003; Evans and Coventry 2006), decision theory (Kahneman 2003) and neuroscience (Bechara *et al.* 2006). Importantly, in these models, if beliefs (equated with outcome expectancies in several major literatures) are addressed, they are part of the explicit system. Both implicit and explicit cognitive processes are important in determining behaviour, but they need to be clearly differentiated in these approaches. Although a term such as 'implicit expectancies' is sometimes used in addiction research, this term may obscure the distinctions just mentioned.

In an alternative framework, consistent with the influential work of Goldman and his colleagues, expectancies can be used to conceptualize and measure one broad underlying process, and they are not necessarily equivalent with beliefs. In this view, one can measure expectancies with either direct or indirect measures. Expectancy is the unifying construct, not association. In the present framework, association or connection is one of the unifying constructs underlying implicit processes, not expectancy, and a hybrid approach that suggests that explicit and implicit processes are essentially the same process is avoided. This is both simplifying and strategic for the present framework. It is simplifying because expectancy has different meanings in different literatures (often involving quite explicit processes, e.g. Bandura 1977), and these meanings often become inconsistent or not applicable when applying basic research on implicit cognition to addictive behaviour. It is strategic because it places more focus on frequently ignored concepts from basic theories of associative memory: more comprehensive application of these concepts may yield important advances in understanding and intervention that might not be uncovered without this emphasis. Readers are referred to Chapter 6 for an alternative framework that focuses on a more general view of expectancy. We believe that multiple approaches with different foci (e.g. associative memory, expectancy) may each lead to important advances if their unique contributions are taken to their limit, as is common in other areas of research. At the same time, commonalities across approaches should continue to be considered and acknowledged.

2.5 Need for integration with dual-process models and interventions

It is beyond the scope of this chapter to attempt a detailed account of dual-process models. However, it is important to note that an integration of connectionist/implicit cognition models with other processes is probably necessary for a complete understanding of addictive behaviour. Dual-process models propose how implicit or automatic cognitions and more controlled, explicit or executive processes may unite to govern behaviour. Several useful dual-process models have been proposed for the addictions; several examples can be found in Wiers and Stacy (2006): a model based on neurobiological learning processes (Yin and Knowlton 2006), a model based on neuropsychological data (Bechara *et al.* 2006), a general dual-process model based on extensive cognitive research in reasoning (Evans 2003) modified to explain pathological gambling (Evans and Coventry 2006), a model based on many findings in social cognition research (Strack and Deutsch 2004) applied to addictive behaviours (Deutsch and Strack 2006), a model based on acute drug effects (Fillmore and Vogel-Sprott 2006) and a model integrating findings from alcohol research (Wiers *et al.* 2006). Common to all these models is that there are *at least* two relatively independent systems involved in addictive behaviours, one based on automatic associative processes in response to drug-related stimuli, and a second lower capacity control system that (if motivation and capacity permit) may steer the person away from the drug-related stimuli (in contrast to the automatic approach reaction triggered by the automatic process). Some models involve more processes (e.g. Bechara *et al.* 2006; Yin and Knowlton 2006) and there are also indications that an automatic aversive reaction may play some role (Wiers *et al.* 2006). However, across the models, one important road to addiction concerns a process in which the automatic appetitive reactions gain relative strength over the controlled processes, over the time the addictive behaviour develops. Further, the same occurs more strongly once a drug-using episode has been initiated, given the strong immediate deleterious effects of drugs on the controlled processes, in the absence of impairing effects on the associative processes (Fillmore and Vogel-Sprott 2006).

From this perspective, individual differences in vulnerability for addictive behaviours can come from different sources, which matches the now common idea that many different genes are important in the vulnerability of addictions (see Chapter 8). Some genes appear to be related to a general risk for addictive behaviours, while others are related to a vulnerability (or an invulnerability) to develop a specific addiction (Goldman and Bergen 1998; McGue 1999). Concerning general factors, individuals differ in their sensitivity to reward (cf. Gray 1990; Bechara *et al.* 2006) as well as in their sensitivity to punishment (Finn *et al.* 1994). There is also evidence that individuals differ in the speed with which they develop sensitization (Robinson and Berridge 2003) (i.e. the rewarding effects of drugs get stronger), once drug use has initiated; there also appear to be genetic predispositions in the development of automatic aversive reactions (Elkins 1986). In addition, for alcohol it has been demonstrated that a low tolerance to alcohol intoxication is a genetic factor in the development of alcoholism, which protects many Asian people (McGue 1999), and the converse is true for individuals with a high tolerance for alcohol, who are at enhanced risk to develop alcoholism (Schuckit 1998). All these factors are likely to exert their influence through the impulsive or associative system in the dual-process models outlined above, because they involve the development of automatic associations with reward or punishment (in which the amygdala plays a major role; Phelps 2005).

Genetic factors play a likely role in individual differences in the ability of the second (reflective or controlled) system, given genetic components in related concepts such as executive functions, working memory and general intelligence (e.g. Plomin et al. 2003). In addition, environmental factors such as socialization, upbringing and social norms are likely to play a role in an individual's ability and motivation to inhibit automatically triggered appetitive processes.

In the development of addictive behaviours, a general problem seems to be that the controlled system overestimates its influence, especially concerning future occasions of control. Again, this principle seems to apply both to development and to behaviour at a specific occasion, once the addictive behaviour has habituated: when the first cigarettes are smoked, one may justifiably feel that there is a relatively large amount of control over that decision. However, this does not imply that control will be as easy later in the development of an addiction, because at that point stronger automatic associative processes play a role than at the moment of initiation. This can be the result of a number of processes: sensitization (Robinson and Berridge 1993, 2003), habituation (Everitt et al. 2001), and other neural adaptations that make the motivational salience of the drug strong in comparison with alternative behavioural options (e.g. adaptations in the striatum; Kalivas and Volkow 2005). The general problem seems to be that the conscious control system largely overestimates its abilities later to inhibit the addictive behaviour once triggered by the adapted brain. The same problem plays a role in the estimated control over appetitive behaviour on an occasion ('I'll have one drink, then go'), especially when one considers that drugs specifically impair the functioning of control systems, while leaving the automatic association systems intact (Fillmore and Vogel-Sprott 2006).

Finally, what could follow from these models regarding interventions (prevention and treatment)? First, it may be useful to assess implicit and explicit cognitions when current interventions are tested (e.g. Cox et al.

2002; Wiers *et al.* 2005). Implicit measures may uniquely predict clinically relevant outcomes such as relapse (Cox *et al.* 2002), and effects of current interventions on implicit versus explicit cognitions appear to be unrelated (i.e. zero correlations; Teachman and Woody 2003; Wiers *et al.* 2005). Including implicit cognition measures in intervention research may provide unique clues regarding the efficacy of (certain aspects of) current interventions, and may be helpful in optimizing the interventions. In addition, perhaps interventions targeting explicit cognitions can modulate the association between implicit cognitions and behaviour (Stacy *et al.* 2004a). Secondly, it may be possible to target implicit alcohol-related cognitions directly. The first studies that have attempted to change an automatic attentional bias for alcohol-related stimuli directly are promising (Wiers *et al.* 2006b). Clearly, this is research in its infancy, but it does provide a good example of a new direction of research inspired by dual-process models of addiction. Thirdly, it may be possible to enhance implicit cognitions that promote behaviours other than drinking or other drug use. A good example concerns implementation intentions (Palfai 2006; Prestwich *et al.* 2006), which may exert their influence in large part through automatic processes.

2.6 Conclusion

Associative memory and connectionist theories are relevant to addictions because they provide global models capable of integrating cognitive, emotional, sensory and motor aspects of addiction within a system that learns from experience, biases decisions, and acquires and perpetuates habits. Thus, such theories provide frameworks from which diverse empirical phenomena obtained from different levels of analysis can be seen in a larger explanatory context. Under such an umbrella, observations at different levels of analysis (e.g. neural versus cognitive) are reflections of a more general architecture that translates experience into cognitions that bias and influence later behaviour. One of the primary reasons for a focus on associative memory theories is that they provide at least several specific propositions that may be fundamental for understanding how these biases operate and how they influence addictive behaviour and its prevention or treatment. Yet, the theories must be extended, at least slightly in some instances or substantially in others, to cover the full range of phenomena uncovered in research on addiction and implicit cognition.

A focus on implicit processes, at least those not requiring deliberation, is relevant to addictive behaviours because they suggest how certain cognitions can spontaneously affect behaviour that is otherwise difficult to understand. The idea that there is a cognitive system, or set of systems, that is not governed by rationality or deliberation provides a parsimonious explanation for behaviour that is counterproductive in the long run. Although several other approaches attempt to explain why people do what they 'know' is bad for them, implicit cognition approaches seem to provide the most comprehensive set of *cognitive* explanations. By proposing that an associative, cognitive system plays a major role in this process, these approaches suggest that interventions, even educational ones, may be able to tap into this system. They also suggest that the processes are just as relevant to prevention as treatment, because of several reasonable assumptions: (1) to be effective, interventions must affect memory; (2) to affect behaviour, memory related to intervention materials and goals must be spontaneously activated, without deliberate retrieval; and (3) memory is associative, and requires links to triggering cues, broadly defined. In this approach, it seems unreasonable to assume that an intervention would affect behaviour without affecting a spontaneous form of memory. Otherwise, one is forced to either assume a model of intervention that relies entirely on explicit memory for learned material or remain unclear about memory assumptions. In our view, it is important to delineate thoroughly the processing assumptions that are thought to affect behaviour, a practice that is rare in intervention research. Only then can alternatives be effectively evaluated.

The present framework can be used to derive either strong or moderate forms of hypotheses for future research, in comparison with some alternatives. Strong forms of associative memory hypotheses would suggest that implicit activation of associations in memory are sufficient to predict a range of behaviours in addiction, such as onset, habit formation and maintenance of intervention effects, without a need for additional variables, such as those involved in executive functions or various decisional or frontal deficits (Volkow and Fowler 2000; Bechara and Damasio 2002). Such deficits may be a product and correlate of addiction, but not a primary cause except perhaps in severe cases of frontal deficit. More moderate forms of hypotheses take into account dual-process approaches, suggesting that a range of other variables are just as important as implicit, associative processes. Some variables may interact with implicit processes in their effects, as in certain dual-process models, or they may have important main effects. A third alternative, possibly the logical extension of strong forms of theories of frontal deficits, is that associative memory plays a minor or insignificant role in comparison with other processes. These alternatives are well worth considering in future research.

Although implicit cognition approaches have shown much promise, there is a long way to go in theory, measurement and especially intervention.

When one considers the evidence, this state of affairs is probably in common across all cognitive models of behaviour. Incorporating non-implicit systems into the mix further complicates things but is certainly important for future research and intervention. Indeed, dual-process models provide an exciting range of possibilities for further development. The present framework suggests that a greater linkage to measures and theories derived from basic cognitive research on both implicit and non-implicit processes may dramatically increase progress in cognitive research on addiction and a range of appetitive behaviours.

Acknowledgements

This chapter was supported by a grant from the National Institute on Drug Abuse, DA16094. The authors thank Jan De Houwer for constructive comments. Any correspondence concerning this chapter should be addressed to Professor Alan Stacy, Department of Preventive Medicine and Institute for Prevention Research, University of Southern California.

References

Ahmed, S.H. and Koob, G.F. (2005). Transition to drug addiction: a negative reinforcement model based on an allostatic decrease in reward function. *Psychopharmacology*, 180, 473–490.

Ames, S.L., Franken, I.H.A. and Coronges, K. (2006). Implicit cognition and drugs of abuse. In R.W. Wiers and A.W. Stacy (ed.), *Handbook of implicit cognition and addiction*. Thousand Oaks, CA: Sage, pp. 363–378.

Anderson, J.R. (1983). *The architecture of cognition*. Cambridge: Harvard University Press.

Arndt, J. and Hirshman, E. (1998). True and false recognition in MINERVA2: explanations from a global matching perspective. *Journal of Memory and Language*, 39, 371–391.

Baker, T.B., Morse, E. and Sherman, J.E. (1986). The motivation to use drugs: a psychobiological analysis of urges. *Nebraska Symposium on Motivation*, 34, 257–323.

Bandura, A. (1977). *Social learning theory*. Oxford: Prentice-Hall.

Bargh, J.A. (2005). Bypassing the will: toward demystifying the nonconscious control of social behavior. In R.R. Hassin, J.S. Uleman and J.A. Bargh (ed.), *The new unconscious*. New York, NY: Oxford University Press, pp. 37–58.

Bargh, J.A., Bond, R.N., Lombardi, W.J. and Tota, M.E. (1986). The additive nature of chronic and temporary sources of construct accessibility. *Journal of Personality and Social Psychology*, 50, 869–878.

Bearden, J.N. and Wallsten, T.S. (2004). MINERVA-DM and subadditive frequency judgments. *Journal of Behavioral Decision Making*, 17, 349–363.

Bechara, A. and Damasio, H. (2002). Decision-making and addiction (part I): impaired activation of somatic states in substance dependent individuals when pondering decisions with negative future consequences. *Neuropsychologia*, 40, 1675–1689.

Bechara, A., Tranel, D., Damasio, H., Adolphs, R., Rockland, C. and Damasio, A.R. (1995). Double dissociation of conditioning and declarative knowledge relative to the amygdala and hippocampus in humans. *Science*, 269, 1115–1118.

Bechara, A., Noel, X. and Crone, E.A. (2006). Loss of willpower: abnormal neural mechanisms of impulse control and decision-making in addiction. In R.W. Wiers and A.W. Stacy (ed.), *Handbook of implicit cognition and addiction.* Thousand Oaks, CA: Sage, pp. 215–232.

Bechtel, W. and Abrahamsen, A. (2002). *Connectionism and the mind: parallel processing, dynamics, and evolution in networks,* 2nd edn. Malden, MA: Blackwell Publishers.

Berridge, K.C. (2001). Reward learning: reinforcement, incentives, and expectations. In D. L. Medin (ed.), *The psychology of learning and motivation: advances in research and theory,* Vol. 40. San Diego, CA: Academic Press, pp. 223–278.

Bolles, R.C. (1975). *Learning theory.* Oxford: Holt, Rinehart and Winston.

Bovasso, G., Szalay, L., Biase, V. and Stanford, M. (1993). A graph theory model of the semantic structure of attitudes. *Journal of Psycholinguistic Research,* 22, 411–425.

Brainerd, C.J. and Reyna, V.F. (1998). When things that were never experienced are easier to 'remember' than things that were. *Psychological Science,* 9, 484–489.

Bruce, G. and Jones, B.T. (2006). Methods, measures, and findings of attentional bias in substance use, abuse, and dependence. In R.W. Wiers and A.W. Stacy (ed.), *Handbook of implicit cognition and addiction.* Thousand Oaks, CA: Sage, pp. 135–149.

Carter, B.L. and Tiffany, S.T. (1999). Meta-analysis of cue-reactivity in addiction research. *Addiction,* 94, 327–340.

Castelli, L., Zogmaister, C., Smith, E.R. and Arcuri, L. (2004). On the automatic evaluation of social exemplars. *Journal of Personality and Social Psychology,* 86, 373–387.

Cox, W.M., Hogan, L.M., Kristian, M.R. and Race, J.H. (2002). Alcohol attentional bias as a predictor of alcohol abusers' treatment outcome. *Drug and Alcohol Dependence,* 68, 237–243.

Cox, W.M., Fadardi, J.S. and Pothos, E.M. (2006). The addiction-Stroop test: theoretical considerations and procedural recommendations. *Psychological Bulletin,* 132, 443–476.

Cramer, P. (1968). *Word association.* Saint Louis, MO: Academic Press.

Cunningham, W.A., Johnson, M.K., Gatenby, J.C., Gore, J.C. and Banaji, M.R. (2003). Neural components of social evaluation. *Journal of Personality and Social Psychology,* 85, 639–649.

Cunningham, W.A., Johnson, M.K., Raye, C.L., Gatenby, J.C., Gore, J.C. and Banaji, M.R. (2004a). Separable neural components in the processing of black and white faces. *Psychological Science,* 15, 806–813.

Cunningham, W.A., Raye, C.L. and Johnson, M.K. (2004b). Implicit and explicit evaluation: fMRI correlates of valence, emotional intensity, and control in the processing of attitudes. *Journal of Cognitive Neuroscience,* 16, 1717–1729.

Curtin, J.J., McCarthy, D.E., Piper, M.E. and Baker, T.B. (2006). Implicit and explicit drug motivational processes: a model of boundary conditions. In R.W. Wiers and A.W. Stacy (ed.), *Handbook of implicit cognition and addiction.* Thousand Oaks, CA: Sage, pp. 233–250.

Dawson, M.R.W. (2004). *Minds and machines: connectionism and psychological modeling.* Malden, MA: Blackwell.

De Houwer, J. (2002). The Implicit Association Test as a tool for studying dysfunctional associations in psychopathology: strengths and limitations. *Journal of Behavior Therapy and Experimental Psychiatry,* 33, 115–133.

De Houwer, J. (2003). The extrinsic affective Simon task. *Experimental Psychology,* 50, 77–85.

De Houwer, J. (2006). What are implicit measures and why are we using them? In R.W. Wiers and A.W. Stacy (ed.), *Handbook of implicit cognition and addiction*. Thousand Oaks, CA: Sage, pp. 11–28.

De Houwer, J., Crombez, G., Koster, E.H.W. and De Beul, N. (2004). Implicit alcohol-related cognitions in a clinical sample of heavy drinkers. *Journal of Behavior Therapy and Experimental Psychiatry*, 35, 275–286.

Deese, J.(1959). Influence of inter-item associative strength upon immediate free recall. *Psychological Reports*, 5, 305–312.

Dehaene, S., Artiges, E., Naccache, L., Martelli, C., Viard, A., Schurhoff, F., Recasens, C., Paillère Martinot, M.L., Leboyer, M. and Martinot, J.-L. (2003). Conscious and subliminal conflicts in normal subjects and patients with schizophrenia: the role of the anterior cingulate. *Proceedings of the National Academy of Sciences of the USA*, 100, 13722–13727.

Deutsch, R. and Strack, F. (2006). Reflective and impulsive determinants of addictive behavior. In R.W. Wiers and A.W. Stacy (ed.), *Handbook of implicit cognition and addiction*. Thousand Oaks, CA: Sage, pp. 45–57.

Di Chiara, G., Bassareo, V., Fenu, S., De Luca, M.A., Spina, L., Cadoni, C., Acquas, E., Carboni, E., Valentini, V. and Lecca, D. (2004). Dopamine and drug addiction: the nucleus accumbens shell connection. *Neuropharmacology*, 47 (Suppl. 1), 227–241.

Dougherty, M.R. P. (2001). Integration of the ecological and error models of overconfidence using a multiple-trace memory model. *Journal of Experimental Psychology: General*, 130, 579–599.

Dougherty, M.R.P., Gettys, C.F. and Ogden, E.E. (1999). MINERVA-DM: a memory processes model for judgments of likelihood. *Psychological Review*, 106, 180–209.

Eiser, J.R., Fazio, R.H., Stafford, T. and Prescott, T.J. (2003). Connectionist simulation of attitude learning: asymmetries in the acquisition of positive and negative evaluations. *Personality and Social Psychology Bulletin*, 29, 1221–1235.

Eldridge, L.L., Masterman, D. and Knowlton, B.J. (2002). Intact implicit habit learning in Alzheimer's disease. *Behavioral Neuroscience*, 116, 722–726.

Elkins, R.L. (1986). Separation of taste-aversion-prone and taste-aversion-resistant rats through selective breeding: implications for individual differences in conditionability and aversion-therapy alcoholism treatment. *Behavioral Neuroscience*, 100, 121–124.

Evans, J.S.B.T. (2003). In two minds: dual-process accounts of reasoning. *Trends in Cognitive Sciences*, 7, 454–459.

Evans, J.S.B.T. and Coventry, K. (2006). A dual process approach to behavioral addiction: the case of gambling. In R.W. Wiers and A.W. Stacy (ed.), *Handbook of implicit cognition and addiction*. Thousand Oaks, CA.Sage, pp. 29–44.

Everitt, B.J. and Robbins, T.W. (2005). Neural systems of reinforcement for drug addiction: from actions to habits to compulsion. *Nature Neuroscience*, 8, 1481–1489.

Everitt, B.J., Dickinson, A. and Robbins, T.W. (2001). The neuropsychological basis of addictive behaviour. *Brain Research Reviews*, 36, 129–138.

Fazio, R.H. (2001). On the automatic activation of associated evaluations: an overview. *Cognition and Emotion*, 15, 115–141.

Fazio, R.H. and Olson, M.A. (2003). Implicit measures in social cognition research: their meaning and uses. *Annual Review of Psychology*, 54, 297–327.

Fazio, R.H., Sanbonmatsu, D.M., Powell, M.C. and Kardes, F.R. (1986). On the automatic activation of attitudes. *Journal of Personality and Social Psychology*, 50, 229–238.

Feldman, J.M. and Lynch, J.G. (1988). Self-generated validity and other effects of measurement on belief, attitude, intention, and behavior. *Journal of Applied Psychology*, 73, 421–435.

Field, M., Mogg, K. and Bradley, B.P. (2004). Cognitive bias and drug craving in recreational cannabis users. *Drug and Alcohol Dependence*, 74, 105.

Field, M., Mogg, K. and Bradley, B.P. (2006). Attention to drug-related cues in drug abuse and addiction: Component processes. In R.W. Wiers and A.W. Stacy (ed.), *Handbook of implicit cognition and addiction*. Thousand Oaks, CA: Sage, pp. 151–164.

Field, M., Mogg, K. and Bradley, B.P. (2005). Craving and cognitive biases for alcohol cues in social drinkers. *Alcohol and Alcoholism*, 40, 504–510.

Fillmore, M.T. and Vogel-Sprott, M. (2006). Acute effects of alcohol and other drugs on automatic and intentional control. In R.W. Wiers and A.W. Stacy (ed.), *Handbook of implicit cognition and addiction*. Thousand Oaks, CA: Sage, pp. 293–306.

Fillmore, M.T., Vogel-Sprott, M. and Gavrilescu, D. (1999). Alcohol effects on intentional behavior: dissociating controlled and automatic influences. *Experimental and Clinical Psychopharmacology*, 7, 372–378.

Finn, P.R., Kessler, D.N. and Hussong, A.M. (1994). Risk for alcoholism and classical conditioning to signals for punishment: evidence for a weak behavioral inhibition system. *Journal of Abnormal Psychology*, 103, 293–301.

Fleischman, D.A., Gabrieli, J.D.E., Reminger, S., Rinaldi, J., Morrell, F. and Wilson, R. (1995). Conceptual priming in perceptual identification for patients with Alzheimer's disease and a patient with right occipital lobectomy. *Neuropsychology*, 9, 187–197.

Franken, I.H. (2003). Drug craving and addiction: integrating psychological and neuropsychopharmacological approaches. *Progress in Neuro-Psychopharmacology and Biological Psychiatry*, 27, 563–579.

Gabrieli, J.D.E. (1998). Cognitive neuroscience of human memory. *Annual Review of Psychology*, 49, 87–115.

Goldman, D. and Bergen, A.W. (1998). General and specific inheritance of substance abuse and alcoholism. *Archives of General Psychiatry*, 55, 964–965.

Goldman, M.S. (2002). Expectancy and risk for alcoholism: the unfortunate exploitation of a fundamental characteristic of neurobehavioral adaptation. *Alcoholism: Clinical and Experimental Research*, 26, 737–746.

Goldman, M.S., Darkes, J. and Del Boca, F.K. (1999). Expectancy mediation of biopsychosocial risk for alcohol use and alcoholism. In Kirsch, I. (ed.), *How expectancies shape experience*. Washington, DC: American Psychological Association, pp. 233–262.

Goldman, M.S., Reich, R.R. and Darkes, J. (2006). Expectancy as a unifying construct in alcohol-related cognition. In R.W. Wiers and A.W. Stacy (ed.), *Handbook of implicit cognition and addiction*. Thousand Oaks, CA: Sage, pp. 105–120.

Gray, J.A. (1990). Brain systems that mediate both emotion and cognition. *Cognition and Emotion*, 4, 269–288.

Greenwald, A.G. and Banaji, M.R. (1995). Implicit social cognition: attitudes, self-esteem, and stereotypes. *Psychological Review*, 102, 4–27.

Greenwald, A.G., McGhee, D.E. and Schwartz, J.L.K. (1998). Measuring individual differences in implicit cognition: the implicit association test. *Journal of Personality and Social Psychology*, **74**, 1464–1480.

Hermans, D. and Van Gucht, D. (2006). Addiction: Integrating learning perspectives and implicit cognition. In R.W. Wiers and A.W. Stacy (ed.), *Handbook of implicit cognition and addiction*. Thousand Oaks, CA: Sage, pp. 483–487.

Hill, A.B. and Paynter, S. (1992). Alcohol dependence and semantic priming of alcohol related words. *Personality and Individual Differences*, **13**, 745–750.

Hintzman, D.L. (1984). MINERVA 2: a simulation model of human memory. *Behavior Research Methods, Instruments and Computers*, **16**, 96–101.

Hintzman, D.L. (1986). 'Schema abstraction' in a multiple-trace memory model. *Psychological Review*, **93**, 411–428.

Hintzman, D.L. (1990). Human learning and memory: connections and dissociations. *Annual Review of Psychology*, **41**, 109–139.

Hintzman, D.L. (1993). Twenty-five years of learning and memory: was the cognitive revolution a mistake? In D.E. Meyer and S. Kornblum (ed.), *Attention and performance 14: synergies in experimental psychology, artificial intelligence, and cognitive neuroscience*. Cambridge, MA: The MIT Press, pp. 359–391.

Hopfield, J.J. and Tank, D.W. (1986). Computing with neural circuits: a model. *Science*, **233**, 625–633.

Houben, K. and Wiers, R.W. (2006). Assessing implicit alcohol associations with the IAT: fact or artifact? *Addictive Behaviors*, **31**, 1346–1362.

Houben, K. and Wiers, R.W. (2006*b*). A test of the salience asymmetry interpretation of the Alcohol-IAT. *Experimental Psychology*, in press.

Houben, K., Wiers, R.W. and Roefs, A. (2006). Reaction time measures of substance-related associations. In R.W. Wiers and A.W. Stacy (ed.), *Handbook of implicit cognition and addiction*. Thousand Oaks, CA: Sage, pp. 91–104.

Huijding, J., de Jong, P.J., Wiers, R.W. and Verkooijen, K. (2005). Implicit and explicit attitudes toward smoking in a smoking and a nonsmoking setting. *Addictive Behaviors*, **30**, 949–961.

Jacoby, L.L. (1998). Invariance in automatic influences of memory: toward a user's guide for the process-dissociation procedure. *Journal of Experimental Psychology: Learning, Memory, and Cognition*, **24**, 3–26.

Jacoby, L.L., Woloshyn, V. and Kelley, C. (1989). Becoming famous without being recognized: unconscious influences of memory produced by dividing attention. *Journal of Experimental Psychology: General*, **118**, 115–125.

Jajodia, A. and Earleywine, M. (2003). Measuring alcohol expectancies with the implicit association test. *Psychology of Addictive Behaviors*, **17**, 126–133.

Johnson, M.K. and Multhaup, K.S. (1992). Emotion and MEM. In S.-A. Christianson (ed.), *The handbook of emotion and memory: research and theory*. Hillsdale, NJ: Lawrence Erlbaum Associates, pp. 33–66.

Jones, B.C., Jones, B.T., Blundell, L. and Bruce, G. (2002). Social users of alcohol and cannabis who detect substance-related changes in a change blindness paradigm report higher levels of use than those detecting substance-neutral changes. *Psychopharmacology*, **165**, 93–96.

Kahneman, D. (2003). A perspective on judgment and choice: mapping bounded rationality. *American Psychologist*, **58**, 697–720.

Kalivas, P.W. and Volkow, N.D. (2005). The neural basis of addiction: a pathology of motivation and choice. *American Journal of Psychiatry*, **162**, 1403–1413.

Kelly, A.B., Masterman, P.W. and Marlatt, G.A. (2005). Alcohol related associative strength and drinking behaviours: concurrent and prospective relationships. *Drug and Alcohol Review*, **24**, 489–498.

Knowlton, B.J., Mangels, J.A. and Squire, L.R. (1996). A neostriatal habit learning system in humans. *Science*, **273**, 1399–1402.

Krank, M.D. and Swift, R. (1994). Unconscious influences of specific memories on alcohol outcome expectancies. *Alcoholism: Clinical and Experimental Research*, **18**, 423.

Krank, M.D. and Wall, A.-M. (2006). Context and retrieval effects on implicit cognition for substance use. In R.W. Wiers and A.W. Stacy (ed.), *Handbook of implicit cognition and addiction*. Thousand Oaks, CA: Sage, pp. 281–292.

Krosnick, J.A. (1999). Survey research. *Annual Review of Psychology*, **50**, 537–567.

Levy, D.A., Stark, C.E.L. and Squire, L.R. (2004). Intact conceptual priming in the absence of declarative memory. *Psychological Science*, **15**, 680–686.

Logan, G.D. (1988). Toward an instance theory of automatization. *Psychological Review*, **95**, 492–527.

Masson, M.E.J. (1995). A distributed memory model of semantic priming. *Journal of Experimental Psychology: Learning, Memory, and Cognition*, **21**, 3–23.

McCusker, C.G. (2001). Cognitive biases and addiction: an evolution in theory and method. *Addiction*, **96**, 47–56.

McEvoy, C.L. and Nelson, D.L. (2006). Measuring, manipulating, and modeling the unconscious influences of prior experience on memory for recent experiences. In R.W. Wiers and A.W. Stacy (ed.), *Handbook of implicit cognition and addiction*. Thousand Oaks, CA: Sage, pp. 59–72.

McEvoy, C.L., Nelson, D.L. and Komatsu, T. (1999). What is the connection between true and false memories? The differential roles of interitem associations in recall and recognition. *Journal of Experimental Psychology. Learning, Memory, and Cognition*, **25**, 1177–1194.

McGue, M. (1999). Behavioral genetic models of alcoholism and drinking. In K.E. Leonard and H.T. Blane (ed.), *Psychological theories of drinking and alcoholism*, 2nd edn. New York, NY: Guilford Press, pp. 372–421.

Mitchell, J.P., Nosek, B.A. and Banaji, M.R. (2003). Contextual variations in implicit evaluation. *Journal of Experimental Psychology: General*, **132**, 455–469.

Mogg, K., Bradley, B.P., Field, M. and De Houwer, J. (2003). Eye movements to smoking-related pictures in smokers: relationship between attentional biases and implicit and explicit measures of stimulus valence. *Addiction*, **98**, 825–836.

Mogg, K., Field, M. and Bradley, B.P. (2005). Attentional and approach biases for smoking cues in smokers: an investigation of competing theoretical views of addiction. *Psychopharmacology*, **180**, 333–341.

Monti, L.A., Gabrieli, J.D.E., Reminger, S.L., Rinaldi, J.A., Wilson, R.S. and Fleischman, D.A. (1996). Differential effects of aging and Alzheimer's disease on conceptual implicit and explicit memory. *Neuropsychology*, **10**, 101–112.

Moody, T.D., Bookheimer, S.Y., Vanek, Z. and Knowlton, B.J. (2004). An implicit learning task activates medial temporal lobe in patients with Parkinson's disease. *Behavioral Neuroscience*, 118, 438–442.

Moors, A. and De Houwer, J.(2006). Automaticity: a theoretical and conceptual analysis. *Psychological Bulletin*, 132, 297–326.

Morris, C.D., Bransford, J.D., and Franks, J.J. (1977). Levels of processing versus transfer appropriate processing. *Journal of Verbal Learning and Verbal Behavior*, 16, 519–533.

Moscovitch, M. (2000). Theories of memory and consciousness. In E. Tulving (ed.), *The Oxford Handbook of Memory*. London: Oxford University Press, pp. 609–625.

Neely, J.H. (1991). Semantic priming effects in visual word recognition: a selective review of current findings and theories. In D. Besner and G.W. Humphreys (ed.), *Basic processes in reading: visual word recognition*. Hillsdale, NJ: Lawrence Erlbaum Associates, pp. 264–336.

Nelson, D.L. and McEvoy, C.L. (2000). What is this thing called frequency? *Memory and Cognition*, 28, 509–522.

Nelson, D.L., McKinney, V.M., Gee, N.R. and Janczura, G.A. (1998). Interpreting the influence of implicitly activated memories on recall and recognition. *Psychological Review*, 105, 299–324.

Nelson, D.L., McEvoy, C.L. and Pointer, L. (2003). Spreading activation or spooky action at a distance? *Journal of Experimental Psychology: Learning, Memory, and Cognition*, 29, 42–52.

Nisbett, R.E. and Wilson, T.D. (1977). Telling more than we can know: verbal reports on mental processes. *Psychological Review*, 84, 231–259.

Nosek, B.A. and Banaji, M.R. (2001). The Go/No-go Association Task. *Social Cognition*, 19, 625–666.

O'Connor, R.M. and Colder, C.R. (2005). Activation of implicit alcohol associations in a college sample: effect of alcohol contextual cue. Paper presented at the Annual Meeting of the Association for Behavioral and Cognitive Therapies, Washington, DC, USA, 2005.

Ostafin, B.D., Palfai, T.P. and Wechsler, C.E. (2003). The accessibility of motivational tendencies toward alcohol: approach, avoidance, and disinhibited drinking. *Experimental and Clinical Psychopharmacology*, 11, 294–301.

Palfai, T.P. (2006). Automatic processes in the self-regulation of addictive behaviors. In R.W. Wiers and A.W. Stacy (ed.), *Handbook of implicit cognition and addiction*. Thousand Oaks, CA: Sage, pp. 411–424.

Palfai, T.P. and Ostafin, B.D. (2003). Alcohol-related motivational tendencies in hazardous drinkers: assessing implicit response tendencies using the modified-IAT. *Behaviour Research and Therapy*, 41, 1149–1162.

Palfai, T.P. and Wood, M.D. (2001). Positive alcohol expectancies and drinking behavior: the influence of expectancy strength and memory accessibility. *Psychology of Addictive Behaviors*, 15, 60–67.

Park, S.M., Gabrieli, J.D.E., Reminger, S.L., Monti, L.A., Fleischman, D.A., Wilson, R.S., Tinklenberg, J.R. and Yesavage, J.A. (1998). Preserved priming across study-test picture transformations in patients with Alzheimer's disease. *Neuropsychology*, 12, 340–352.

Phelps, E.E. (2005). The interaction of emotion and cognition: the relation between the human amygdala and cognitive awareness. In R.R. Hassin, J.S. Uleman and J.A. Bargh (ed.), *The new unconscious*. New York, NY: Oxford University Press, pp. 61–76.

Phillips, P.E., Stuber, G.D., Heien, M.L., Wightman, R.M. and Carelli, R.M. (2003). Subsecond dopamine release promotes cocaine seeking. *Nature*, **422**, 614–618.

Plaut, D.C. and Booth, J.R. (2000). Individual and developmental differences in semantic priming: empirical and computational support for a single-mechanism account of lexical processing. *Psychological Review*, **107**, 786–823.

Plomin, R., DeFries, J.C., Craig, I.W. and McGuffin, P. (2003). Behavioral genetics. In R. Plomin, J.C. DeFries, I.W. Craig and P.McGuffin (ed.), *Behavioral genetics in the postgenomic era*. Washington, DC: American Psychological Association, pp. 3–15.

Posner, M.I. and Snyder, C.R. (1975). Attention and cognitive control. In R.L. Solso (ed.), *Information processing and cognition: the Loyola symposium*. Hillsdale, NJ: Lawrence Erlbaum Associates, pp. 55–85.

Prestwich, A., Conner, M. and Lawton, R. (2006). Implementation intentions: can they be used to prevent and treat addiction? In R.W. Wiers and A.W. Stacy (ed.), *Handbook of implicit cognition and addiction*. Thousand Oaks, CA: Sage, pp. 455–470.

Queller, S. and Smith, E.R. (2002). Subtyping versus bookkeeping in stereotype learning and change: connectionist simulations and empirical findings. *Journal of Personality and Social Psychology*, **82**, 300–313.

Rather, B.C., Goldman, M.S., Roehrich, L. and Brannick, M. (1992). Empirical modeling of an alcohol expectancy memory network using multidimensional scaling. *Journal of Abnormal Psychology*, **101**, 174–183.

Reich, R.R., Goldman, M.S. and Noll, J.A. (2004). Using the false memory paradigm to test two key elements of alcohol expectancy theory. *Experimental and Clinical Psychopharmacology*, **12**, 102–110.

Rescorla, R.A. (2003). More rapid associative change with retraining than with initial training. *Journal of Experimental Psychology: Animal Behavior Processes*, **29**, 251–260.

Richardson-Klavehn, A. and Gardiner, J.M. (1996). Cross-modality priming in stem completion reflects conscious memory, but not voluntary memory. *Psychonomic Bulletin and Review*, **3**, 7.

Robinson, T.E. and Berridge, K.C. (1993). The neural basis of drug craving: an incentive-sensitization theory of addiction. *Brain Research Reviews*, **18**, 247–291.

Robinson, T.E. and Berridge, K.C. (2003). Addiction. *Annual Review of Psychology*, **54**, 25–53.

Roediger, H.L. (1990). Implicit memory: retention without remembering. *American Psychologist*, **45**, 1043–1056.

Roediger, H.L. and McDermott, K.B. (1995). Creating false memories: remembering words not presented in lists. *Journal of Experimental Psychology: Learning, Memory, and Cognition*, **21**, 803–814.

Roediger, H.L., Gallo, D.A. and Geraci, L. (2002). Processing approaches to cognition: the impetus from the levels-of-processing framework. *Memory*, **10**, 319–332.

Roehrich, L. and Goldman, M.S. (1995). Implicit priming of alcohol expectancy memory processes and subsequent drinking behavior. *Experimental and Clinical Psychopharmacology*, **3**, 402–410.

Rothermund, K. and Wentura, D. (2004). Underlying processes in the Implicit Association Test: dissociating salience from associations. *Journal of Experimental Psychology: General*, **133**, 139–165.

Rotter, J.B. (1954). *Social learning and clinical psychology*. Oxford, Prentice-Hall.

Rumelhart, D.E., Hinton, G.E. and Williams, R.J. (1986). Learning representations by back-propagating errors. *Nature*, 323, 533–536.

Sayette, M.A. (1993). An appraisal-disruption model of alcohol's effects on stress responses in social drinkers. *Psychological Bulletin*, 114, 459–476.

Sayette, M.A., Martin, C.S., Perrott, M.A. and Wertz, J.M. (2001). Parental alcoholism and the effects of alcohol on mediated semantic priming. *Experimental and Clinical Psychopharmacology*, 9, 409–417.

Schacter, D.L. and Badgaiyan, R.D. (2001). Neuroimaging of priming: new perspectives on implicit and explicit memory. *Current Directions in Psychological Science*, 10, 4.

Schott, B.H., Henson, R.N., Richardson-Klavehn, A., Becker, C., Thoma, V., Heinze, H.-J. and Düzel, E. (2005). Redefining implicit and explicit memory: the functional neuroanatomy of priming, remembering, and control of retrieval. *Proceedings of the National Academy of Sciences of the USA*, 102, 1257–1262.

Schuckit, M.A. (1998). Biological, psychological and environmental predictors of the alcoholism risk: a longitudinal study. *Journal of Studies on Alcohol*, 59, 485–494.

Sherman, S.J., Rose, J.S., Koch, K., Presson, C.C. and Chassin, L. (2003). Implicit and explicit attitudes toward cigarette smoking: the effects of context and motivation. *Journal of Social and Clinical Psychology*, 22, 13–39.

Smith, E.R. and DeCoster, J. (1998). Knowledge acquisition, accessibility, and use in person perception and stereotyping: simulation with a recurrent connectionist network. *Journal of Personality and Social Psychology*, 74, 21–35.

Smith, E.R. and Zarate, M.A. (1992). Exemplar-based model of social judgment. *Psychological Review*, 99, 3–21.

Spaniol, J. and Bayen, U.J. (2005). Aging and conditional probability judgments: a global matching approach. *Psychology and Aging*, 20, 165–181.

Squire, L.R. (2004). Memory systems of the brain: a brief history and current perspective. *Neurobiology of Learning and Memory*, 82, 171–177.

Stacy, A.W. (1994). Evidence of implicit memory activation of alcohol concepts: toward parallel memory models of addiction. Paper presented at the Annual Meeting of the Research Society on Alcoholism, Maui HI, USA, 1994.

Stacy, A.W. (1997). Memory activation and expectancy as prospective predictors of alcohol and marijuana use. *Journal of Abnormal Psychology*, 106, 61–73.

Stacy, A.W. and Ames, S.L. (2000). Implict cogition theory in drug use and driving-under-the-influence interventions. In S. Sussman (ed.), *Handbook of program development for health behavior research and practice*. Thousand Oaks, CA: Sage, pp. 107–130.

Stacy, A.W., Widaman, K.F. and Marlatt, G.A. (1990). Expectancy models of alcohol use. *Journal of Personality and Social Psychology*, 58, 918–928.

Stacy, A.W., Ames, S.L. and Knowlton, B.J. (2004a). Neurologically plausible distinctions in cognition relevant to drug use etiology and prevention. *Substance Use and Misuse*, 39, 1571–1623.

Stacy, A.W., Ames, S.L. and Leigh, B.C. (2004b). An implicit cognition assessment aproach to relapse, secondary prevention, and media effects. *Cognitive and Behavioral Practice*, 11, 139–149.

Stacy, A.W., Ames, S.L. and Grenard, J.(2006a). Word association tests of associative memory and implicit processes: theoretical and assessment issues. In R.W. Wiers and

A. W. Stacy (ed.), *Handbook on implicit cognition and addiction*. Thousand Oaks, CA: Sage, pp. 75–90.

Stacy, A.W., Ames, S.L., Ullman, J.B., Zogg, J.B. and Leigh, B.C. (2006b). Spontaneous cognition and HIV risk behavior. *Psychology of Addictive Behaviors*, 20, 196–206.

Stewart, S.H., Hall, E., Wilkie, H. and Birch, C. (2002). Affective priming of alcohol schema in coping and enhancement motivated drinkers. *Cognitive Behaviour Therapy*, 31, 68–80.

Strack, F. and Deutsch, R. (2004). Reflective and impulsive determinants of social behavior. *Personality and Social Psychology Review*, 8, 220–247.

Swanson, J.E., Rudman, L.A. and Greenwald, A.G. (2001). Using the Implicit Association Test to investigate attitude–behaviour consistency for stigmatised behaviour. *Cognition and Emotion*, 15, 207–230.

Szalay, L.B., Strohl, J.B. and Doherty, K.T. (1999). *Psychoenvironmental forces in substance abuse prevention*. Dordrecht, The Netherlands: Kluwer Academic Publishers.

Teachman, B.A. and Woody, S.R. (2003). Automatic processing in spider phobia: implicit fear associations over the course of treatment. *Journal of Abnormal Psychology*, 112, 100–109.

Thush, C. and Wiers, R.W. (2006). Explicit and implicit alcohol-related cognitions and the prediction of current and future drinking in adolescents. Submitted.

Tiffany, S.T. (1990). A cognitive model of drug urges and drug-use behavior: role of automatic and nonautomatic processes. *Psychological Review*, 97, 147–168.

Tobias, B.A., Kihlstrom, J.F. and Schacter, D.L. (1992). Emotion and implicit memory. In S.-A. Christianson (ed.), *The handbook of emotion and memory: research and theory*. Hillsdale, NJ: Lawrence Erlbaum Associates, pp. 67–92.

Tolman, E.C. (1932). *Purposive behavior in animals and men*. Oxford: Appleton-Century.

Tulving, E. (1983). *Elements of episodic memory*. New York, NY: Oxford University Press.

Vaidya, C.J., Gabrieli, J.D.E., Monti, L.A., Tinklenberg, J.R. and Yesavage, J.A. (1999). Dissociation between two forms of conceptual priming in Alzheimer's disease. *Neuropsychology*, 13, 516–524.

Volkow, N.D. and Fowler, J.S. (2000). Addiction, a disease of compulsion and drive: involvement of the orbitofrontal cortex. *Cerebral Cortex*, 10, 318–325.

Warren, H.C. (1921). *A history of the association psychology*. New York, NY: Scribner.

Waters, A.J. and Sayette, M.A. (2006). Implicit cognition and tobacco addiction. In R.W.Wiers and A.W. Stacy (ed.), *Handbook of implicit cognition and addiction*. Thousand Oaks, CA: Sage, pp. 309–338.

Waters, A.J., Shiffman, S., Bradley, B.P. and Mogg, K. (2003). Attentional shifts to smoking cues in smokers. *Addiction*, 98, 1409–1417.

Weingardt, K.R., Stacy, A.W. and Leigh, B.C. (1996). Automatic activation of alcohol concepts in response to positive outcomes of alcohol use. *Alcoholism: Clinical and Experimental Research*, 20, 25–30.

White, N.M. (1996). Addictive drugs as reinforcers: multiple partial actions on memory systems. *Addiction*, 91, 921–949.

Wiers, R.W. and Stacy, A.W. (ed.) (2006). *Handbook of implicit cognition and addiction*. Thousand Oaks, CA: Sage.

Wiers, R.W., Stacy, A.W., Ames, S.L., Noll, J.A., Sayette, M.A., Zack, M. and Krank, M. (2002a). Implicit and explicit alcohol-related cognitions. *Alcoholism: Clinical and Experimental Research*, 26, 129–137.

Wiers, R.W., Van Woerden, N., Smulders, F.T.Y. and De Jong, P.J. (2002*b*). Implicit and explicit alcohol-related cognitions in heavy and light drinkers. *Journal of Abnormal Psychology*, 111, 648–658.

Wiers, R.W., de Jong, P.J., Havermans, R. and Jelicic, M. (2004). How to change implicit drug use-related cognitions in prevention: a transdisciplinary integration of findings from experimental psychopathology, social cognition, memory, and experimental learning psychology. *Substance Use and Misuse*, 39, 1625–1684.

Wiers, R.W., Van de Luitgaarden, J., Van den Wildenberg, E. and Smulders, F.T. (2005). Challenging implicit and explicit alcohol-related cognitions in young heavy drinkers. *Addiction*, 100, 806–819.

Wiers, R.W., Houben, K., Smulders, F.T.Y., Conrod, P.J. and Jones, B. (2006*a*). To drink or not to drink: the role of automatic and controlled cognitive processes in the etiology of alcohol-related problems. In R.W. Wiers and A.W. Stacy (ed.), *Handbook of implicit cognition and addiction*. Thousand Oaks, CA: Sage, pp. 339–362.

Wiers, R.W., Cox, W.M., Field, M., Fadardi, J.S., Palfai, T.P., Tim, S. *et al.* (2006*b*). The search for new ways to change implicit alcohol-related cognitions in heavy drinkers. *Alcoholism, Clinical and Experimental Research*, in press.

Yin, H.H. and Knowlton, B.J. (2006). Addiction and learning in the brain. In R.W. Wiers and A.W. Stacy (ed.), *Handbook of implicit cognition and addiction*. Thousand Oaks, CA: Sage, pp. 167–184.

Zack, M., Toneatto, T. and MacLeod, C.M. (1999). Implicit activation of alcohol concepts by negative affective cues distinguishes between problem drinkers with high and low psychiatric distress. *Journal of Abnormal Psychology*, 108, 518–531.

Zack, M., Poulos, C.X., Fragopoulos, F. and MacLeod, C.M. (2003). Effects of negative and positive mood phrases on priming of alcohol words in young drinkers with high and low anxiety sensitivity. *Experimental and Clinical Psychopharmacology*, 11, 176–185.

3

Attentional biases in drug abuse and addiction: cognitive mechanisms, causes, consequences and implications

Matt Field

Summary

Drug abuse and addiction are associated with biases in selective attention for drug-associated stimuli. This chapter reviews this literature and discusses it within existing theoretical frameworks. Although the existence of attentional biases is well documented, a variety of different paradigms (that may tap different mechanisms) have been used, leaving the cognitive and attentional processes involved in attentional biases poorly understood and in need of clarification. Consistent with some theoretical predictions, the evidence suggests that attentional biases operate in early stages of attentional processing and thus they may be 'automatic'. Attentional biases are closely associated with subjective drug craving, and recent research suggests that this relationship may be bidirectional in nature: elevated drug craving may make drug-related cues more salient, but pronounced attentional biases may promote further increases in craving. Theoretical predictions that attentional biases are ultimately caused by classical conditioning mechanisms, and the relationships between attentional biases and drug-use behaviours at different stages of addiction, are also discussed.

3.1 Introduction

In drug users, drug-associated environmental stimuli exert powerful influences on affect, cognition and behaviour. For example, when tobacco smokers are presented with smoking-related cues, such as a lit cigarette, alterations in mood, physiology and tobacco-seeking behaviours are typically observed (for a review, see Carter and Tiffany 1999). In recent years, the focus of research has shifted to the study of cognitive responses to drug-related cues. In this chapter, I will review evidence relating to cognitive processing biases for drug cues, specifically biases in selective attention.

This evidence will be discussed in the context of recent theories of motivation and addiction, which suggest that biases in selective attention reflect an enhanced incentive-motivational state that could mediate drug-seeking behaviour.

3.2 Theoretical background

Historically, theories of addiction have emphasized either the positive incentive (e.g. Stewart *et al.* 1984) or the negative reinforcing (e.g. Wikler 1948) properties of drugs of abuse, with more recent theories stressing the importance of cognitive factors (e.g. Ryan 2002*a*; Cox and Klinger 2004). Contemporary theoretical models of addiction predict the presence of attentional biases in addiction, and these will be briefly reviewed here.

Many theorists argue that addiction is maintained by the positive incentive properties of drugs and drug-related cues. For example, Stewart *et al.* (1984) proposed that, through a classical conditioning process, drug-related cues become associated with the positive reinforcing properties of drugs and, as a consequence, those cues acquire conditioned incentive properties, thereby drawing the individual towards the drug, resulting in drug taking. According to Robinson and Berridge (1993, 2000), stimuli associated with drug taking acquire motivational salience and, as a result, become highly attractive, wanted and they 'grab' attention. According to those authors, the incentive-salience mechanism that underlies this process is mediated by dopamine neurotransmission in the mesolimbic dopamine system, which plays a key role in maintaining drug-seeking behaviour. Thus, the extent to which drug cues capture attention should provide a direct index of the extent to which this incentive-salience mechanism is being activated by such cues. Moreover, according to this model, the attentional bias for drug cues is mediated primarily by automatic processes, which operate in early aspects of stimulus processing and do not depend on intentional, controlled strategies.

Negative reinforcement models of addiction suggest that compulsive drug seeking is maintained because drugs alleviate the negative mood states which occur during drug withdrawal. According to Baker and co-workers (Baker *et al.* 2004*a,b*), the strong links between negative affect and the motivation to smoke may occur because negative affect 'inflate[s] the incentive value of smoking cues' (Baker *et al.* 2004*a*, p. 483). Baker *et al.*'s (2004*b*) model predicts that, when negative affect increases (e.g. as a consequence of drug withdrawal or environmental stressors), drug-related cues become more salient. Thus, this model implies that attentional biases for drug cues may be enhanced by high negative affect, given that attention seems more likely to be allocated preferentially to information that is highly salient for the individual.

Other theories of drug dependence suggest that an attentional bias for drug cues arises primarily from effortful and resource-demanding processes. For example, Tiffany (1990) proposed that drug use is largely controlled by automatized action schemata. Thus, drug-taking behaviours tend to be relatively fast and occur automatically, often without awareness. However, if drug-taking behaviours are impeded, then more effortful, resource-demanding processes are recruited to overcome the obstacle. According to Tiffany's model, the latter, non-automatic processes underlie drug urges and elicit an accompanying attentional bias, which directs processing resources away from ongoing activities towards the goal of drug consumption.

Cox and Klinger's (2004) motivational model of alcohol use (see Chapter 4) suggests that alcohol abusers who wish to discontinue drinking will have a greater chance of success if they are able to engage in non alcohol-related activities and concerns, rather than alcohol-related concerns. Therefore, the degree to which addicts get easily distracted by (or have an attentional bias for) stimuli that relate to drug use, rather than stimuli that are related to other concerns, should be directly associated with the risk of relapse. McCusker (2001) and Sayette et al. (2000) propose that attentional biases provide a cognitive measure of an individual's motivation to use drugs (see Chapter 5). Ryan (2002a) and Franken (2003) developed this argument by suggesting that attentional bias plays an important role as a mediating link between the perception of drug cues, and drug-seeking behaviours in response to those stimuli. Specifically, like Robinson and Berridge (1993), Franken argues that presentation of drug cues increases the release of dopamine in the nucleus accumbens and associated structures in the mesolimbic dopamine system. This results in 'motor preparation and a hyperattentive state towards drug-related stimuli that, ultimately, promotes further craving and relapse' (p. 563). In fact, many of the theories discussed above predict that attentional biases for drug-related cues will be related to indices of drug-use motivation, such as drug craving, and this prediction will be addressed in this chapter.

Finally, it is important to note that cognitive processing biases are likely to be a feature of all motivational states, and they should not be limited to pathological motivational states such as those seen in addiction. For example, according to the 'elaborated intrusion' theory of desire (Kavanagh et al. 2005), drive states (e.g. hunger, cigarette craving) can be initially triggered by internal stimuli (such as sensations that the stomach is empty or the effects of physiological withdrawal from nicotine), or external stimuli (such as the sight of a pizza or the smell of a lit cigarette). However, once elicited, these subjective desires are then 'elaborated' as individuals tend to ruminate on their desire for the drug (or food), and focus their attention on drug-related (or food-related) cues in the environment. This cognitive elaboration then increases the strength of the desire, which increases elaboration on drug- or food-related cues, and so on, until ultimately the substance (e.g. food, tobacco) is sought out and consumed. Studies which demonstrate that hungry individuals have attentional biases for food-related environmental cues (e.g. Mogg *et al.* 1998) are entirely consistent with this model, and they suggest that the literature demonstrating attentional biases in addiction, which is reviewed below, should be considered in this wider context of general motivational states.

In summary, a number of theories of addiction predict the presence of an 'attentional bias' for drug-related cues in addiction, and they also predict that the attentional bias is 'automatic', i.e. it occurs at an early stage of cognitive processing, possibly before the drug user is aware of the stimuli. A common theme in these theories is the suggestion that subjective drug craving should be correlated with measures of attentional bias. In the past 10–15 years a considerable body of evidence has accumulated which enables us to evaluate these predictions, and this literature will be discussed in the following sections.

3.3 Evidence for attentional biases in addiction
3.3.1 The modified Stroop task

The modified Stroop task has been widely used to demonstrate cognitive biases in mood disorders. Stroop tasks involve presenting words to participants who have to name the colour of the ink that the word is written in whilst ignoring the semantic content of the word. Numerous studies have demonstrated that participants are slower to colour-name individual words or word lists if those words are semantically related to their current concerns, so, for example, anxious patients are slower to name the colours of threatening words (e.g. 'cancer', 'collapse') than matched neutral words, but this difference is not seen in non-anxious controls (for a review, see Mogg and Bradley 1998). Gross *et al.* (1993) obtained colour-naming times for lists of smoking-related (e.g. 'cigarette') and neutral (e.g. 'tablespoon') words from two groups of smokers: one group who were instructed to abstain from smoking for 12 h before completing the task, and another group who were instructed to smoke normally beforehand. They demonstrated that the abstinent smokers were slower to name the colours of the smoking-related words than the control words, but the reverse pattern was seen in the non-abstinent smokers.

Subsequently there have been numerous further reports of slower colour-naming times for smoking-related than neutral control words in adult smokers (Rusted *et al.* 2000; Waters and Feyerabend 2000; Wertz and Sayette 2001; Mogg and Bradley 2002; Munafò *et al.* 2003; Waters *et al.* 2003*a*; although see Johnsen *et al.* 1997, for a failure to replicate these findings), and one report of the same phenomenon in adolescent smokers (Zack *et al.* 2001). Only one of these studies compared smokers with a non-smoker control group, and these results demonstrated that interference effects were specific to smokers, as they were not seen in non-smokers (Munafò *et al.* 2003). The other studies did not include a suitable control group, which complicates interpretation of the results from those studies (for a discussion, see Robbins and Ehrman 2004).

Colour-naming interference for drug and alcohol-related words has also been investigated in other addictions. Copersino *et al.* (2004) demonstrated that cocaine-dependent patients were significantly slower to colour-name cocaine-related words, relative to non-addict controls, although this attentional bias was also absent in participants with co-morbid cocaine dependence and schizophrenia, which may be attributable to general cognitive deficits in the schizophrenic group. Hester *et al.* (2006) also demonstrated that active cocaine users were slower to name the colour of cocaine-related word and pictorial stimuli, but these biases were absent in non-users. Franken *et al.* (2000*a*) presented heroin-related and matched control words to heroin addicts and controls, and found slower colour-naming times for the heroin-related compared with the matched control words in the addict group only. In a subsequent study, Franken *et al.* (2004) demonstrated that the attentional bias in heroin addicts was eliminated after administration of haloperidol, a dopamine antagonist. Field (2005) demonstrated that cannabis users who met criteria for cannabis dependence (based on their responses on a self-report questionnaire) had a significant attentional bias for cannabis-related words, but non-dependent cannabis users did not have this attentional bias, although a control group of non-users of cannabis was not included in that study.

Johnsen *et al.* (1994) reported that inpatient alcoholics but not non-alcoholic controls were significantly slower to colour-name alcohol-related words than matched control words. Stetter *et al.* (1995), Stormark *et al.* (2000), Sharma *et al.* (2001) and Lusher *et al.* (2004) subsequently replicated these findings. Numerous other studies have demonstrated significant attentional biases for alcohol-related words in alcoholics, although interpretation of some of these results is complicated due to the failure to find differences in attentional bias between alcoholic and non-alcoholic controls (see Stetter *et al.* 1994; Bauer and Cox 1998; Duka *et al.* 2002; Ryan 2002*b*). Cox *et al.* (2002) demonstrated that Stroop interference effects were specific to alcoholics who were subsequently unable to maintain abstinence after treatment for alcoholism, but were not seen in alcoholics who did manage to maintain abstinence, or in non-alcoholic control participants. These interesting results will be discussed in more detail in Section 3.5.

Attentional bias for alcohol words has also been demonstrated in non-dependent drinkers, although biases appear to be linked to the level of alcohol consumption. For example, Sharma *et al.* (2001) demonstrated that heavy, but not light, non-dependent drinkers were slower to colour-name alcohol-related than matched neutral words. Bruce and Jones (2004) demonstrated comparable interference effects on colour naming when alcohol-related and matched control pictures, rather than words, were used, in heavy compared with light social drinkers. Cox *et al.* (1999) did not report all appropriate statistical comparisons, but it would appear from the figure in that paper that only heavy, but not light social drinkers who had been exposed to an alcohol cue rather than a neutral cue exhibited a significant attentional bias for alcohol-related words. These results were subsequently replicated using an alternative statistical methodology by Cox *et al.* (2003).

There is some argument as to whether the modified Stroop task actually measures attentional bias. Colour-naming interference may occur because processing resources (i.e. selective attention) are allocated to the semantic content of the word rather than its colour; limitations in parallel processing capacity mean that colour-naming times are consequently slowed (Williams *et al.* 1996). However, other explanations of the Stroop effect centre on the fact that the interference may be occurring at the response selection stage rather than at the attentional input stage (see Mogg and Bradley 1998). For example, smokers may see the word 'cigarette' which may readily activate the 'smoking schema' in these participants and they may be inclined to say the word 'cigarette' instead of the correct response (i.e. to name the colour of the word); this would generate conflict at the response selection stage and should either increase response latencies (while the correct response is selected) or increase the number of errors (if the wrong response is selected). Alternatively, with regard to the emotional Stroop, anxiety-relevant words may induce a temporary state of anxiety, particularly in anxious patients, a state which could lead to a general slowdown in all other activities, including colour naming (for empirical demonstrations, see Algom *et al.* 2004). In the same way, drug-relevant words may elicit drug-relevant thoughts (e.g. craving for the drug), which could interfere with all other cognitive processing, including colour naming.

A further issue is that the modified Stroop task may be sensitive to an individual's current concerns and level of expertise with the stimuli used, rather than any underlying pathology (for a discussion, see Cox *et al.* 2006), and the relevance of this to an understanding of Stroop effects in addiction is demonstrated by a study by Ryan (2002*b*). He reported that both alcoholics, and treatment workers at a clinic for the treatment of alcoholism, had a significant attentional bias for alcohol-related words, which raises concerns that Stroop effects in addiction research may reflect the fact that addicts are simply more experienced with drug-related words. Thus, the modified Stroop task provides evidence that is suggestive of an attentional bias for drug-relevant words in addicts, but problems with interpretation of results from this task make it desirable to use alternative methods to assess the deployment of attention.

3.3.2 Dual-task procedures

The rationale behind the use of dual-task procedures is that the response latency to react to a (usually auditory) probe stimulus is dependent on the available attentional resources; therefore, if attentional resources are involved in ongoing processing of another stimulus, response latencies to the probe will be slowed. Sayette and Hufford (1994) reported two experiments in which smokers responded to an auditory probe under two conditions: whilst holding a lit cigarette and whilst holding a roll of tape. In both experiments, the participants were slower to respond to the probe when holding the cigarette than when holding the tape, which suggests that attentional resources were diverted to the former stimulus more so than to the latter. Furthermore, there was a trend for this effect to be larger in nicotine-deprived rather than non-deprived participants, and the magnitude of the effect was correlated with variations in the level of subjective craving. Baxter and Hinson (2001) replicated this finding in groups of experienced and inexperienced smokers, although the mediating role of nicotine deprivation was not investigated in that study. Juliano and Brandon (1998) reported a similar finding, although this was restricted to a subgroup of smokers who believed that they would be unable to smoke after the session, which may implicate frustration at being unable to obtain the drug as being partially responsible for the 'attentional bias' seen when the dual task paradigm is used. Sayette *et al.* (1994) measured reaction times to auditory probes in male alcoholics while they held alcoholic and non-alcoholic beverages. As predicted, response times were slower during the alcohol beverage condition. Finally, a recent report by Waters and Green (2004) used a novel dual-task procedure to demonstrate that abstinent alcoholics, but not controls, were significantly impaired on a cognitive task when alcohol-related words were presented in the periphery of a visual display, compared with when neutral words were presented in the periphery.

Data from experiments that have used dual-task procedures complement the data from modified Stroop tasks as the dual-task data suggest that visual, tactile and olfactory stimuli associated with drug use can command attentional resources, just as drug-associated words can in the modified Stroop. However, like the Stroop task, dual-task procedures suffer from problems when attempting to interpret the results. In particular, slowed reaction times to an auditory probe while in the presence of a drug-relevant stimulus may arise because the presence of the drug-relevant stimulus causes the addict to experience drug craving, and it could be this resultant drug craving, not the allocation of selective attention to the drug stimulus *per se*, that causes the reduction in available attentional capacity. Indeed, the majority of experiments reported above were designed with the specific aim of testing a specific prediction made by Tiffany's (1990) theory of addiction: that drug craving is a conscious, non-automatic process that uses attentional resources. These studies sought to evoke craving in addicts, by exposing them to drug-related stimuli, and then measured the degree to which attentional capacity was compromised by craving. The evidence discussed above is entirely consistent with Tiffany's formulation that drug craving utilizes attentional resources, but is also consistent with the notion that presentation of drug-relevant stimuli attracts attentional resources. Hence this paradigm is unable to differentiate between attention being directed at drug-relevant stimuli, or the attention-consuming craving that results when drug-relevant stimuli are presented.

3.3.3 Visual probe tasks/attentional cueing tasks

Unlike the modified Stroop and dual-task procedures, described above, the visual probe task and its variants are fairly direct measurements of the allocation of visuo-spatial attention. In a typical visual probe task, two stimuli are presented side by side on a computer screen. On critical trials, one stimulus may be a drug-related picture and the other may be a picture with similar perceptual characteristics, but lacking any drug-related content. Immediately after the pictures disappear, a small probe stimulus appears in the location of one of the pictures, and participants are required to press a key as quickly as possible in response to the probe. The rationale behind the task is that people respond faster to stimuli that appear in an attended, rather than unattended, region of a visual display (Posner *et al.* 1980). Thus one can infer the deployment of attention to the different pictures from the reaction times to probes, with faster reaction times to probes replacing drug-related stimuli, indicating an attentional bias for those stimuli. In the anxiety literature, numerous studies have used this task to demonstrate that highly anxious subjects have an attentional bias for threatening stimuli, but subjects with low levels of anxiety do not (for a review, see Mogg and Bradley 1998).

In the past few years, the visual probe task has been used to demonstrate an attentional bias for smoking-related pictures in smokers (Mogg and Bradley 2002, Waters *et al.* 2003*b*; Field *et al.* 2004*a*, 2005*a*; Mogg *et al.* 2005) which is specific to smokers and is not seen in non-smokers (Ehrman *et al.* 2002; Bradley *et al.* 2003, 2004; Hogarth *et al.* 2003*a*; Mogg *et al.* 2003). Attentional bias for heroin-related pictures has also been demonstrated in opiate addicts but not non-addict controls, using this task (Lubman *et al.* 2000), and a recent report demonstrated that individuals who consumed high amounts of caffeinated drinks had significant attentional biases for caffeine-related words, but these biases were absent in moderate or non-consumers of caffeine (Yeomans *et al.* 2005).

Like the modified Stroop, the visual probe task has also been used to demonstrate attentional biases in recreational, non-dependent drug users. Townshend and Duka (2001) and Field *et al.* (2004*b*) demonstrated that heavy social drinkers had a significantly larger attentional bias for alcohol-related pictures than light social drinkers, and Field *et al.* (2004*c*) demonstrated that recreational cannabis users with high levels of cannabis craving had an attentional bias for cannabis-related words. Similar results were reported in a sample of social drinkers by Field *et al.* (2005*b*), who demonstrated potentiated attentional biases in social drinkers who reported a high level of alcohol craving before completing the task, compared with social drinkers who had low levels of alcohol craving. Finally, one study demonstrated that, in social drinkers, the visual probe task index of attentional bias for alcohol cues was potentiated by a priming dose of alcohol (Duka and Townshend 2004).

Stormark *et al.* (1997) and Franken *et al.* (2000*b*) used conceptually similar 'attentional cueing tasks' to assess attentional biases in alcoholics and cocaine addicts, respectively. In the tasks used by these authors, drug-related and matched control words were presented individually (i.e. not in pairs) on the left or right of a computer screen. After offset of the words, a probe stimulus appeared either in the same hemifield in which the word had been presented, or in the opposite hemifield, and latencies to respond to the probes were recorded. Stormark *et al.* (1997) reported that alcoholics, but not non-alcoholic controls, were significantly slower to respond to probes that appeared in the opposite hemifield to that cued by the alcohol word rather than to probes that appeared in the same hemifield as that cued by the alcohol word, which is consistent with an attentional bias for alcohol words. In the Franken *et al.* (2000*b*) study, there were no within-subject differences in reaction times to probes that appeared in the opposite hemifield to that cued by the cocaine-related word. However, it was reported that craving for cocaine, obsessive thoughts about cocaine and feelings of reduced control over cocaine use were correlated with reaction times to probes that appeared in the opposite hemifield to that cued by the cocaine-related words (reaction times to probes that appeared in the *same* hemifield as the drug-related word were not reported by Franken *et al.* 2000*b*). Therefore, this study is suggestive of a relationship between attentional bias for cocaine-related words, and psychological aspects of cocaine addiction, including craving.

3.3.4 Eye movement monitoring

Mogg *et al.* (2003) measured the eye movements of smokers and non-smokers whilst they completed a visual probe task with smoking-related and matched control pictures (presented for 2000 ms). Results revealed that smokers' initial fixations were of significantly longer duration when they were directed at the smoking compared with the control pictures, with no difference in the non-smokers, which suggests increased maintenance of attention on smoking-related pictures. Furthermore, smokers, but not non-smokers, directed a higher proportion of their initial fixations to the smoking compared with the control pictures, which was interpreted as increased initial orienting to the smoking pictures. The finding that smokers tend to maintain their gaze for longer on smoking-related pictures has been replicated in several subsequent studies (Field *et al.* 2004*a*, 2005*a*; Mogg *et al.* 2005), although the finding of a bias to direct the initial fixation towards smoking pictures was not replicated in all of those studies. A recent (currently unpublished) study from my laboratory demonstrated comparable results in regular cannabis users, who maintained their gaze on cannabis-related pictures for longer than on matched control images, whereas non-users of cannabis spent a comparable amount of time directing their gaze at cannabis-related and control images.

3.3.5 Other paradigms

Jones *et al.* (2002, 2003) used the 'change blindness' or 'flicker' paradigm to explore attentional biases in social users of alcohol and cannabis (see Chapter 7). During the task, two near-identical photographs, which contained both drug-related and neutral objects, were repeatedly presented in quick succession. The photographs differed with regard to two small elements: one drug-related and one unrelated to drug use. So, for example, one of the photographs contained both alcohol (e.g. a bottle of wine, a glass of beer) and non-alcohol (e.g. a mug, a videotape) elements and some of these elements were subtly different in the two different photographs (e.g. the wine bottle and the videotape were rotated by 90°). Participants watched the computer display 'flicker' as the two photographs were sequentially presented until they could identify one of the differences between the two photographs. Results indicated that social users of alcohol with high levels of alcohol consumption were more likely to detect the alcohol-related change than the non alcohol-related change, relative to social users with low levels of consumption. A comparable relationship between consumption and the ability to detect drug-related changes was also seen in recreational cannabis users. Furthermore, heavy users of alcohol and cannabis were quicker to detect alcohol- and cannabis-related changes, respectively, compared with light users of alcohol and cannabis. Therefore, these results would appear consistent with the notion that social users of a drug who report high levels of consumption are more likely to notice a drug-related feature of the environment than a feature of the environment that is not drug related, which can be interpreted as an attentional bias.

3.4 Are attentional biases 'automatic'?

According to several theorists (e.g. Robinson and Berridge 1993), attentional biases for drug-related cues operate automatically and at very early stages of cognitive processing, such that addicts may shift their attention to drug-related cues before those cues have crossed the threshold of conscious awareness. As reviewed elsewhere (see Field *et al.* 2006), several studies have used subliminal presentation paradigms to probe for biased attentional processing of drug cues that were presented below the threshold of conscious awareness (Franken *et al.* 2000*a*; Mogg and Bradley 2002; Munafò *et al.* 2003; Bradley *et al.* 2004). In each of those studies, there was no evidence for any significant attentional bias for drug-related cues when those cues were presented below the threshold of conscious awareness, although all of those studies reported attentional bias for drug-related stimuli that were presented under supraliminal conditions (i.e. above the threshold of conscious awareness). Therefore, the available evidence suggests that attentional biases for drug-related cues do not occur before those stimuli have crossed the threshold of conscious awareness. However, a recent report by Ingjaldsonn *et al.* (2003) suggests that subliminally presented alcohol-related stimuli can provoke subjective and physiological reactions in alcoholics, although this is clearly separable from biases in selective attention.

Other evidence suggests that attentional biases may be under the control of 'automatic' cognitive processes, although it is important to note that this can be distinguished from processes that operate in the absence of conscious awareness. Current models of visuo-spatial attention emphasize the distinctions between the orienting versus the maintenance of attention (e.g. Allport 1989;

LaBerge 1995). The orienting (or shifting) of attention can occur fairly rapidly and automatically, and it may function to programme shifts in eye movements (Jonides 1981; Kowler 1995). This can be contrasted with the maintenance of attention, which according to Laberge (1995) is more likely to be influenced by strategic cognitive processes and motivational states. In the context of visual probe and attentional cueing tasks (described above), drug-related pictures and words can be presented very briefly (e.g. 200 ms or less in some studies) or for much longer durations (e.g. 2000 ms in some studies). If drug-related stimuli are presented for 200 ms or less, then any observed attentional biases are likely to represent biases in the orienting of attention, and are thus likely to be under the control of rapid, automatic processes (Egeth and Yantis 1997). In dependent populations, there is evidence of attentional biases for drug-related stimuli that are presented for durations of 200 ms or shorter. Using an attentional cueing task, Stormark *et al.* (1997) found that alcoholics, but not non-alcoholic controls, had a significant attentional bias for alcohol-related words presented for 100 ms. Bradley *et al.* (2004) used a conventional visual probe task and demonstrated that regular smokers, but not non-smokers, were faster to respond to probes that replaced smoking-related rather than control pictures, and this attentional bias was not significantly different when picture pairs were presented for 200 and 2000 ms. However, Field *et al.* (2004c) found no significant difference between heavy and light social drinkers in attentional bias for alcohol-related pictures presented for 200 ms, even though the heavy drinkers had significantly larger attentional biases than the light drinkers when the pictures were presented for longer durations (500 and 2000 ms). Also, Franken *et al.* (2000b) used an attentional cueing task and found no absolute evidence of attentional bias for cocaine-related words presented for 100 ms in cocaine addicts, although this attentional bias measure was correlated with subjective craving in this group. Therefore, the available evidence suggests that attentional biases *can* be fairly rapid and 'automatic', although the variety of paradigms that have been employed (e.g. attentional cueing, visual probe) and populations that have been studied (e.g. heavy drinkers versus alcoholics) may have contributed to the lack of consistency across studies.

3.5 Distal and proximal causes of attentional biases in addiction

3.5.1 Attentional biases may result from a classical conditioning process

Classical conditioning processes may explain the development of attentional biases for drug-related stimuli. Drug self-administration should support classical conditioning because drugs function as unconditioned stimuli (US) which produce unconditioned responses (UR); therefore, they should become associated with environmental stimuli that are contiguous and contingent with these effects as conditioned stimuli (CS), which would in turn come to elicit conditioned responses (CR) in the addict, such as physiological changes and a desire for the drug. One consequence of classical conditioning is an orienting of attention to conditioned stimuli after they come to predict unconditioned stimuli (Bindra 1974; Mackintosh 1975). Therefore, as drug-related cues are paired with the effects of abused drugs in the brain, those cues should 'grab the attention' (see Robinson and Berridge 1993, 2000).

While the numerous demonstrations of attentional bias for naturalistic drug-related stimuli that have been described in this review are consistent with the involvement of a classical conditioning process (because drug-related cues are, by nature, associated with the effects of drugs), it is impossible to state for certain that a classical conditioning process was responsible for the effects reported in those studies. If it could be demonstrated in controlled laboratory conditions that arbitrary environmental stimuli that have been paired with drugs of abuse will 'grab the attention', this would support a classical conditioning explanation. As discussed below, numerous authors have attempted to demonstrate this.

Arbitrary stimuli that are repeatedly paired with the effects of drugs of abuse can elicit conditioned responses that are comparable with those elicited by 'naturalistic' drug-related stimuli, such as the sight of lit cigarettes or bottles of alcohol. For example, Lazev et al. (1999) trained cigarette smokers to associate a novel stimulus with the effects of cigarette smoking, and reported that exposure to that stimulus evoked higher levels of subjective craving than a stimulus that had been explicitly unpaired with smoking (see also Foltin and Haney 2000; Field and Duka 2001). There are also demonstrations that such stimuli can function as incentive stimuli that will command the attention and will be preferred over stimuli that have not been previously paired with drug intake. For example, Mucha et al. (1998) demonstrated that, among smokers who had previously experienced discriminative conditioning sessions in which one auditory stimulus (CS+) was paired with smoking and another stimulus (CS−) was paired with the absence of smoking, a significant majority of participants chose to experience the CS+ rather than the CS−. Similarly, Foltin and Haney (2000) demonstrated that cocaine addicts preferred to experience a compound audiovisual stimulus that had been previously paired with cocaine rather than a stimulus that had been previously unpaired with cocaine. Field and Duka (2002) paired visuo-gustatory compound stimuli with low doses of alcohol, and matched stimuli with placebo, in a group of social drinkers and subsequently observed the shifting of gaze when those stimuli were placed in front of the subjects. They reported that subjects shifted their gaze towards the alcohol-related stimulus more often than to the control stimulus, even though they were not consciously aware of the relationship between the stimuli and prior alcohol administration and, at the time of testing, they knew that no alcohol was available.

In the studies described above, the measures of attentional bias that were used were at best very crude measures of attention, and some measures may even be described as measures of overt stimulus preference, rather than the deployment of attention. Therefore, there is a need to employ more sophisticated measures of attentional deployment in order to clarify the nature of attentional biases to drug-paired CS+. In two recent studies, Hogarth *et al.* (2003*b*, 2005) exposed smokers to a discriminative conditioning procedure in which one arbitrary stimulus (CS+) was paired with the availability of smoking and another stimulus (CS−) was explicitly unpaired with the availability of smoking. After conditioning trials, the allocation of attention to both stimuli was assessed with a visual probe task, in which both stimuli were presented simultaneously on a computer screen. In both studies, participants' response latencies were significantly faster to probes that replaced CS+ than CS−, which is consistent with an attentional bias for CS+. However, in both studies, the attentional bias was not particularly robust: in the first study (Hogarth *et al.* 2003*b*), the bias was only evident after repeated unreinforced stimulus presentations, and in the second study (Hogarth *et al.* 2005), the bias was only seen in a subset of participants who were unaware of the experimental contingencies between the CS+ and the availability of smoking. It is clear that researchers need to make use of more sophisticated techniques to infer the deployment of attention to drug-paired conditioned stimuli, such as the visual probe task and eye movement recording, in order to clarify any similarities in attentional bias for 'naturalistic' drug-paired cues, and for conditioned stimuli that have been paired with drug taking in the laboratory. Nonetheless, the few studies that currently exist are encouraging in that they suggest that classical conditioning could explain the development of attentional biases in addiction.

3.5.2 Attentional biases for drug cues are closely associated with subjective craving

According to Robinson and Berridge (1993), 'craving' can be defined as the subjective component of the incentive-motivational state underlying drug-seeking behaviour, with attentional biases providing a cognitive index of this motivational state. Subjective craving should therefore be positively associated with attentional biases. Furthermore, manipulations that increase the activation of this incentive-motivational state should have parallel effects on subjective craving, and attentional biases for drug-related cues.

In smokers, craving increases with increasing periods of nicotine deprivation (Schuh and Stitzer 1995), and nicotine deprivation enhances attentional biases in smokers (Gross et al. 1993; Waters and Feyerabend 2000; Field et al. 2004a). An additional unpublished study from my laboratory demonstrated that, in regular smokers, exposure to smoking-related cues (holding a lit cigarette) led to increased subjective craving and a corresponding increase in Stroop interference for smoking-related words, relative to a control manipulation with no smoking cues. Moreover, several studies have demonstrated positive correlations between cigarette craving and various measures of attentional bias, independently of any experimental manipulation of nicotine deprivation (Sayette and Hufford 1994; Zack et al. 2001; Mogg and Bradley 2002; Mogg et al. 2003, 2005; Waters et al. 2003b). Associations between craving and attentional bias have also been reported in alcoholics (Sayette et al. 1994), cocaine addicts (Rosse et al. 1993, 1997; Franken et al. 2000b; Copersino et al. 2004), and recreational users of cannabis and alcohol (Field et al. 2004b,c, 2005b; Field 2005). In addition, as described in the preceding sections, when the motivation for alcohol is experimentally increased, for example by exposure to alcohol-related cues or by the administration of a priming dose of alcohol, attentional biases for alcohol cues also increase (Cox et al. 1999, 2003; Duka and Townshend 2004). On the other hand, numerous investigations of attentional bias in addiction either did not measure subjective craving, or did not report correlations between craving and attentional bias measures, while other studies have reported that such correlations were not statistically significant (e.g. Ehrman et al. 2002). Nonetheless, the widely replicated association between craving and attentional bias for drug cues across different drug classes suggests that the association is reasonably robust, especially considering that studies differ widely in the tasks used to measure attentional bias, and in the self-report measures of craving that were used (e.g. single item visual analogue scales versus multifactorial questionnaires).

Studies which monitor eye movements to smoking-related cues in smokers suggest that craving may be selectively associated with biases in the maintenance, rather than the initial shift, of overt attention to drug-related cues. Mogg et al. (2003) found that subjective cigarette craving was correlated with biases in the maintenance of gaze on smoking-related pictures, but not with the bias to shift gaze to smoking-related pictures. This demonstration of an association between subjective cigarette craving and the maintenance, but not the shifting, of gaze was subsequently replicated (Mogg *et al.* 2005). In two further studies, craving was experimentally manipulated by imposing 12 h of nicotine deprivation (Field *et al.* 2004*a*), or by administering a moderate dose of alcohol (Field *et al.* 2005*a*). In both of these studies, the manipulations were effective, as subjective cigarette craving was elevated by nicotine deprivation and by alcohol. Importantly, in both of these studies, the manipulations were selectively associated with an increased bias in the maintenance of gaze on smoking-related pictures, but they were not associated with an increased bias in the shifting of gaze. Therefore, the available evidence suggests that, in cigarette smokers at least, subjective craving is selectively associated with biased maintenance of attention on drug-related cues, but not with biases in the orienting of attention to those cues. Comparable associations were reported by Rosse *et al.* (1993, 1997), who demonstrated that the number of attentional fixations made to a cocaine-related cue was positively correlated with subjective cocaine craving, in a sample of cocaine addicts.

These demonstrations of a selective association between subjective drug craving and biased maintenance of attention to drug cues are perhaps not surprising. Drug craving is a subjectively experienced motivational state, and the maintenance of attention is thought to be under the influence of strategic cognitive processes and motivational variables (LaBerge 1995). However, the initial orienting of attention may occur more automatically and may be strongly influenced by simple stimulus features, such as brightness and colour (Egeth and Yantis 1997). Models of addiction which predict associations between subjective craving and measures of the salience of drug-related cues (e.g. Robinson and Berridge 1993) may require modification in the light of these findings, which suggest that measures of the maintenance of attention (or perhaps *rumination* on drug-related cues) are selectively associated with subjective craving, but measures of the initial orienting or *shifting* of attention are not so closely associated with craving.

Like Franken (2003), I suggest that subjective craving and attentional bias (particularly biases in the maintenance of attention) have a reciprocal causal relationship. As discussed above, variables that increase drug craving (e.g. nicotine deprivation in smokers, alcohol priming or alcohol cue exposure in social drinkers) produce increases in attentional bias. Recent results also indicate that attentional bias can influence subjective drug craving. Field and Eastwood (2005) experimentally manipulated attentional bias for alcohol-related cues in two groups of heavy social drinkers. Using a modified visual probe task, one group of participants ('attend alcohol') were trained to direct their attention towards alcohol-related cues over repeated trials, and another group ('avoid alcohol) were trained to direct their attention away from alcohol cues. After this period of attentional training, the attend alcohol group showed a significant increase in attentional bias for alcohol cues, relative to baseline, but the avoid alcohol group showed a significant reduction in attentional bias for alcohol cues, relative to baseline. Furthermore, the subjective urge to drink alcohol significantly increased after attentional training in the attend alcohol group, again relative to baseline, although there was no significant change in the avoid alcohol group. In addition, after attentional training, the attend alcohol group consumed more beer than the avoid alcohol group during an experimental 'taste test', which suggests that the attentional training had effects on both subjective and behavioural indices of the motivation to drink alcohol. These results are entirely consistent with Franken's (2003) model of the role of attentional bias in drug-seeking behaviours, as they suggest that an increase in attentional bias can play a causal role in such behaviours. They are also consistent with Kavanagh *et al.*'s (2005) more general motivational model, as they suggest that rumination on (or the focusing of selective attention on) alcohol-related stimuli can increase the desire for alcohol. This provides a mechanism to explain the nature of the association between drug abuse, craving and attentional bias. Drug-related cues may acquire incentive-motivational properties in drug users through an incentive learning process such as that proposed by Robinson and Berridge (1993) and Franken (2003). However, once acquired, attentional biases for drug-related cues lead addicts to increase their exposure to those cues, which increases their craving (for a meta-analysis of the effects of drug cue exposure on subjective craving, see Carter and Tiffany 1999). Eventually, subjective craving reaches a threshold level which ultimately leads to drug-seeking and drug-taking behaviours. The potential clinical implications of these models and findings are the focus of investigation by several research groups, as discussed in Section 3.6, below.

3.5.3 The presence of cognitive–behavioural biases is moderated by the severity of dependence

According to incentive sensitization theory (Robinson and Berridge 1993), cognitive indices of a sensitized incentive-motivational state should increase after each exposure to drugs of abuse, as individuals become more dependent on a given drug. Therefore, one may expect attentional biases to be positively associated with the level of drug intake, or with the severity of dependence. Research with recreational cannabis users and social drinkers appears consistent with this prediction. In cannabis users, attentional biases for cannabis-related words are positively correlated with the frequency of cannabis use (Jones et al. 2003; Field et al. 2004c; Field 2005), and with the severity of cannabis dependence (Field 2005). In social drinkers, attentional biases are positively associated with the amount of alcohol that is regularly consumed (Townshend and Duka 2001; Jones et al. 2003; Field et al. 2004b; although see Field et al. 2005b, for a failure to replicate this association), and with the severity of alcohol-related problems as assessed by the Alcohol Use Disorders Identification Test (AUDIT; see Sharma et al. 2001).

In tobacco smokers, some modified Stroop studies suggest that attentional bias is greater in heavy (rather than light) smokers (e.g. Zack et al. 2001; Mogg and Bradley 2002), although this has not been found consistently (e.g. Waters and Feyerabend 2000; Munafò et al. 2003; Waters et al. 2003a). On the other hand, some studies using the visual probe task found that the attentional bias correlated negatively with daily smoking rate, suggesting a greater attentional bias in lighter smokers (Bradley et al. 2003, Experiment 1; Hogarth et al. 2003; Waters et al. 2003b). However, this relationship between indices of nicotine dependence and attentional bias has not been consistently found in studies that used the visual probe task (e.g. Mogg and Bradley 2002). Furthermore, Mogg et al. (2005) measured smokers' eye movements while they viewed smoking-related and control scenes in a visual probe task. A key aim was to examine whether biases in eye movements to smoking cues would be enhanced, or attenuated, as a function of the level of nicotine dependence. Based on participants' responses on the modified Fagerstrom Tolerance Questionnaire (mFTQ; Prokhorov et al. 1996), participants were split into those with a low level of nicotine dependence and those with a moderate level of dependence. Results indicated that attentional biases for smoking-related cues (as indicated by longer duration of gaze on smoking cues) were significantly greater in the low dependence group, compared with the moderate dependence group.

How can the latter, seemingly counterintuitive results, be interpreted? It is important to consider theories of addiction which emphasize the role of habit, rather than incentive processes, in maintaining drug-taking behaviour. According to Tiffany's (1990) model, drug-taking behaviour is under the control of automatized action schemata and is largely habitual (i.e. requiring little conscious control), unless the behaviour is blocked or interrupted. More recently, theorists have suggested that incentive and habit processes may play differential roles in the acquisition and maintenance of drug dependence. There is evidence that, during the early stages of stimulus–response learning, incentive learning factors primarily control operant responding but, after an operant response has been learned and repeatedly performed, the behaviour becomes automatically initiated by triggering stimuli (the development of 'habit'; Dickinson and Balleine 1995). With respect to smoking, Di Chiara (2000) proposed that, in the early stages of nicotine dependence, smoking behaviour is controlled by incentive learning processes and that, as a result of dopamine release, smoking-related cues acquire positive motivational properties. However, after extensive experience of smoking, incentive learning processes no longer play a primary role in determining smoking behaviour, as there is a 'switch of responding' from incentive responding to a mode of habit-based responding. Thus, in smokers with higher levels of dependence, smoking behaviour is more likely to become a 'habit' that is initiated automatically by drug-related stimuli in the absence of incentive-motivational processes (see also Everitt *et al.* 2001). As habit becomes a more influential determinant of drug-seeking behaviour, the strength of incentive-motivational processes may actually diminish. These theoretical views may account for the observed demonstrations of an inverse relationship between frequency of smoking, or dependence, and attentional bias for smoking-related cues in tobacco smokers

3.5.4 A conflict may operate between cue-elicited attentional biases and cognitive avoidance strategies

It is helpful to make a distinction between: (1) cognitive responses elicited by drug-related cues (e.g. attentional biases), and associated increases in subjective craving; and (2) effortful regulation of those cognitions, such as strategic processes which are aimed at avoidance of drug cues, in order to reduce subjective craving.

Much research has examined attentional biases in smokers, commonly students who have relatively low to moderate levels of nicotine dependence [e.g. participants in the study reported by Bradley *et al.* (2004) smoked 15 cigarettes per day, on average]. Typically, these are individuals who do not regard their smoking behaviour as problematic, and they are not actively seeking treatment for nicotine addiction. Thus, while there may be common features between the cognitive–behavioural biases shown in addiction (as predicted by a variety of models), there may also be some important differences between relatively lightly addicted individuals (e.g. student smokers) and more severely dependent individuals, such as alcoholics or opiate addicts who are undergoing treatment. A key difference may be in the extent to which they use cognitive avoidance strategies.

Some recent research on alcoholics in treatment suggests that these individuals actually *avoid* attentional processing of alcohol-related cues. For example, Stormark *et al.* (1997) demonstrated a significant attentional bias for alcohol-related words in alcoholics (who were attending treatment) when those words were presented very briefly (100 ms) in the context of an attentional cueing task. However, in another experimental condition, the words were presented for 500 ms. In this condition, alcoholics demonstrated significant attentional avoidance of the alcohol-related words. A further study by Townshend and Duka (2003) also demonstrated attentional avoidance of alcohol-related pictures in a sample of inpatient alcoholics, when those pictures were presented for 500 ms in the context of a visual probe task. Using the modified Stroop task, Cox *et al.* (2002) demonstrated that alcoholics who successfully completed treatment and maintained abstinence had a reduced attentional bias for alcohol-related words at the end of treatment compared with at the beginning, whereas no such reduction in attentional bias was observed in a subset of patients who did not successfully complete the treatment programme (and presumably relapsed). These results may suggest that active avoidance of drug-related stimuli may actually be a useful strategy for treatment-seeking drug users to employ, as it may increase the chances of future abstinence. Interventions which aim directly to manipulate and reduce attentional biases as a form of treatment for addictive disorders are briefly discussed below.

3.6 Treatment implications

So far, research into attentional biases has mainly focused on evaluating hypotheses that are derived from theories of addiction which predict the existence of these biases, and elucidating the underlying cognitive mechanisms. There has been little research relating to the implications of this research for the treatment of addictions. However, this is a potentially important area for further development (see Chapters 9 and 10). For example, recent research suggests that an attentional bias for drug cues may be of clinical significance in predicting treatment outcome (Waters *et al.* 2003*a*). As discussed above, cognitive avoidance of drug-related cues may be a useful strategy for addicts in treatment to employ, and this suggestion is consistent with a recent theoretical model. According to Franken (2003), attentional biases for drug cues may be an important determinant of drug craving and drug-seeking behaviour. Therefore, if drug users could be trained to allocate their attention away from drug-related cues, this may in turn reduce their craving and drug-seeking behaviour (for a discussion, see Wiers *et al.* 2004; De Jong *et al.*

2006; for some promising preliminary findings, see Wiers *et al.* 2006). Such attentional training could be extended to 'high-risk' situations, thereby reducing craving and the risk of relapse to drug taking in these situations. According to Franken's model, such an approach may not prevent drug-related cues from attracting attentional resources in the first instance, but it may prevent this attentional bias from growing in strength and provoking further increases in drug craving.

3.7 **Concluding comments**

In this chapter I have summarized research which indicates that, across a variety of experimental psychology paradigms, drug abuse and addiction are associated with biases in selective attention for drug-related stimuli, as predicted by many contemporary models of addiction and motivation. These 'attentional biases' may be influenced by aspects of cognition that are fairly rapid and automatic, although there is no evidence that attentional biases are sufficiently 'automatic' to occur in the absence of conscious awareness. Recent research suggests that these biases may be the consequence of a classical conditioning process, although, once established, they may both be caused by, and may provoke further increases in, subjective drug craving. It is also apparent that these attentional biases can be moderated by variables such as the severity of dependence, and future research efforts should focus on identifying how these biases can be directly and indirectly modified, as the theoretical and clinical implications could be highly significant.

Acknowledgements

Much of my research is supported by research grants from the British Academy (reference LRG-37196) and the University of Liverpool. I thank Karin Mogg and Brendan Bradley for their helpful comments on an early version of this chapter. Any correspondence concerning this chapter should be addressed to Dr Matt Field, School of Psychology, University of Liverpool, Eleanor Rathbone Building, Liverpool, L69 7ZA.

References

Algom, D., Chajut, E. and Lev, S. (2004). A rational look at the emotional Stroop phenomenon: a generic slowdown, not a Stroop effect. *Journal of Experimental Psychology: General*, 133, 323–338.

Allport, A. (1989). Visual attention. In M.I. Posner (ed.), *Foundations of cognitive science*. Cambridge, MA: MIT Press, pp. 631–682.

Baker, T.B., Brandon, T.H. and Chassin, L. (2004a). Motivational influences on cigarette smoking. *Annual Review of Psychology*, 55, 463–491.

Baker, T.B., Piper, M.E., McCarthy, D.E., Majeskie, M.R. and Fiore, M.C. (2004b). Addiction motivation reformulated: an affective processing model of negative reinforcement. *Psychological Review*, 111, 33–51.

Bauer, D. and Cox, W.M. (1998). Alcohol-related words are distracting to both alcohol abusers and non-abusers in the Stroop colour-naming task. *Addiction*, 93, 1539–1542.

Baxter, B.W. and Hinson, R.E. (2001). Is smoking automatic? Demands of smoking behavior on attentional resources. *Journal of Abnormal Psychology*, 110, 59–66.

Bindra, D. (1974). A motivational view of learning, performance, and behaviour modification. *Psychological Review*, 81, 199–213.

Bradley, B.P., Mogg, K., Wright, T. and Field, M. (2003). Attentional bias in drug dependence: vigilance for cigarette-related cues in smokers. *Psychology of Addictive Behaviors*, 17, 66–72.

Bradley, B.P., Field, M., Mogg, K. and De Houwer, J. (2004). Attentional and evaluative biases for smoking cues in nicotine dependence: component processes of biases in visual orienting. *Behavioural Pharmacology*, 15, 29–36.

Bruce, G. and Jones, B.T. (2004). A pictorial Stroop paradigm reveals an alcohol attentional bias in heavier compared to lighter social drinkers. *Journal of Psychopharmacology*, 18, 527–533.

Carter, B.L. and Tiffany, S.T. (1999). Meta-analysis of cue reactivity in addiction research. *Addiction*, 94, 327–340.

Copersino, M.L., Serper, M.R., Vadhan, N., Goldberg, B.R., Richarme, D., Chou, J.C.Y., Stitzer, M. and Cancro, R. (2004). Cocaine craving and attentional bias in cocaine-dependent schizophrenic patients. *Psychiatry Research*, 128, 209–218.

Cox, W.M. and Klinger, E. (2004). A motivational model of alcohol use: determinants of use and change. In W.M. Cox and E. Klinger (ed.), *Handbook of motivational counseling: concepts, approaches, and assessment*. Chichester, UK: Wiley, pp. 121–138.

Cox, W.M., Yeates, G.N. and Regan, C.M. (1999). Effects of alcohol cues on cognitive processing in heavy and light drinkers. *Drug and Alcohol Dependence*, 55, 85–89.

Cox, W.M., Hogan, L.M., Kristian, M.R. and Race, J.H. (2002). Alcohol attentional bias as a predictor of alcohol abusers' treatment outcome. *Drug and Alcohol Dependence*, 68, 237–243.

Cox, W.M., Brown, M.A. and Rowlands, L.J. (2003). The effects of alcohol cue exposure on non-dependent drinkers' attentional bias for alcohol-related stimuli. *Alcohol and Alcoholism*, 38, 45–49.

Cox, W.M., Fadardi, J.S. and Pothos, E.M. (2006). The Addiction-Stroop test: theoretical considerations and procedural recommendations. *Psychological Bulletin*, 132, 443–476.

De Jong, P.J., Kindt, M. and Roefs, A. (2006). Relevance of research on experimental psychopathology to substance misuse. In R. Wiers and A.S tacy (ed.), *Handbook of implicit cognition and addiction*. Thousand Oaks, CA: Sage, 425–437.

Di Chiara, G. (2000). Role of dopamine in the behavioural actions of nicotine related to addiction. *European Journal of Pharmacology*, 393, 295–314.

Dickinson, A. and Balleine, B. (1995). Motivational control of instrumental action. *Current Directions in Psychological Science*, 4, 162–167.

Duka, T. and Townshend, J.M. (2004). The priming effect of alcohol pre-load on attentional bias to alcohol-related stimuli. *Psychopharmacology (Berlin)*, 176, 353–361.

Duka, T., Townshend, J.M., Collier, K. and Stephens, D.N. (2002). Kindling of withdrawal: a study of craving and anxiety after multiple detoxifications in alcohol inpatients. *Alcoholism: Clinical and Experimental Research*, 36, 785–795.

Egeth, H.E. and Yantis, S. (1997). Visual attention: control, representation, and time-course. *Annual Review of Psychology*, 48, 269–297.

Ehrman, R.N., Robbins, S.J., Bromwell, M.A., Lankford, M.E., Monterosso, J.R. and O'Brien, C.P. (2002). Comparing attentional bias to smoking cues in current smokers, former smokers, and non-smokers using a dot-probe task. *Drug and Alcohol Dependence*, 67, 185–191.

Everitt, B.J., Dickinson, A. and Robbins, T.W. (2001). The neuropsychological basis of addictive behaviour. *Brain Research Reviews*, 36, 129–138.

Field, M. (2005). Cannabis 'dependence' and attentional bias for cannabis-related words. *Behavioural Pharmacology*, 16, 473–476.

Field, M. and Duka, T. (2001). Smoking expectancy mediates the conditioned responses to arbitrary smoking cues. *Behavioural Pharmacology*, 12, 183–194.

Field, M. and Duka, T. (2002). Cues paired with a low dose of alcohol acquire conditioned incentive properties in social drinkers. *Psychopharmacology (Berlin)*, 159, 325–334.

Field, M. and Eastwood, B. (2005). Experimental manipulation of attentional bias increases the motivation to drink alcohol. *Psychopharmacology (Berlin)*, 183, 350–357.

Field, M., Mogg, K. and Bradley, B.P. (2004a). Eye movements to smoking-related cues: effects of nicotine deprivation. *Psychopharmacology (Berlin)*, 173, 116–123.

Field, M., Mogg, K., Zetteler, J. and Bradley, B.P. (2004b). Attentional biases for alcohol cues in heavy and light social drinkers: the roles of initial orienting and maintained attention. *Psychopharmacology (Berlin)*, 176, 88–93.

Field, M., Mogg, K. and Bradley, B.P. (2004c). Cognitive bias and drug craving in recreational cannabis users. *Drug and Alcohol Dependence*, 74, 105–111.

Field, M., Mogg., K. and Bradley, B.P. (2005a). Alcohol increases cognitive biases for smoking cues in smokers. *Psychopharmacology (Berlin)*, 180, 63–72.

Field, M., Mogg, K. and Bradley, B.P. (2005b). Craving and cognitive biases for alcohol cues in social drinkers. *Alcohol and Alcoholism*, 40, 504–510.

Field, M., Mogg., K. and Bradley, B.P. (2006). Attention to drug-related cues in drug abuse and addiction: component processes. In R. Wiers and A. Stacy (ed.), *Handbook of implicit cognition and addiction*. Thousand Oaks, CA: Sage, pp. 151–164.

Foltin, R.W. and Haney, M. (2000). Conditioned effects of environmental stimuli paired with smoked cocaine in humans. *Psychopharmacology (Berlin)*, 135, 82–92.

Franken, I.H.A. (2003). Drug craving and addiction: integrating psychological and neuro-psychopharmacological approaches. *Progress in Neuro-Psychopharmacology and Biological Psychiatry*, 27, 563–579.

Franken, I.H.A., Kroon, L.Y., Wiers, R.W. and Jansen, A. (2000a). Selective cognitive processing of drug cues in heroin dependence. *Journal of Psychopharmacology*, 14, 395–400.

Franken, I.H.A., Kroon, L.Y , and Hendriks, V.M. (2000b). Influence of individual differences in craving and obsessive cocaine thoughts on attentional processes in cocaine abuse patients. *Addictive Behaviors*, 25, 99–102.

Franken, I.H.A., Hendriks, V.M., Stam, C.J. and Van Den Brink, W. (2004). A role for dopamine in the processing of drug cues in heroin dependent patients. *European Neuropsychopharmacology*, 14, 503–508.

Gross, T.M., Jarvik, M.E. and Rosenblatt, M.R (1993). Nicotine abstinence produces context-specific Stroop interference. *Psychopharmacology (Berlin)*, 110, 333–336.

Hester, R., Dixon, V. and Garavan, H. (2006). A consistent attentional bias for drug-related material in active cocaine users across word and picture versions of the emotional Stroop task. *Drug and Alcohol Dependence*, **81**, 251–257.

Hogarth, L.C., Mogg, K., Bradley, B.P., Duka, T. and Dickinson, T. (2003*a*). Attentional orienting towards smoking-related stimuli. *Behavioural Pharmacology*, **14**, 153–160.

Hogarth, L.C., Dickinson, A. and Duka, T. (2003*b*). Discriminative stimuli that control instrumental tobacco-seeking by human smokers also command selective attention. *Psychopharmacology (Berlin)*, **168**, 435–445.

Hogarth, L., Dickinson, A. and Duka, T. (2005). Explicit knowledge of stimulus–outcome contingencies and stimulus control of selective attention and instrumental action in human smoking behaviour. *Psychopharmacology (Berlin)*, **177**, 428–437.

Ingjaldsson, J.T., Thayer, J.F. and Laberg, J.C. (2004). Craving for alcohol and pre-attentive processing of alcohol stimuli. *International Journal of Psychophysiology*, **49**, 29–39.

Johnsen, B.H., Laberg, J.C., Cox, W.M., Vaksdal, A. and Hugdahl, K. (1994). Alcoholic subjects' attentional bias in the processing of alcohol-related words. *Psychology of Addictive Behaviors*, **8**, 111–115.

Johnsen, B.H., Thayer, J.F., Laberg, J.C. and Asbjornsen, A.E. (1997). Attentional bias in active smokers, abstinent smokers, and non-smokers. *Addictive Behaviors*, **22**, 813–817.

Jones, B.C., Jones, B.T., Blundell, L. and Bruce, G. (2002). Social users of alcohol and cannabis who detect substance-related changes in a change blindness paradigm report higher levels of use than those detecting substance-neutral changes. *Psychopharmacology (Berlin)*, **165**, 93–96.

Jones, B.T., Jones, B.C., Smith, H. and Copley, N. (2003). A flicker paradigm for inducing change blindness reveals alcohol and cannabis information processing biases in social users. *Addiction*, **98**, 235–244.

Jonides, J. (1981). Voluntary versus automatic control over the mind's eye movements. In J. Long and A. Baddeley (ed.), *Attention and performance IX*. Hillsdale, NJ: Erlbaum, pp. 187–203.

Juliano, L.M. and Brandon, T.H. (1998). Reactivity to instructed smoking availability and environmental cues: evidence with urge and reaction time. *Experimental and Clinical Psychopharmacology*, **6**, 45–53.

Kavanagh, D.J., Andrade, J. and May, J. (2005). Imaginary relish and exquisite torture: the elaborated intrusion theory of desire. *Psychological Review*, **112**, 446–467.

Kowler, E. (1995). Eye movements. In S.M. Kosslyn and D.M. Osheron (ed.), *Visual cognition*. Cambridge, MA: Harvard University Press, pp. 215–265.

LaBerge, D. (1995). *Attentional processing*. Cambridge, MA: Harvard.

Lazev, A.B., Herzog, T.A. and Brandon, T.H. (1999). Classical conditioning of environmental cues to cigarette smoking. *Experimental and Clinical Psychopharmacology*, **7**, 56–63.

Lubman, D.I., Peters, L.A., Mogg, K., Bradley, B.P. and Deakin, J.F.W. (2000). Attentional bias for drug cues in opiate dependence. *Psychological Medicine*, **30**, 169–175.

Lusher, J., Chandler, C. and Ball, D. (2004). Alcohol dependence and the alcohol Stroop paradigm: evidence and issues. *Drug and Alcohol Dependence*, **75**, 225–231.

Mackintosh, N.J. (1975). A theory of attention: variations in the associability of stimuli with reinforcement. *Psychological Review*, **82**, 276–298.

McCusker, C.G. (2001). Cognitive biases and addiction: an evolution in theory and method. *Addiction*, **96**, 47–56.

Mogg, K. and Bradley, B.P. (1998). A cognitive-motivational analysis of anxiety. *Behaviour Research and Therapy*, **36**, 809–848.

Mogg, K. and Bradley, B.P. (2002). Selective processing of smoking-related cues in smokers: manipulation of deprivation level and comparison of three measures of processing bias. *Journal of Psychopharmacology*, **16**, 385–392.

Mogg, K., Bradley, B.P., Hyare, H. and Lee, S. (1998). Selective attention to food-related stimuli in hunger: are attentional biases specific to emotional and psychopathological states, or are they also found in normal drive states? *Behavior Research and Therapy*, **36**, 227–237.

Mogg, K., Bradley, B.P., Field, M. and De Houwer, J. (2003). Eye movements to smoking-related pictures in smokers: relationship between attentional biases and implicit and explicit measures of stimulus valence. *Addiction*, **98**, 825–836.

Mogg, K., Field, M. and Bradley, B.P. (2005). Attentional and evaluative biases for smoking cues in smokers: an investigation of competing theoretical views of addiction. *Psychopharmacology (Berlin)*, **180**, 333–341.

Mucha, R.F., Pauli, P. and Angrilli, A. (1998). Conditioned responses elicited by experimentally produced cues for smoking. *Canadian Journal of Physiology and Pharmacology*, **76**, 259–268.

Munafò, M., Mogg, K., Roberts, S., Bradley, B.P. and Murphy, M. (2003). Attentional bias in cigarette smokers, ex-smokers and never-smokers on the modified Stroop task. *Journal of Psychopharmacology*, **17**, 311–317.

Posner, M.I., Snyder, C.R. and Davidson, B.J. (1980). Attention and the detection of signals. *Journal of Experimental Psychology: General*, **109**, 160–174.

Prokhorov, A.V., Pallonen, U.E., Fava, J.L., Ding, L. and Niaura, R. (1996). Measuring nicotine dependence among high-risk adolescent smokers. *Addictive Behaviors*, **21**, 117–127.

Robbins, S.J. and Ehrman, R.N. (2004). The role of attentional bias in substance abuse. *Behavioral and Cognitive Neuroscience Reviews*, **3**, 243–260.

Robinson, T.E. and Berridge, K.C. (1993). The neural basis of drug craving: an incentive-sensitisation theory of addiction. *Brain Research Reviews*, **18**, 247–291.

Robinson, T.E. and Berridge, K.C. (2000). The psychology and neurobiology of addiction: an incentive-sensitization view. *Addiction*, **95** (Suppl. 2), S91–S117.

Rosse, R.B., Miller, M.W., Hess, A.L., Alim, T.N. and Deutsch, S.I. (1993). Measures of visual scanning as a predictor of cocaine cravings and urges. *Biological Psychiatry*, **33**, 554–556.

Rosse, R.B., Johri, S., Kendrick, K., Hess, A.L., Alim, T.N., Miller, M. and Deutsch, S.I. (1997). Preattentive and attentive eye movements during visual scanning of a cocaine cue: correlation with intensity of cocaine cravings. *Journal of Neuropsychiatry and Clinical Neurosciences*, **9**, 91–93.

Rusted, J.M., Caulfield, D., King, L. and Goode, A. (2000). Moving out of the laboratory: does nicotine improve everyday attention? *Behavioural Pharmacology*, **11**, 621–629.

Ryan, F. (2002a). Detected, selected, and sometimes neglected: cognitive processing of cues in addiction. *Experimental and Clinical Psychopharmacology*, **10**, 67–76.

Ryan, F. (2002*b*). Attentional bias and alcohol dependence: a controlled study using the modified Stroop paradigm. *Addictive Behaviors*, 27, 471–482.

Sayette, M.A. and Hufford, M.R. (1994). Effects of cue exposure and deprivation on cognitive resources in smokers. *Journal of Abnormal Psychology*, 103, 812–818.

Sayette, M.A., Monti, P.M., Rohsenow, D.J., Gulliver, S.B., Colby, S.M., Sirota, A.D., Niaura, R. and Abrams, D.B. (1994). The effects of cue exposure on reaction time in male alcoholics. *Journal of Studies on Alcohol*, 55, 629–633.

Sayette, M., Shiffman S., Tiffany, S.T., Niaura, R.S., Martin, C.S. and Shadel, W.G. (2000). The measurement of drug craving. *Addiction*, 95 (Suppl. 2), S189–S210.

Schuh, K.J. and. Stitzer, M.L. (1995). Desire to smoke during spaced smoking intervals. *Psychopharmacology (Berlin)*, 120, 289–295.

Sharma, D., Albery, I.P. and Cook, C. (2001). Selective attentional bias to alcohol related stimuli in problem drinkers and non-problem drinkers. *Addiction*, 96, 285–295.

Stetter, F., Chaluppa, C., Ackermann, K., Straube, E.R. and Mann, K. (1994). Alcoholics' selective processing of alcohol related words and cognitive performance on a Stroop task. *European Psychiatry*, 9, 71–76.

Stetter, F., Ackermann, K., Bizer, A., Straube, E.R. and Mann, K. (1995). Effects of disease-related cues in alcoholic inpatients: results of a controlled 'alcohol Stroop' study. *Alcoholism: Clinical and Experimental Research*, 19, 593–599.

Stewart, J., de Wit, H. and Eikelboom, R. (1984). The role of conditioned and unconditioned drug effects in the self-administration of opiates and stimulants. *Psychological Review*, 91, 251–268.

Stormark, K.M., Field, N.P., Hugdahl, K. and Horowitz, M. (1997). Selective processing of visual alcohol cues in abstinent alcoholics: an approach–avoidance conflict. *Addictive Behaviors*, 22, 509–519.

Stormark, K.M., Laberg, J.C., Nordby, H. and Hugdahl, K. (2000). Alcoholics' selective attention to alcohol stimuli: automated processing? *Journal of Studies on Alcohol*, 61, 18–23.

Tiffany, S.T. (1990). A cognitive model of drug urges and drug-use behaviour: role of automatic and nonautomatic processes. *Psychological Review*, 97, 147–168.

Townshend, J.M. and Duka, T. (2003). Avoidance of alcohol-related cues by alcoholic inpatients undergoing detoxification. *Behavioural Pharmacology*, 14 (Suppl. 1), S69.

Townshend, J.M. and Duka, T. (2001). Attentional bias associated with alcohol cues: differences between heavy and occasional social drinkers. *Psychopharmacology (Berlin)*, 157, 67–74.

Waters, A.J. and Feyerabend, C. (2000). Determinants and effects of attentional bias in smokers. *Psychology of Addictive Behaviors*, 14, 111–120.

Waters, A.J., Shiffman, S., Sayette, M.A., Paty, J.A., Gwaltney, C.J. and Balabanis, M.H. (2003*a*). Attentional bias predicts outcome in smoking cessation. *Health Psychology*, 22, 378–387.

Waters, A.J., Shiffman, S., Bradley, B.P. and Mogg, K. (2003*b*). Attentional shifts to smoking cues in smokers. *Addiction*, 98, 1409–1417.

Waters, H. and Green, M.W. (2004). A demonstration of attentional bias, using a novel dual-task paradigm, towards clinically salient material in recovering alcohol abuse patients? *Psychological Medicine*, 33, 491–498.

Wertz, J.M. and Sayette, M.A. (2001). Effects of smoking opportunity on attentional bias in smokers. *Psychology of Addictive Behaviors*, 15, 268–271.

Wiers, R.W., de Jong, P.J., Havermans, R. and Jelicic, M. (2004). How to change implicit drug use-related cognitions in prevention: a transdisciplinary integration of findings from experimental psychopathology, social cognition, memory, and experimental learning psychology. *Substance Use and Misuse*, 39, 1625–1684.

Wiers, R.W., Cox, W.M., Field, M., Fadardi, J.S., Palfai, T.P., Schoenmakers, T. and Stacy, A.W. (2006). The search for new ways to change alcohol-related cognitions in heavy drinkers. *Alcoholism: Clinical and Experimental Research*, 30, 320–331.

Wikler, A. (1948). Recent progress in research on the neurophysiological basis of morphine addiction. *American Journal of Psychiatry*, 105, 329–338.

Williams, J.M.G., Mathews, A. and Macleod, C. (1996). The emotional Stroop task and psychopathology. *Psychological Bulletin*, 120, 3–25.

Yeomans, M.R., Javaherian, S., Tovey, H.M. and Stafford, L.D. (2005). Attentional bias for caffeine-related stimuli in high but not moderate or non-caffeine consumers. *Psychopharmacology (Berlin)*, 181, 477–485.

Zack, M., Belsito, L., Scher, R., Eissenberg, T. and Corrigall, W.A. (2001). Effects of abstinence and smoking on information processing in adolescent smokers. *Psychopharmacology (Berlin)*, 153, 249–257.

4

Motivational basis of cognitive determinants of addiction

W. Miles Cox, Eric Klinger and
Javad Salehi Fadardi

4.1 Introduction

The concept *motivation* accounts for why an individual selects, moves toward and persists in trying to achieve specific goals. Although motivation is generally recognized as related to goal-directed behaviour, different motivational psychologists emphasize different aspects of that behaviour. In our view, motivation is 'the internal states of the organism that lead to the instigation, persistence, energy, and direction of behaviour towards a goal' (Klinger and Cox 2004, pp. 4–5). That is, when a human or an animal attempts to reach a particular goal, internal processes (i.e. motivation) instigate and energize the behaviour and keep it persistently directed toward the goal. What is the nature of these internal processes?

We use the term *current concern* (Klinger 1975, 1977; Klinger and Cox 2004) to refer to the internal processes that provide the neural substrate for the striving to achieve a goal. Each concern begins when the individual first makes a commitment—consciously or non-consciously—to pursue a particular goal. A current concern, as a motivational state, has several characteristics. First, there is a separate current concern corresponding to each goal pursuit. Therefore, current concerns can be compatible or incompatible with each other. As the number of incompatible current concerns increases, so does the amount of goal conflict and negative affect due to the person's inability to resolve the conflict (e.g. between drinking alcohol heavily and working productively). Secondly, a current concern is a *time-binding* motivational entity, and may act at a conscious and at a non-conscious level. Although people are commonly aware of most of their goals, the concern itself is latent, i.e. implicit, non-conscious. That is, a current concern continues until the goal is reached or the person gives up the pursuit, regardless of momentary awareness of the presence of the concern. Thirdly, a current concern is an active entity, which coordinates motivational processes toward achieving a goal. These processes are adaptive, helping to keep the individual on track until the goal is successfully reached or abandoned. Fourthly, the *goal-lurking* feature of current concerns has the effect of potentiating responses to goal-related cues, including attentional, emotional, covert mental and overt motor responses. The goal lurking can become automatic and eventually separate from its source; this phenomenon explains the uncontrollability of many thoughts and behaviours. The current concern thereby focuses the person's cognitive processes on the goal pursuit, including what he or she thinks about, the stimuli attended to and the information stored and recalled.

The following sections explain how people's current concerns and the way that they attempt to achieve their goals influence their emotional regulation, and, in turn, their motivation to use addictive substances. The emotional, cognitive and behavioural consequences of having current concerns about substance use are also explained.

4.2 A motivational model of alcohol and other drug use

4.2.1 Substance use as a goal and the role of anticipated emotional payoff

For a person who voluntarily and knowingly drinks alcohol or uses another addictive drug, the substance use is a goal that the person is motivated to pursue, however ambivalently; such a person has a current concern for drinking or using. When the use reaches an addictive level, the motivation can be perplexing in view of the overwhelming negative consequences. To account for the motivation, we proposed the motivational model (Cox and Klinger 1988, 1990, 2004a), which takes into account an array of variables that interact with one another to determine the degree to which a person is motivated to drink alcohol or to use another drug. The model shows how each of various kinds of variables affects the motivational pathway to alcohol or drug use.

In general, people are motivated to obtain positive incentives (objects or events that they expect will make them feel good, or enhance their positive affect) and to get rid of negative incentives (objects or events that they expect will make them feel bad, or enhance their negative affect). The crucial role of expected emotional payoff in goal choice is argued extensively elsewhere (e.g. Klinger 1977, 1996; Mellers 2000, Klinger and Cox 2004). Thus, to be motivated to drink alcohol or use another drug, a person must expect that doing so will bring about net positive affective changes. For example, people might expect that drinking alcohol will make them feel happier or more optimistic, or that it will help them gain access to other kinds of incentives that will make them feel good (e.g. social companionship). On the other hand, people might expect that drinking alcohol will help to reduce their anxiety or other negative feelings, or that it will alleviate other kinds of negative incentives that cause them distress (e.g. physical pain). Needless to say, people's beliefs and expectations do not necessarily match reality, and the long-term adverse consequences of abusing alcohol or other substances might well be overshadowed by the perceived short-term benefits of using them.

4.2.2 Influences on expected outcomes of drinking alcohol or using other drugs

A variety of biological, psychological and socio-cultural influences give rise to these expectations.

With regard to biological factors, the manner in which each person's body metabolizes alcohol determines the degree to which that person experiences drinking as positive or negative. For example, when some people drink, their bodies accumulate the toxic metabolite acetaldehyde, which causes malaise. Such a person would form expectations of negative affective changes from drinking alcohol—expectations that would reduce the person's motivation to drink.

One example of the contribution of psychological factors to the motivation to drink, or not to do so, is the drinker's personality characteristics. For example, some personality characteristics (e.g. impulsivity, reward seeking or low tolerance for frustration) increase the chances that the drinker will emphasize the immediate, positive benefits of drinking and discount the long-range negative consequences.

In some cultures, individuals are reinforced for drinking heavily; in other cultures, people are reinforced for drinking moderately or not at all. People living in the former kind of culture develop expectations of positive affective changes from social support if they drink heavily. These expectations, in turn, increase the people's motivation to overindulge in alcohol.

A prominent feature of the motivational model of alcohol and other substance use is the theoretical interactions among the factors described above that it brings together. Interactions among these factors determine both a person's expectations of affective changes from drinking and his or her expectations about achieving substance-unrelated goals. The interactions among the factors determine individual differences in the motivation to drink or not drink.

4.2.3 Motivational structure

Motivational structure is a product of the interaction among the different factors in the model. It refers to a person's set of goals and his or her manner of relating to them. It includes (1) major goals and minor ones; and (2) both the goal of drinking alcohol or using other drugs and goals to resolve concerns in other areas of one's life.

Some motivational structures could be regarded as *adaptive* in that they increase the person's chances of finding successful resolutions of substance-unrelated goal strivings, thereby reducing their motivation to manipulate their emotions through alcohol or other drug use. Other motivational structures are *maladaptive* in that they reduce the chances of the person's successfully reaching goals and increase the motivation to induce affective changes chemically. Because motivational structure is an important determinant of addictive behaviours, we next discuss the components of motivational structure and how they are assessed.

4.3 Assessing motivational structure: the Motivational Structure Questionnaire and Personal Concerns Inventory

Various techniques have been used to assess motivational structure (see Klinger 1987), the most recent examples of which are the Motivational Structure Questionnaire (MSQ) and Personal Concerns Inventory (PCI). The MSQ is longer and more comprehensive; the PCI is an abridged, more user-friendly version of the MSQ. Both instruments combine idiographic and nomothetic assessment of motivation. The assessment is idiographic in that respondents name and describe their concerns and goals in an open-ended, highly individualized manner. The assessment is nomothetic in that all respondents use the same rating scales to characterize their goal strivings in various life areas. These ratings, in turn, allow comparisons to be made across different goals and individuals.

The MSQ and the PCI begin by showing respondents a list of life areas (e.g. *Home and Housekeeping, Money and Finances, Relationships*) and asking them to think about the concerns that they might have that are related to each area. It is explained to them that a concern can either be about something negative that they want to get rid of, prevent or avoid, or it can be about something positive that they want to achieve. In other words, a 'concern' does not necessarily refer to a problem. Next, respondents are asked to write a brief description of (1) each concern in each of the selected areas; and (2) their goal for resolving each concern. Because respondents describe the concerns and goals in their own words, the descriptions vary greatly from one respondent to another.

As mentioned, rating scales are used for the nomothetic part of the assessment; on each of them, respondents rate each goal that they have described.

Some of the rating scales are common to both the MSQ and PCI; they are the *Commitment, Happiness, Unhappiness, Chances of Success* and *Goal Distance* scales. Each of the first four of these is a scale whose endpoints range from the least amount of the characteristic (e.g. 'no happiness at all') to the maximum amount of the characteristic (e.g. 'the most happiness I can imagine feeling'). The Goal Distance scale asks respondents to estimate the amount of time that will elapse from the present to the time in the future when they will reach the goal for which they are striving.

A motivational profile can be drawn for each respondent who completes the MSQ or the PCI. Typically, the indices plotted on the profile are the ratings from each scale averaged across all of the respondent's goals. As mentioned, one way to classify different profiles is on the basis of how motivationally adaptive or maladaptive they are. An individual with adaptive motivation would probably be one who actively pursues goals, feels strongly committed to them, expects to feel strong happiness if the goals are attained and believes that the goals are attainable in a reasonable period of time. An individual with maladaptive motivation is one who is low in one or more of the essential components of motivation, or there is incompatibility among the scales. For example, he or she might not feel strongly committed to goals, might not be actively involved in pursuing them or does not expect strong emotional benefits if the goals are achieved. An equally maladaptive motivational profile might be that of a person who reports strong commitment to achieving goals but does not expect strong emotional reactions to either achieving or not achieving goals.

The prior research supports our theoretical premise that people's motivation to drink alcohol or use other drugs is negatively related to their ability to obtain emotional satisfaction from other (i.e. substance-unrelated) areas of their life, including the satisfaction that comes from having adaptive motivational patterns. We next briefly review this evidence.

4.4 The effect on substance use of having attractive non-substance incentives

The value that individuals place on drinking alcohol or using other drugs occurs in the context of other incentives in their lives. Both laboratory and applied research has shown that as the availability and the accessibility of other attractive incentives increase, drinkers' motivation to drink alcohol decreases (Tucker *et al.* 1995; Vuchinich and Tucker 1996, 1998; Fadardi 2003; Correia 2004). As explained earlier, people with an adaptive motivational profile should theoretically be better able to achieve their goals than are those with a maladaptive profile and are, thereby, emotionally more satisfied and less motivated to use addictive substances. In addition, when excessive drinkers are able to reduce their drinking, they are more likely to continue to drink moderately (or not at all), if they can find meaningful incentives to pursue and enjoy in other areas of their life (e.g. Moos *et al.* 1990; Tucker *et al.* 2002*a,b*). A similar result has been found in reducing other drug use (Iguchi *et al.* 1997). This effect may be of particular importance when an individual's substance use becomes problematic. University students who have encountered drink-related problems, such as effects on their relationships or work, seem better able to reduce their drinking if they have the motivational resources to facilitate the pursuit of important life goals (Cox *et al.* 2002*a*). Finally, community reinforcement and other contingency management approaches that reinforce healthy alternatives to alcohol or other drug use have been highly successful in helping addicted individuals to reduce substance use and improve the quality of other areas of their lives (Wong *et al.* 2004).

4.5 Effects of current concerns on cognitive processing

As mentioned earlier, the theory of current concerns (e.g. Klinger 1977, 1996; Klinger and Cox 2004) posits that when a person initiates a goal pursuit, and the current concern that corresponds to it is set into motion, goal-related stimuli influence the person's cognitive activity. Current concerns appear to affect all major classes of cognition: perception, attention, recall, thought content and dream content, and the effects appear to be automatic. There is strong evidence to support this premise.

4.5.1 Effects on attention, recall and thought content: the general case

Initial investigations of this theory asked participants to listen to series of two different, simultaneous, 15 min narratives on audiotape, one narrative to each ear. The narratives were in each case drawn from the same author, but at particular time points a few words were altered so that participants heard passages in one ear that were associated with their own concerns, and, simultaneously, passages in the other ear that were related to another's concerns. Participants spent significantly more time listening to passages associated with their own concerns than to the others, recalled those passages much more often and had thought content that (by ratings of blind judges) was much more often related to them than to the simultaneous passages that were related to someone else's concerns (Klinger 1978).

   One might question whether these effects can truly be attributed to the cues rather than simply reflecting a greater likelihood of people having memories and thoughts related to their concerns. To examine this possibility, recalled passages and reported thoughts were compared with cues that participants had not yet heard. The results indicated clearly that the effects were specific to the particular cues, i.e. the tape-recorded material that participants recalled and the thoughts they reported having had resembled the concern-related cues they had just heard much more than they resembled the concern-unrelated cues they had just heard; but when the recalled material and the thought reports were compared with cues that came only later in the session (which participants had therefore not yet heard), there was little resemblance to those cues, regardless of whether they were concern-related cues or not. That makes it clear that it was the interaction of cue content with participants' individual concerns that produced the effects on recall and thought content. Thus, the concerns that underlie goal pursuits potentiate cognitive processing of goal-related cues on the three levels of attention, recall and thought content.

4.5.2 Automaticity of the effects

Subsequent studies of both waking and sleeping participants indicated that these effects are apparently non-conscious and automatic, rather than controlled and attributable to a conscious process, such as deliberately focusing on concern-related stimuli. In fact, concern-related stimuli seem to impose an extra cognitive processing load even when they are peripheral and participants are consciously ignoring them (Young 1987); when asked to judge as quickly as possible whether a string of letters on a screen constitutes a word, these apparently irrelevant distractor stimuli nevertheless slowed the lexical decisions about the target words much more to the extent that they were related to the participants' concerns.

Effects of concerns on cognition have been shown using yet another class of cognitive method: Stroop and quasi-Stroop procedures. These procedures present words on a screen (or card, in earlier versions) and instruct participants to name the font (or ink) colour of the words as quickly as possible. Distractor stimuli that slow the colour-naming response have presumably received greater processing priority than the colour-naming task. Participants in these experiments name font colours more slowly when the words are related to one of their own concerns than when they are not (Riemann and McNally 1995; Riemann et al. 1995; Cox et al. 2000; de Jong-Meyer et al. 2002; Carrigan et al. 2004), indicating that the concern-related content of these words automatically receives priority in processing. Because of the millisecond range within which such a bias occurs, the effect is presumably automatic rather than consciously controlled.

Even when people are asleep, concern-related stimuli influence dream content much more reliably than do other stimuli (Hoelscher *et al.* 1981; Nikles *et al.* 1998). Taken together, these results confirm that the effects of concern-related cues on cognitive processing are substantially automatic and probably inexorable.

4.5.3 Effects on action

These automatic effects of goal-related cues, even non-conscious ones, appear to spill over onto goal-directed actions. A series of investigations (Chartrand and Bargh 1996, 2002; Bargh *et al.* 2001) has shown that priming cues related to particular goals influence how participants perform on subsequent laboratory tasks. For example, when participants performed a first task that included unobtrusively embedded words related to achievement (versus receiving achievement-unrelated words), they performed better on a different second task, persisted longer and were more likely to resume it if interrupted (Bargh *et al.* 2001). This was true even though no participant knew the true connection between the first and second tasks; this means that the effect was probably non-conscious and hence automatic. Thus, non-conscious cues can affect performance in ways similar to the established effects (e.g. Locke 1968, 2001) of setting conscious performance goals for oneself. Priming cues related to cooperation also had a comparable effect on participants' cooperative behaviour (Chartrand and Bargh 1996), showing that these effects are not restricted to just one kind of behaviour.

4.5.4 Effects on attention to substance-related cues

When people resort to chemical substances to manipulate their moods and emotions, use and abuse of these substances become personal goals. Pursuing them instates similar processes that accompany other goal pursuits. People explicitly or implicitly become committed to these pursuits, and the resulting concerns enlist cognitive processing in the same ways. In addition, chemical incentives (e.g. alcohol, heroin) that are strongly valued and used frequently come adversely to affect brain processes involved in evaluating, selecting and pursuing incentives, such that an addicted brain tends to ignore other rewards and instead reacts strongly to chemical incentives and cues associated with them (e.g. Comings and Blum 2000; Cardinal and Everitt 2004). Theoretically, heavy drinkers, for example, will give processing priority to alcohol cues and hence will be more likely to notice, dwell on, store and think about them than would otherwise be the case.

That is the theoretical view, and it has gained strong empirical support. A variety of paradigms have been used to study cognitive processes associated with alcohol and other drug use (see, for example, Stacy 1997; Pothos and Cox 2002; Wiers *et al.* 2002; Hogarth *et al.* 2003; Jones *et al.* 2003; Waters and Green 2003). However, the addiction-Stroop is the paradigm that has been most widely used (Cox *et al.* 2006). The addiction-Stroop paradigm uses stimuli (typically words) from two semantic categories, addiction related and emotionally neutral. Each stimulus is presented in a different colour of ink (typically red, yellow, blue or green); the participant's task is to name the colour of ink as quickly and accurately as possible while ignoring the meaning of the stimuli. Addiction-related attentional interference is indexed as the difference in participants' mean time to respond to the addiction-related stimuli and their mean time to respond to the neutral stimuli. It indicates the degree to which performance on the task is impaired presumably as a result of participants' focusing their attention away from the task-relevant dimension of the stimuli (colour relatedness) and toward the dimension that should be ignored (addiction relatedness).

Four kinds of addictive behaviours—alcohol abuse, smoking, illicit drug abuse and gambling—have been studied with the Stroop paradigm. However, the majority of studies (88 per cent) have been related to either alcohol abuse or smoking (Cox *et al.* 2006). A recent meta-analysis of the alcohol-related and smoking-related Stroop studies (Cox *et al.* 2006) reached the following conclusions: drinkers show greater alcohol attentional bias than non-drinkers, and drinkers are ordered as follows from greatest to least attentional bias: alcohol abusers, heavy drinkers and light drinkers. Likewise, smokers show greater smoking-related attentional bias than non-smokers, and heavy smokers show greater bias than light smokers; moreover, when current smokers are required to remain abstinent for experimental purposes (from 2 to 12 h), their attentional bias increases. This research is described in detail elsewhere in this monograph (for example, see Chapter 3).

The slowing of the colour-naming response suggests that the brain gives processing priority to the concern-related features of the stimulus words, including to substance-related cues in the case of substance users, which delays the processing of other features and therefore slows judgements about these other features. Because of the very narrow millisecond range of this attentional bias for substance-related cues, it is almost certainly automatic and hence outside conscious control.

Waters and Feyerabend (2000) argued that attentional bias for addiction-related stimuli is important for three reasons. First, attentional bias causes substance users more easily to become aware of addiction-related stimuli in their environment than other people. Addicted individuals' sensitized brains should continue to influence their attentional system, even after they have quit, thus increasing their susceptibility to relapse. In fact, research has shown that alcohol abusers who have undergone treatment (Cox *et al.* 2002*b*) and smokers who have quit (Waters *et al.* 2003) show increases in addiction-related attentional bias, and the increases predict subsequent relapse. Secondly, the automatic responses to addiction-related stimuli can elicit conditioned responses such as withdrawal or compensatory responses, and these might increase the desire to drink or use, or they might invoke automatic patterns of behaviour that lead to substance use (see Niaura *et al.* 1988; Tiffany 1990; Feldtkeller *et al.* 2001; Baker *et al.* 2004). Thirdly, attentional bias for addiction-related stimuli is undesirable for abstainers. It disturbs their mood and interferes with their thought processes and daily activities (e.g. Forgas 1995; Bradley *et al.* 2001).

In conclusion, having goals related to addictive substances (e.g. to drink alcohol or to smoke cigarettes) sensitizes a person to attend to the relevant addiction-related stimuli. Moreover, the strength of the current concern about using the substance (as inferred from the amount that the person habitually uses) is directly related to the degree of the addiction-related attentional bias.

4.6 **Cognitive–motivational interventions**

As we saw, according to the motivational model (Cox and Klinger 1988, 2004*a*), there are two major cognitive–motivational determinants of drinking and drug use: (1) users' motivational structure, which prevents them from focusing on and successfully achieving healthy, adaptive goal pursuits as an alternative to drinking alcohol or using other drugs; and (2) users' attentional bias for addiction-related stimuli, which reflects their preoccupation with use. Paradoxically, users' attempts to reduce their use *increase* their attentional bias for addiction-related stimuli, which in turn intensifies their desire to use (e.g. Cox *et al.* 2002*b*). These mechanisms can overshadow the person's intentions not to use a drug, leading to a vicious cycle of repeated relapses. This means that interventions for substance abuse should address users' sensitized attentional system. There is a recent trend among cognitively oriented addiction researchers to develop interventions to reduce abusers' attentional bias for addiction-related stimuli (cf. Wiers *et al.* 2006)

In an attempt to reduce alcohol-specific attentional bias, we developed the Addiction Attentional Control Training Programme (AACTP: Fadardi 2003; Wiers *et al.* 2006). The AACTP trains excessive drinkers and other substance abusers to overcome their automatic attentional distraction for addiction-related stimuli. It does so in three phases. First, it assesses users' addiction-related attentional bias and provides them with feedback about the results of the assessment. This process helps them to understand the meaning and consequences of the distraction and motivates them to engage in the training programme. Secondly, the AACTP helps drinkers and drug users to set challenging but realistic goals for controlling their distractions during the training programme. Thirdly, the AACTP continually evaluates participants' progress while they take part in the programme and provides them with immediate feedback about their success (or lack of it).

In previous research to evaluate the effectiveness of the AACTP with excessive drinkers (Fadardi 2003; Wiers *et al.* 2006), participants showed progressive reductions in their alcohol-specific attentional bias. The alcohol specificity of the programme was established by finding no generalization of the effects of the training to other appetitive, personally relevant stimuli. Moreover, a group of student heavy drinkers who underwent the training showed increases in their readiness to reduce their alcohol consumption. Fadardi (2003) showed that detoxified alcohol abusers who were trained with the AACTP reduced their alcohol attentional bias, and their attention to other life concerns was enhanced. Recently, a group of excessive drinkers recruited from the community were trained with the AACTP, and their alcohol consumption was measured at baseline and at follow-up. Preliminary analysis of the results suggests that both participants' alcohol attentional bias and their alcohol consumption decreased as a result of the AACTP training.

Although the AACTP is beneficial in helping alcohol abusers to reduce their distraction for alcohol stimuli, re-directing their attention to other productive alternatives is not a part of the training. Nevertheless, the reduced distractibility for alcohol should help abusers to shift their resources toward more adaptive goal pursuits. The effects of the AACTP cannot be expected to endure, especially if the former drinker lacks the motivational skills necessary to find emotional satisfaction without alcohol. Such a person needs help to re-direct his or her attention toward other goal pursuits, so that alternative, healthy incentives can replace the alcohol. We expect that the greatest success in helping problem drinkers to control their attention and reduce their drinking will be achieved with a combination of attentional and motivational interventions. A current study from our laboratory on excessive drinkers is testing this prediction.

Accordingly, we have developed another brief intervention—called Brief Personal Concerns Counseling (PCC-B)—to help substance abusers to modify their maladaptive motivational structure. PCC-B starts by administering the PCI (Cox and Klinger 2004*b*), in order to identify participants' maladaptive motivational structure that interferes with their ability to pursue meaningful, healthy goals as an alternative to drinking alcohol or using other drugs. The motivational difficulties might include: (1) an inability to identify interim steps necessary for the achievement of long-range goals; (2) undue pessimism about goal attainments; and (3) a lack of anticipated emotional gratification from goal attainments. In short, respondents' goals might be too unrealistic and too remote for much pleasure to be derived from them. Alternatively, they might be emotionally blasé and put forth little effort to reach their goals. Such patterns make them susceptible to self-defeating behaviours such as substance abuse.

The aim of the PCC-B is to help excessive drinkers or other drug abusers to lead a satisfying life without the need for using addictive substances. PCC-B is delivered in the following steps: first, the rationale for PCC-B is made explicit to participants. They are given a clear explanation for the reciprocal relationship between the satisfactions and frustrations that people experience from other areas of their life and their need to use alcohol to regulate their positive and negative affect. Various factors can influence a person's life satisfaction, such as their feelings of control, knowing what to do to achieve goals and how realistic their goals are. Secondly, in order to personalize the principles discussed in step 1, the results from the PCI taken at baseline are depicted graphically and the meaning of them interpreted. The feedback includes concrete details about the person's motivational patterns that interfere with achieving sustained emotional satisfaction from goal pursuits, thus making it less likely that the person will be able to reduce drinking or other drug use. Thirdly, participants are helped to develop skills needed for resolving important concerns by teaching them motivational enhancement techniques and exercises. The techniques include: (1) re-evaluating their sources of emotional satisfaction; (2) shifting toward an appetitive motivational style and away from an avoidant one; (3) constructing goal ladders to facilitate mastery of small, interim steps that underlie long-range achievements; and (4) developing skills needed to cope with negative affect and enhance positive affect. The person is also given worksheets detailing the principles underlying successful goal achievements and homework assignments to practise the newly learned skills. Fourthly, a summary is prepared of the participant's progress and his or her plan for continuing the application of the motivational principles to his or her prioritized goals during the next 3 and 6 months.

The effectiveness of PCC-B is currently being evaluated. However, previous evaluations of an extended version of the technique called Systematic Motivational Counseling (SMC: Cox and Klinger 2004c) have shown that it reduces alcohol consumption, and the reduction is mediated by improvements in alcohol abusers' maladaptive motivation (e.g. Cox *et al.* 2003; Miranti and Heinemann 2004). Theoretically, the two major variables in the cognitive–motivational model that influence drinking decisions—alcohol attentional bias and motivational structure—are important keys for maximizing success in changing abusive drinking patterns. Focusing on their combined and interacting influences is expected considerably to enhance our ability to change excessive drinking behaviour.

Acknowledgements

Any correspondence concerning this chapter should be addressed to Professor W. Miles Cox, School of Psychology, University of Wales, Bangor, LL57 2AS.

References

Baker, T.B., Brandon, T.H. and Chassin, L. (2004). Motivational influences on cigarette smoking. *Annual Review of Psychology*, 55, 463–491.

Bargh, J.A., Gollwitzer, P.M., Lee-Chai, A., Barndollar, K. and Trötschel, R. (2001). The automated will: nonconscious activation and pursuit of behavioral goals. *Journal of Personality and Social Psychology*, 81, 1014–1027.

Bradley, M.M., Codispoti, M., Cuthbert, B.N. and Lang, P.J. (2001). Emotion and motivation I: defensive and appetitive reactions in picture processing. *Emotion*, 1, 276–298.

Cardinal, R.N. and Everitt, B.J. (2004). Neural and psychological mechanisms underlying appetitive learning: links to drug addiction. *Current Opinions on Neurobiology*, 14, 156–162.

Carrigan, M.H., Drobes, D.J. and Randall, C.L. (2004). Attentional bias and drinking to cope with social anxiety. *Psychology of Addictive Behaviors*, 18, 374–380.

Chartrand, T.L. and Bargh, J.A. (1996). Automatic activation of impression formation and memorization goals: nonconscious goal priming reproduces effects of explicit task instructions. *Journal of Personality and Social Psychology*, 71, 464–478.

Chartrand, T.L. and Bargh, J.A. (2002). Nonconscious motivations: their activation, operation, and consequences. I.A. Tesser and D.A. Stapel (ed.), *Self and motivation: emerging psychological perspectives*. Washington, DC.American Psychological Association, pp. 13–41.

Comings, D.E. and Blum, K. (2000). Reward deficiency syndrome: genetic aspects of behavioral disorders. *Progress in Brain Research*, 126, 325–341.

Correia, C.J. (2004). Behavioral economics: basic concepts and clinical applications. In W.M. Cox and E.Klinger (ed.), *Handbook of motivational counseling: concepts, approaches, and assessment*. Chichester, UK: Wiley, pp. 49–64.

Cox, W.M. and Klinger, E. (1988). A motivational model of alcohol use. *Journal of Abnormal Psychology*, 97, 168–180.

Cox, W.M. and Klinger, E. (1990). Incentive motivation, affective change, and alcohol use: a model. In W.M. Cox (ed.), *Why people drink: parameters of alcohol as a reinforcer*. New York, NY: Amereon Press, pp. 291–314.

Cox, W.M. and Klinger, E. (2004a). A motivational model of alcohol use: determinants of use and change. In W.M. Cox and E. Klinger (ed.), *Handbook of motivational counseling: concepts, approaches, and assessment*. Chichester, UK: Wiley, pp. 121–138.

Cox, W.M. and Klinger, E. (2004b). Measuring motivation: the Motivational Structure Questionnaire and Personal Concerns Inventory. In W.M. Cox and E. Klinger (ed.), *Handbook of motivational counseling: concepts, approaches, and assessment*. Chichester, UK: Wiley, pp. 141–175.

Cox, W.M. and Klinger, E. (2004c). Systematic motivational counseling: the Motivational Structure Questionnaire in action. In W.M. Cox and E. Klinger (ed.), *Handbook of motivational counseling: concepts, approaches, and assessment*. Chichester, UK: Wiley, pp. 217–239.

Cox, W.M., Blount, J.P. and Rozak, A.M. (2000). Alcohol abusers' and nonabusers' distraction by alcohol and concern-related stimuli. *American Journal of Drug and Alcohol Abuse*, **26**, 489–495.

Cox, W.M., Schippers, G.M., Klinger, E., Skutle, A., Stuchlíková, I., Man, F., King, A.L. and Inderhaug, I. (2002a). Motivational structure and alcohol use of university students with consistency across four nations. *Journal of Studies on Alcohol*, **63**, 280–285.

Cox, W.M., Hogan, L.M., Kristian, M.R. and Race, J.H. (2002b). Alcohol attentional bias as a predictor of alcohol abusers' treatment outcome. *Drug and Alcohol Dependence*, **68**, 237–243.

Cox, W.M., Heinemann, A.W., Miranti, S.V., Schmidt, M., Klinger, E. and Blount, J.P. (2003). Outcomes of systematic motivational counseling for substance use following traumatic brain injury. *Journal of Addictive Diseases*, **22**, 93–110.

Cox, W.M., Fadardi, J.S. and Pothos, E.M. (2006). The addiction-Stroop test: theoretical considerations and procedural recommendations. *Psychological Bulletin*, **132**(3), 443–476.

de Jong-Meyer, R., Eickelmann, S., Lindenmeyer, J. and Gerlach, A.L. (2002). Selective attention of alcohol addicts to alcohol related and current concern related stimulus words: no effects of exposure in the Stroop paradigm/Selektive Aufmerksamkeit Alkoholabhaeniger bei alkoholbezogenen und motivational bedeutsamen Stimulusworten: Keine Expositionseffekte im Stroop-Paradigma. *Sucht: Zeitschrift für Wissenschaft und Praxis*, **48**, 422–430.

Fadardi, J.S. (2003). Cognitive–motivational determinants of attentional bias for alcohol-related stimuli: implications for an attentional-control training programme. Unpublished PhD thesis. University of Wales, Bangor, UK.

Feldtkeller, B., Weinstein, A., Cox, W.M. and Nutt, D. (2001). Effects of contextual priming on reactions to craving and withdrawal stimuli in alcohol-dependent participants. *Experimental and Clinical Psychopharmacology*, **9**, 343–351.

Forgas, J.P. (1995). Mood and judgments: the affect infusion model (AIM). *Psychological Bulletin*, **117**, 39–66.

Hoelscher, T.J., Klinger, E. and Barta, S.G. (1981). Incorporation of concern- and nonconcern-related verbal stimuli into dream content. *Journal of Abnormal Psychology*, **49**, 88–91.

Hogarth, L., Dickinson, A. and Duka, T. (2003). Discriminative stimuli that control instrumental tobacco-seeking by human smokers also command selective attention. *Psychopharmacology (Berlin)*, **168**, 435–445.

Iguchi, M.Y., Belding, M.A., Morral, A.R., Lamb, R.J. and Husband, S.D. (1997). Reinforcing operants other than abstinence in drug abuse treatment: an effective alternative for reducing drug use. *Journal of Consulting and Clinical Psychology*, **65**, 421–428.

Jones, B.T., Jones, B.C., Smith, H. and Copely, N. (2003). A flicker paradigm for inducing change blindness reveals alcohol and cannabis information processing biases in social users. *Addiction*, **98**, 235–244.

Klinger, E. (1977). *Meaning and void: inner experience and the incentives in people's lives.* Minneapolis, MN: University of Minnesota Press.

Klinger, E. (1987). The interview questionnaire technique: reliability and validity of a mixed idiographic-nomothetic measure of motivation. In J.N. Butcher and C.D. Spielberger (ed.), *Advances in personality assessment*, Vol. 6. Hillsdale, NJ: Erlbaum, pp. 31–48.

Klinger, E. (1996). Emotional influences on cognitive processing, with implications for theories of both. I.P. Gollwitzer and J.A. Bargh (ed.), *The psychology of action: linking cognition and motivation to behavior*. New York, NY: Guilford Press. pp. 168–189.

Klinger, E. (1975). Consequences of commitment to and disengagement from incentives. *Psychological Review*, **82**, 1–25.

Klinger, E. (1978). Modes of normal conscious flow. In K.S. Pope and J.L. Singer (ed.), *The stream of consciousness: scientific investigations into the flow of human experience*. New York, NY: Plenum, pp. 225–258.

Klinger, E. and Cox, W.M. (2004). Motivation and the theory of current concerns. In W.M. Cox and E. Klinger (ed.), *Handbook of motivational counseling: concepts, approaches, and assessment*. Chichester, UK: Wiley, pp. 3–29.

Locke, E.A. (1968). Toward a theory of task motivation and incentives. *Organizational Behavior and Human Performance*, **3**, 157–189.

Locke, E.A. (2001). Motivation by goal setting. In R.T. Golembiewski (ed.), *Handbook of organizational behavior*, 2nd. edn. New York, NY: Marcel Dekker, pp. 43–56.

Mellers, B.A. (2000). Choice and the relative pleasure of consequences. *Psychological Bulletin*, **126**, 910–924.

Miranti, S.V. and Heinemann, A.W. (2004). Systematic motivational counseling in rehabilitation settings. In W.M. Cox and E. Klinger (ed.), *Handbook of motivational counseling: concepts, approaches, and assessment*. Chichester, UK: Wiley, pp. 302–317.

Moos, R.H., Finney, J.W. and Cronkite, R.C. (1990). *Alcoholism treatment: context, process, and outcome*. New York, NY: Oxford University Press.

Niaura, R.S., Rohsenow, D.J., Binkoff, J.A., Monti, P.M., Pedraza, M. and Abrams, D.B. (1988). Relevance of cue reactivity to understanding alcohol and smoking relapse. *Journal of Abnormal Psychology*, **97**, 133–152.

Nikles, C.D. II, Brecht, D.L., Klinger, E. and Bursell, A.L. (1998). The effects of current-concern and nonconcern related waking suggestions on nocturnal dream content. *Journal of Personality and Social Psychology*, **75**, 242–255.

Pothos, E.M. and Cox, W.M. (2002). Cognitive bias for alcohol-related information in inferential processes. *Drug and Alcohol Dependence*, **66**, 235–241.

Riemann, B.C. and McNally, R.J. (1995). Cognitive processing of personally-relevant information. *Cognition and Emotion*, **9**, 325–340.

Riemann, B.C., Amir, N. and Louro, C.E. (1995). Cognitive processing of personally relevant information in panic disorder. Unpublished manuscript.

Sharma, D., Albery, I.P. and Cook, C. (2001). Selective attentional bias to alcohol related stimuli in problem drinkers and non-problem drinkers. *Addiction*, **96**, 285–295.

Stacy, A.W. (1997). Memory activation and expectancy as prospective predictors of alcohol and marijuana use. *Journal of Abnormal Psychology*, **106**, 61–73.

Tiffany, S.T. (1990). A cognitive model of drug urges and drug-use behavior: role of automatic and nonautomatic processes. *Psychological Review*, **97**, 147–168.

Tucker, J.A., Vuchinich, R.E. and Pukish, M.M. (1995). Molar environmental contexts surrounding recovery from alcohol problems by treated and untreated problem drinkers. *Experimental and Clinical Psychopharmacology*, **3**, 195–204.

Tucker, J.A., Vuchinich, R.E. and Rippens, P.D. (2002*a*). Environmental contexts surrounding resolution of drinking problems among problem drinkers with different help-seeking experiences. *Journal of Studies on Alcohol*, **63**, 334–341.

Tucker, J.A., Vuchinich, R.E. and Rippens, P.D. (2002*b*). Predicting natural resolution of alcohol-related problems: a prospective behavioral economic analysis. *Experimental and Clinical Psychopharmacology*, **10**, 248–257.

Vuchinich, R.E. and Tucker, J.A. (1996). Alcoholic relapse, life events, and behavioral theories of choice: a prospective analysis. *Experimental and Clinical Psychopharmacology*, **4**, 19–28.

Vuchinich, R.E. and Tucker, J.A. (1998). Choice, behavioral economics, and addictive behavior patterns. In W.R. Miller and N. Heather (ed.), *Treating addictive behaviors*, 2nd edn. New York, NY: Plenum Press, pp. 93–104.

Waters, A.J. and Feyerabend, C. (2000). Determinants and effects of attentional bias in smokers. *Psychology of Addictive Behaviors*, **14**, 111–120.

Waters, A.J., Shiffman, S., Sayette, M.A., Paty, J.A., Gwaltney, C.J. and Balabanis, M.H. (2003). Attentional bias predicts outcome in smoking cessation. *Health Psychology*, **22**, 378–387.

Waters, H. and Green, M.W. (2003). A demonstration of attentional bias, using a novel dual task paradigm, towards clinically salient material in recovering alcohol abuse patients? *Psychological Medicine*, **33**, 491–498.

Wiers, R.W., van Woerden, N., Smulders, F.T. and de Jong, P.J. (2002). Implicit and explicit alcohol-related cognitions in heavy and light drinkers. *Journal of Abnormal Psychology*, **111**, 648–658.

Wiers, R.W., Cox, W.M., Field, M., Fadardi, J.W., Palfai, T., Schoenmakers, T. and Stacy, A. (2006). *The search for new ways to change implicit alcohol-related cognitions in heavy drinkers. Alcoholism: Clinical and Experimental Research*, **30**(2), 320–331.

Williams, J.M.G., Mathews, A. and MacLeod, C. (1996). The emotional Stroop task and psychopathology. *Psychological Bulletin*, **120**, 3–24.

Wong, C. J., Jones, H. E. & Stitzer, M. L. (2004). Community Reinforcement Appproach and contingency management interventions for substance abuse. In W. M. Co & E. Klinger (Eds.), *Handbook of motivational counselling: concepts, approaches, and assessment* (pp. 421–437). Chichester, United Kingdom: Wiely.

Young, J. (1987). The role of selective attention in the attitude–behavior relationship. Unpublished PhD thesis. University of Minnesota, USA.

5

Towards understanding loss of control: an automatic network theory of addictive behaviours

Chris McCusker

5.1 Introduction

The highest possible stage in moral culture is when we recognise that we ought to control our thoughts

Darwin (1871). *The descent of man*

The hallmark of addictive behaviours is undoubtedly loss of control. Most diagnostic descriptions of addiction-related syndromes make reference to the individual ingesting larger amounts of the substance, and at more regular intervals, than intended (Edwards and Gross 1976; American Psychiatric Association 1994). Related features include repeated relapses—at odds with abstention intentions—and the apparent inability to curtail use of the substance despite increasing costs to physical, psychological and social functioning. Confronted with such a self-defeating and irrational pattern of behaviour, it is unsurprising that early theories of addictive behaviours made reference to a 'disease' of the will (e.g. Jellinek 1960).

A disease implies some internal, pathological, agent is at play, largely outside the individual's control. As such, this could plausibly explain loss of control features described above. So seductive have disease theories been for explaining loss of control that such notions have been generalized in various ways to other excessive behaviours, from substance abuse to eating and sexual disorders, and have entered both popular and therapeutic discourses as a sign of enlightened understanding and treatment (Heather and Robertson 1997).

However plausible and metaphorically useful the disease notion has been, its explanatory potential for understanding loss of control proved to be empirically flawed. Theoretically tautologous, driven by social rather than scientific imperatives, and paradigmatically undermined by a return to controlled drinking in those diagnosed with the 'disease' of 'alcoholism' (Marlatt 1983; Heather and Robertson 1997), disease notions became increasingly assailed through the

1960s and 1970s by evidence that addictive behaviours were subject to the scientific laws of learning theory. Predominant was the negative reinforcement hypothesis. This suggested that drug usage was a generalized response to aversive stimulation such as stress, negative emotional states or indeed drug withdrawal itself (Bandura 1969; Wikler 1980; Cappell and Greely 1987). By the 1980s, expectancy constructs, again seen as scientifically grounded (Jones *et al.* 2001), but which allowed for indirect as well as direct experience to shape behavioural preferences, were integrated within increasingly dominant social learning theories of addictive behaviours (e.g. Marlatt and Gordon 1985). These have grown increasingly to influence treatment models which espouse the credibility of a grounding in psychological science (Monti *et al.* 1989).

Whilst acknowledging the empirical basis to social learning theory accounts of addictive behaviour, the present chapter will argue that these models remain fundamentally limited—especially in accounting for the loss of control features of addictive behaviours described above. *Automaticity* is posited as a central theoretical concept which, it will be argued, has been neglected by social learning models and which needs greater prominence in theories of addictive behaviours. Research which has pursued the relevance of automatic processes in addictive behaviours is discussed.

Non-volitional physiological responses to drug-related stimuli became a first focus of attention through the 1980s and 1990s. These phenomena offered a promising window to automatic processes, outside of the individual's control, which might help explain the loss of control features in addiction and the desynchrony between cognitive intentions to remain abstinent and behavioural patterns of ongoing drug usage. Despite the promise, this research vein was ultimately compromised by the classical conditioning models used to explain it and drive therapeutic implications.

In the past decade, an emergent body of work has drawn attention to automatic processes of a cognitive nature. This work has been contextualized within the principles and paradigms of cognitive science, rather than social learning theory. These automatic cognitive processes are described and their parameters discussed. A theoretical formulation, which can incorporate the automatic physiological phenomena indicated above, is delineated and relevance for understanding loss of control considered.

5.2 Cognitive social learning theory: advent of the expectancy construct

Recent psychological approaches to treating addictive behaviours have generally been grounded in cognitive social learning theory (e.g. Marlatt and

Gordon 1985, Monti *et al.* 1989). Initial behavioural formulations emphasized drug use as an instrumentally learned response, which functionally served to alleviate negative affective states including drug withdrawal (Wikler 1980; Cappell and Greely 1987). More recent formulations suggest that the individual may also respond to external or internal cues that such negative affect is imminent and use the drug to avert such a state occurring (Baker *et al.* 2004). Other behavioural models emphasized appetitive conditioning—either through primary reinforcement processes (Stewart *et al.* 1984) or through secondary conditioning to other reinforcers (McCusker and Bell 1988).

Perhaps in response to the determinism of the behaviourist and psychodynamic movements, and certainly in the light of evidence that direct experience was not necessary to shape behavioural preferences and responses, the emergence of cognitive social learning theory in the 1960s emphasized personal expectancies and beliefs (Bandura 1977). In the addiction arena, it was initially emphasized that the shaping of drug-using behaviours could occur through 'differential reinforcement *and* modelling experiences' (Bandura 1969, p. 536). Evidence grew that whilst the actual stress-relieving and hedonistic properties of dominant drugs of abuse such as alcohol might be equivocal, the key factors were perceptions and expectations that such was so (Falk 1983). The socio-cultural influences of family, peer group and popular culture came to be seen as at least as important as direct experience in the creation of internalized expectancies for drug effects (Biddle *et al.* 1980). However, whether through direct experience or observational learning, the key construct in cognitive theories of addiction came to be the 'expectancy' (Jones *et al.* 2001; Leigh and Stacy 2004).

Jones *et al.* (2001) argue that the expectancy construct is the most scientifically grounded in current cognitive theories of drug-use motivation. Expectancies are defined by these authors as 'structures in long-term memory that have impact on cognitive processes governing current and future consumption' (p. 59). They represent assumptions and beliefs about the outcomes of drug-using behaviours (e.g. 'Drinking will relax me...Smoking is cool and sophisticated...'); the variability in the extent and strength of these expectancies across individuals is generally thought to explain the variability in drug-using behaviours therein. Indeed, biased expectancies (inflated positive and minimized negative) are thought to underlie addictive phenomenology in much the same way that 'dysfunctional assumptions' about the self, the world and the future are thought to underpin mood disorders (Beck 1976; Beck *et al.* 1993). Cognitive appraisals related to low self-efficacy, or the ability to cope with 'high-risk' situations without the drug, have been a theoretical adjunct in some models (e.g. Marlatt and Gordon 1985; Monti *et al.* 1989), but the critical ingredient appears to be the activation of the drug-related expectancy network, whether in the context of stress or not.

In support of this position, numerous studies have charted an association between the extent of positive and negative drug-related beliefs and drug-using behaviours in both cross-sectional and longitudinal designs (Brown *et al.* 1987; Schafer and Brown 1991; Stacy *et al.* 1991; Sher *et al.* 1996). Alcohol expectancies modulate from predominantly negative to positive from childhood to adolescence, in a way which is associated with the onset of drinking behaviour, and expectancies of negative outcomes increase with age and drinking experience as negative experience is accrued (McMahon *et al.* 1994; Leigh and Stacy 2004). The extent of positive expectancies has been shown to differentiate heavy from light drinkers (Southwick *et al.* 1981) and, as highlighted in Fig. 5.1, the degree of the disjunction between negative

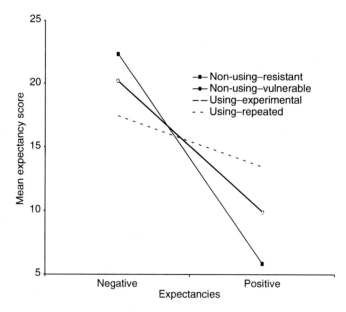

Fig. 5.1 Mean expectancy scores on positive and negative dimensions across four groups of teenagers. One group had never used illegal drugs and their behavioural intentions suggested they were resistant to the idea (non-using–resistant); one group had never used but indicated they would like to try at least one illegal drug listed (non-using–vulnerable); a third group had used at least one illegal drug but only 'once or twice' (using–experimental); a fourth group used at least one illegal drug 'sometimes or regularly' (using–repeated). From McCusker *et al.* (1995). **NB** The non-using–vulnerable and the using-experimental group lines overlap on the above figure.

and positive expectancies was shown to observe a linear relationship across four groups of teenagers on a behavioural continuum of 'non-using–resistant', 'non-using–vulnerable', 'using–experimental' and 'using–repeated' illegal drug users (McCusker *et al.* 1995). Finally, expectancies, together with the subjectively weighted 'values' attached to those expectancies, have been shown to predict post-treatment outcomes (Connors *et al.* 1993; Jones and McMahon 1994).

Given this evidence base, therapies in the cognitive–behavioural tradition have aimed to restructure the expectancy network assumed to be of motivational significance. Interventions have therefore included interpersonal skills training to help the individual cope with stressful situations without the drug, and mood management therapies have been used to eliminate the need for the drug as mood modulator (Marlatt and Gordon 1985; Monti *et al.* 1989). These strategies have been complemented with cognitive and motivational techniques which challenge positive outcome expectancies, increase the value ascribed to negative expectations and bolster self-efficacious beliefs about coping and attaining drug control or abstinence (Miller and Rollnick 1984; Marlatt and Gordon 1985; Monti *et al.* 1989; Beck *et al.* 1993). Whilst such interventions have enjoyed some success (Monti *et al.* 1989), fundamental problems remain.

5.3 Anomalies and limitations in traditional cognitive and expectancy theories

The association between expectancies and drug-using behaviours has been statistically significant, as detailed above. However, expectancies have been shown to be a function of demographic and behavioural factors such as age, gender and previous drug use. When studies control for these factors, the additional predictive utility of expectancies for behaviour in regression models has been as little as 1–2 per cent (Leigh 1989; Carey 1995).

Cognitive restructuring, as noted above, is the logical intervention which follows from expectancy theory. However, psychoeducational initiatives driven by assumptions that promoting negative drug expectancies would counter rising patterns of illegal drug use in young people have had equivocal, and even counterproductive, success (Ho 1992; Plant and Plant 1992). Where changes in behaviour have been found, it is not at all clear that this has been accompanied, let alone precipitated, by expectancy change (Jones *et al.* 2001). Furthermore, high relapse rates still bedevil intervention programmes which assume that bolstering coping and self-control strategies will reduce the value ascribed to positive drug outcomes in high-risk situations (Davis and Glaros 1986).

Conceptually, the research outlined above is largely based on correlational designs. Even in longitudinal studies, it has not been clearly established that expectancies precede patterns of addictive behaviours. It is entirely possible that expectancies expressed by these research samples are *post hoc* rationalizations of drug-using behaviours (McCusker 2001). Thus, if I drink heavily, dissonance might be created if I did not agree with the statement that 'drinking is relaxing' and where behavioural change is difficult, attitudes are moderated to ensure cognitive–behavioural consistency (Festinger 1962). The stereotype of the addict 'in denial' at a pre-contemplative stage of behavioural change (Prochaska and DiClemente 1982) becomes more understandable when gauged in this way. Expectancy change may accompany or bolster behavioural change, but it is not at all certain that it is the crucial precipitant of such.

In addition to the use of correlational designs with equivocal significance, the studies summarized above have emphasized patterns of between-group differences (e.g. heavy versus light drinkers or users versus non-users of a drug). As argued in a previous paper (McCusker 2001), insufficient attention has been given to *within*-group differences in the balance of positive versus negative outcome expectancies. When reported, these have often thrown up anomalous findings for expectancy theory. Smokers, for example, may endorse more positive and fewer negative outcome expectancies than non-smokers, but within smoking samples the difference between positive and negative expectancies appears to be at best non-significant and, at worst (for expectancy theory), negative beliefs outweigh the positive (Hill and Grey 1983; Litz *et al.* 1987; Sutton *et al.* 1990; Leung and McCusker 1999). Although the disjunction between negative and positive drug-related expectancies reduced in line with vulnerability to, and actual, drug-using behaviours, all groups in the McCusker *et al.* (1995) study, from 'resistant' non-users to 'repeated' users, manifested a greater number of negative than positive expectancies *per se* (see Fig. 5.1). Such findings must be at odds with the predicted relationships between cognitive structures and behavioural outcomes outlined in expectancy theory.

Related, cognitive theories of health behaviours (e.g. Ajzen 1991) would suggest that expectancies, when integrated with other cognitive factors (e.g. subjective expected utility estimates—or the 'value' attached to outcome expectancies), mediate 'behavioural intention', which has superior predictive utility for ongoing behaviour than expectancies or attitudes alone (Conner and Sparks 1996). Whilst integrating value estimates with expectancies has proven useful in addiction research (Jones *et al.* 2001), the essential relationship between behavioural intentions and ongoing behaviours appears fundamentally compromised in addictive behaviours. Given the relatively greater number of negative versus positive expectancies held by smokers, as described above, it is not surprising that at any given time 50 per cent *intend* to quit (Ho 1992). However, many fewer actually do quit and a very significant proportion of those who do (up to 90 per cent) will have returned to smoking within a year (Cohen 1989; Hughes 1992). Indeed, as argued previously (McCusker 2001), and indicated above, desynchrony between expressed intentions to abstain from, or moderate, drug usage and ongoing behaviours to the contrary essentially represents the loss of control hallmark of addictive behaviours. This suggests there may be processes governing the behaviour which are outside of conscious awareness and volitional control.

5.4 Autonomic cue-reactivity phenomena: initial indices of automatic processes

In the 1980s, research interest grew in the finding that detoxified abusers of alcohol and heroin manifested autonomic reactivity to environmental stimuli previously associated with their drug usage (Hinson and Siegel 1982; Pomerleau *et al.* 1983; Monti *et al.* 1987). These findings were variously interpreted as conditioned withdrawal (Wikler 1965), conditioned compensatory responses (Hinson and Siegel 1982) or conditioned appetitive responses (Stewart *et al.* 1984). Consistent across theories was a classical conditioning mechanism of action, whereby cues associated with previous drug use came automatically to trigger drug-related physiological responses, which were proposed to be the biological basis of craving and subsequent relapse. Autonomic cue reactivity was indeed shown to differentiate problem from non-problem drinkers, be associated with craving and predict latency to relapse in a more robust way than results of cognitive self-reports (Pomerleau *et al.* 1983; Monti *et al.* 1987; McCusker and Brown 1991; Rohsenow *et al.* 1992). More recent formulations suggest that conditioned responses may occur to internal as well as external cues, that negative affect will occur if the drug is not ingested and that relapse can be triggered without measurable or observable reactivity taking place (Baker *et al.* 2004). Since individuals can no more control these automatic autonomic responses to drug-related cues than Pavlov's dogs could control their conditioned salivation, here perhaps was a theoretical candidate to explain the loss of control phenomenon so poorly explicated by traditional social learning theories. Such hopes, however, proved to be premature.

First, drug-using episodes or relapses do not appear to be necessarily preceded by craving or negative affect (McAuliffe 1982). Secondly, the utility of cue reactivity in differentiating problem versus non-problem drinking status, or indeed its association with degree of dependence, was inconsistent and equivocal across studies (Childress *et al.* 1986; Meyer 1988; McCusker and Brown 1991). Thirdly, cue-exposure treatments, the logical implication of conditioning explanations of cue-reactivity phenomena, did not appear to have benefits over and above standard cognitive–behavioural treatment packages (Dawe *et al.* 1993). Finally, and perhaps of most conceptual damage to these theories, cue reactivity did not appear sufficient in itself to trigger craving. Rather, cognitive factors, such as self-efficacy (Niaura 2000), or the degree to which cue reactivity promoted the subjective experience of anxiety in the individual exposed to drug cues (McCusker and Brown 1991), appeared to mediate the relationship between autonomic reactivity and craving. Moreover, that cue exposure could apparently extinguish autonomic reactivity in response to drug-related stimuli, but not the subjective experience of craving, which it putatively precipitated (McCusker and Brown 1995), suggested that for all their promise as the automatic agents of loss of control, the story was likely to be more complex than conditioning models suggested. It appeared that cognitive processes would need further attention and integration.

Some have attempted this integration by essentially patching together the two social learning and conditioning theories reviewed above. In essence, autonomic cue reactivity, especially in so far as it undermines an already fragile self-efficacy, may give impetus to the sorts of drug expectancies outlined earlier and precipitate further drug-using behaviours (Beck *et al.* 1993; Drummond *et al.* 1995; Niaura 2000). Although this appears theoretically parsimonious, the cognitive dimension of expectancies remains unchanged and thus the anomalies described above remain. However, an evolution in the theoretical frameworks and methods of cognitive analysis has indicated another way forward (McCusker 2001).

5.5 Rethinking methods of cognitive assessment

In previous studies (Leung and McCusker 1999; McCusker 2001) it has been argued that the methods employed by research grounded in cognitive social learning models of addiction may have some fundamental limitations of a conceptual nature, which significantly muddy the picture painted. Most research in this genre has purported to measure the extent and strength of drug-related cognitions by asking individuals to endorse, or rate the strength to which they endorse, propositions related to drug use (e.g. 'Alcohol is relaxing ... smoking is cool and sophisticated ...'). Supposedly, such methods are evaluating the content of the individual's long-term memory with respect to the propositions stored there about drug use (Ingram 1990). However, this method may be as useful—and as flawed—as basing a neuropsychological assessment of an individual's memory on their self-report of all the different aspects of their memory functioning. Certainly, some insights may be apparent. However, when memory is most severely disturbed, awareness and insight may be grossly impaired. Investigating cognitive structures in memory related to drug use, using what is essentially a cued judgement protocol, and without paying sufficient attention to the cognitive processes involved in information encoding, storage and retrieval, may come to be seen as a naïve period in the history of addiction research (Leung and McCusker 1999).

General problems with these methods have long been known (Nisbett and Wilson 1977), and limitations in their use in addiction research have been recently described (Lowman et al. 2000; McCusker 2001). Perhaps most fundamental is the tendency for retrieval of information from memory to be highly cue and situation dependent. Thus, information in memory may be best retrieved under conditions comparable with that in which it was initially encoded (Tulving and Thomson 1973). Propositions accessed in the typical research, or indeed clinical, context may bear little resemblance to the sorts of propositions or constructs activated within drug-related contexts. Both sets of propositions may be 'true' for the espouser and, rather than represent a contradiction, this may highlight what Mischel (2004) has long noted about the situational specificity of cognitive dimensions of motivation. Cooney et al.'s (1987) findings of a change in reported alcohol expectancies pre- and post-alcohol cue exposure attest to this inference.

Other problems with using inventories of expectancies include the fact that the responses to earlier probe items may actually prime or otherwise bias responses to later items, e.g. in an effort to be internally consistent (Bargh et al. 1986). As noted earlier, dissonance processes may mean that expectancy endorsements are made to be consistent with current behaviour and context, rather than being precursors to that behaviour. Relatedly, demand characteristics and social conformity effects (e.g. 'Since it is a known fact that smoking causes cancer I would be a fool not so say that I agree with this statement') may also mediate inventory responses without telling us anything about whether these are indeed the propositions which are personally relevant and accessible from memory in drug-using contexts.

Emergent ideas, that to understand cognitive features of addictive behaviours we need to consider how constructs and paradigms from cognitive science can help reveal the propositions available and accessible from memory (Leung and McCusker 1999), have prompted a second wave, or evolution, in the methods used to study cognitive structures and processes related to addictive behaviours (McCusker 2001). These methods have not relied solely on what individuals, on directed introspection, say about their expectancies. Rather, inferences are made about underlying cognitive processes and structures based on behavioural responses. Such methods, which have a long history in cognitive neuroscience, include memory, priming, reaction time, perceptual judgement and other tasks as described below. These tasks incorporate drug-related stimuli as material within their paradigms, but do not ask the individual to introspect about reasons for behaviour or make judgements about expectancies. In this way, the problems which have beset traditional methods of cognitive enquiry outlined above (dissonance and demand processes, etc.), and which undoubtedly contributed to the anomalies in traditional cognitive models of addictive behaviours outlined earlier, may be avoided. From these paradigms, cognitive processes and structures of an 'automatic' nature have been revealed.

5.6 Automatic cognitive processes in addictive behaviours

Although theoretically and empirically incomplete, cue-reactivity research established two important conclusions. First, this research established that *automatic* drug-related responses existed and, although problematic for the original conditioning theory, subsequent research indicated there could exist a desynchrony between such automatic processes and non-automatic cognitive reports of craving and expectancy. Secondly, there did appear to be some indication that such automatic processes had greater predictive significance for key features of the addictive behaviour, such as latency to relapse (Rohsenow *et al.* 1992), than non-automatic cognitive reports.

Tiffany's (1990) review suggested a conceptual way forward. His thesis suggested that cue reactivity may be simply a physiological index of other automatic processes of a cognitive nature. Traditional cognitive features such as craving and expectancy reports, he suggested, were to some extent only revealed when an automatic cognitive–behaviour sequence was interrupted (e.g. by attempts at abstinence or obstacles to drug availability). Automatic and non-automatic processes were seen as at least partially independent and, although both could mediate the addictive behaviour, an emphasis was given to automatic processes. Tiffany (1990) and Tiffany and Carter (1998) likened addictive behaviours to other over-rehearsed, skilled, behaviours (such as piano playing or driving a car). Although initially guided by intentional, non-automatic, cognitive processes, with time such behaviours become increasingly mediated by cognitive processes and judgements which are largely outside of conscious awareness and control. Indeed, such a proposition, that the control of behavioural preferences and patterns shifts in cognitive control from intentional to automatic processes, has been more widely applied (Bargh and Chartrand 1999). These authors present evidence for such across multifarious social behaviours and suggest that this reduces the demand on limited cognitive resources, thus conferring evolutionary advantage (Bargh and Chartrand 1999). In addiction research, however, if such were true, then asking the drug user to introspect about the expectancies and judgements underlying their drug-using behaviour may be as useful, and as hopelessly limited, as the professional golfer simply 'telling' the weekend player how to execute the perfect drive.

Cognitive science has defined automatic processes as fast, autonomous and without intention, difficult to control, effortless and involving little conscious awareness (Posner and Snyder 1975; Shiffrin and Schneider 1977; Logan 1988). A number of studies across a range of addictive behaviours have now suggested that cognitive processes related to addictive behaviours may operate in such an automatic fashion.

A well-utilized protocol has involved the modified Stroop paradigm (Stroop 1935). In this task, the goal-directed behaviour and cognitive intention is to name the varying colours of ink in which a series of words are printed. However, when the word stimuli are names of conflicting colours (e.g. 'blue' printed in red ink), an interference effect is manifested and colour-naming response times increase. It has been interpreted that reading, and the processing of the semantic meaning of stimuli, is an automatic process which interferes with the non-automatic perceptual task of colour naming (Stroop 1935; MacLeod 1991). Using modified Stroop paradigms, it has been repeatedly demonstrated that problem drinkers (Johnsen et al. 1994; Cox et al. 1999), smokers (Gross et al. 1993) and heroin users (Franken et al. 2000) will manifest a selective interference effect for stimuli related to their addictive behaviour in comparison with control stimuli. The same effect has been reported with compulsive gamblers (McCusker and Gettings 1997; see Fig. 5.2) and individuals with eating disorders (Cooper et al. 1992). Moreover, the strength of this interference has been associated with the degree of dependence (Ryan 2002) and treatment outcome (Cox et al. 2002). This research is described in detail elsewhere in this monograph (see Chapter 3).

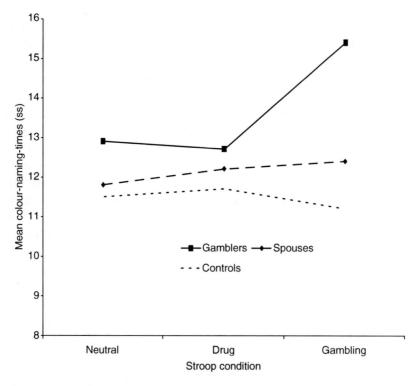

Fig. 5.2 Mean colour-naming times on modified card Stroops of neutral, drug and gambling stimuli for gamblers, their spouses and an additional control group (McCusker and Gettings 1997).

The automaticity of this effect has generally been inferred from the fact that interference is affected on an intentional, goal-directed, behaviour (i.e. colour naming). The attentional bias has also been demonstrated in such studies even when conscious perception of the salient word does not occur (e.g. on masked Stroop tasks: Franken 2003). The Stroop interference effect in smokers has been associated, moreover, with evoked potentials (N100 waves) associated with early, pre-conscious, selection in attention (Leung 2000).

Studies using other paradigms have also suggested that automatic cognitive responses occur to addiction-related stimuli. Ehrman *et al.* (2002) used a dot-probe task to demonstrate that reaction times to the detection of the spatial location of dots on a computer task were selectively faster in smokers when a smoking-related image primed the dot location. Task-irrelevant auditory distractions selectively impaired performance times on an experimental task in smokers and drinkers when the distracting auditory stimuli were related to the addictive behaviour in question (Sayette and Hufford 1994; Sayette *et al.* 1994). These paradigms suggest that addiction-related information automatically 'grabs' cognitive resources in a way which, in the first example, facilitated non-automatic, intentional, cognitive activity, and, in the second example, compromised it.

These cognitive processes related to addictive behaviours appear to meet the criteria for automaticity outlined above. They appear to be operating in a way which is at odds with intention and compromise goal-directed behaviour of a conscious nature. Implications for understanding loss of control features in addictive behaviours start to occur. If drug-related stimuli can automatically 'grab' cognitive processing resources, then cognitive efficiency for intentional activity (e.g. searching for and rehearsing information to motivate abstinence) will be compromised—perhaps severely so if emotional distress is also high (see Section 5.8). An apparent 'loss of control' may follow. However, it is suggested that there is a further dimension to automatic cognitive processing. The processes highlighted by the studies described in this section largely relate to attention for addiction-related information. Whilst this in itself may compromise behavioural control, as described, it is suggested that by automatically diverting attentional resources to addiction-related stimuli, behaviour-motivating propositions (i.e. expectancies) automatically associated with these stimuli are also triggered, which further potentiates loss of control. The differences between automatic addiction-related propositions and those elucidated by the traditional self-report or inventory approach to assessment will be discussed in the next section.

5.7 Implicit memory structures and processes in addictive behaviours

At best, traditional methods of cognitive assessment only elucidate those drug-related propositions which are 'available' in long-term memory (Leung and McCusker 1999) and, as described above, problems beset even this assumption. What they do not do is reveal whether these are the propositions which are most automatically 'accessible' from long-term memory in drug-related contexts (Leung and McCusker 1999; Palfai and Wood 2001). Measuring *accessibility* of drug-related information from memory will distinguish between those automatically triggered propositions and those which are the products of elaborative search processes, which may themselves be influenced by extraneous aspects of the research context (McCusker 2001). Reaction time and memory paradigms have been used as alternative methods of cognitive enquiry.

Response time paradigms have highlighted that *accessibility* versus *availability* of drug-related propositions has been a neglected parameter of cognitive models of addictive behaviours. Timed, free association, memory paradigms are commonly used to measure the associative strength in semantic memory between concepts (e.g. Balota and Lorch 1986; McNamara 1992; Nelson *et al.* 1993). Thus, the greater the associative strength between two concepts, the more active, and accessible from memory, one becomes when the other is presented. Such tasks may, therefore, reveal automatically triggered associations rather than those associations which are only retrieved from more elaborative search processes (Mandler 1980; Diamond and Rozin 1984; Graf and Mandler 1984). Using such a paradigm, Leung and McCusker (1999) showed that smokers generated more smoking-related associations *per se* than non-smokers, consistent with more extensive cognitive architecture related to smoking. Problematic for social learning theory, however, was that *both* smokers and non-smokers generated more negative than positive smoking associations across a free association time period. However, whilst the ratio of positive/negative associations remained constant across the total time period for non-smokers, smokers generated proportionally more of their positive associations in the early time interval and proportionally more of their negative associations in the later time period. This suggested that whilst smokers may very well have a greater number of negative smoking expectancies 'available' to them in long-term memory, positive propositions are relatively more accessible than negative.

Similarly, Stacy and his colleagues have argued that as stimulus–outcome associations become repetitively rehearsed in practice, related conceptual units come automatically to trigger each other in memory (Stacy 1997). This research group have used word or picture association paradigms to measure feelings or events which are spontaneously triggered by the stimulus. Stimuli used have included positive outcomes (e.g. 'feeling good'), ambiguous words (e.g. 'draft') or pictures and open-ended phrase completions tasks (e.g. 'When I feel happy I just want to: _____'), and participants are instructed to respond with the first word or phrase which comes to mind (Stacy *et al.* 2004). Being demand free, since they are not asking the individual to introspect directly about their beliefs about dug use, the authors argue that such methods access the processes and content of implicit cognition related to drug use, i.e. automatic associations which act without conscious awareness or intention. The nature and extent of drug-related associations and responses revealed by such methods have been shown to be a stronger predictor of drinking behaviour than demographics, personal traits (e.g. sensation seeking), and outcome expectancies elicited by traditional, explicit, methods (Weingardt *et al.* 1996; Stacy 1997). Similar effects have been demonstrated for marijuana users (Stacy *et al.* 1996; Ames *et al.* 2005).

Combining explicit measures of expectancy associations with the response times taken to endorse these associations, Armstrong (1997) found that the latter differentiated drinking status more than endorsed expectancies *per se*. In the task used, sentence stems such as 'When I drink alcohol I feel . . .' or (in the control condition) 'When I drive a car I feel . . .' were presented on a computer screen, following which various single word affective associations briefly flashed up (e.g. 'good . . . relaxed . . . tense'). Participants had to press key pad buttons to agree or disagree with these associations 'as quickly as possible'. Whilst heavy social drinkers endorsed more positive alcohol associations than light drinkers, the within-group patterns were the same. Perhaps influenced by subcultural norms about alcohol (groups were drawn from student populations), both groups endorsed more positive than negative alcohol associations, as summarized in Fig. 5.3. However, when response times to make endorsements were examined, *within-* as well as between-group differences became obvious. Thus heavy drinkers endorsed positive associations significantly faster than light drinkers *and* significantly faster than they endorsed negative associations (see Fig. 5.3). The reverse pattern was true for light drinkers. Although they endorsed more positive than negative alcohol associations, mean response times for the negative associations were faster than for the positive. It was concluded that positive associations were more available to both heavy and light drinkers. However, whereas these were also more accessible from memory in heavy drinkers, negative and behaviourally inhibiting propositions were more accessible for light drinkers. The pattern of results in this study suggested that the cognitive feature which best differentiates excessive from moderate alcohol consumption is the accessibility, rather than the availability, of alcohol propositions in memory.

As summarized above, a feature of automatic cognitive processing is that it does not require effort or intention. Incidental memory procedures, which ask participants to recall information to which they had been previously exposed but not directed to memorize, may reflect automatic processing of the information. Whilst positive propositions related to smoking may not be preferentially endorsed by smokers, both Litz *et al.* (1987) and Leung (2000) observed an incidental memory bias for positive versus negative smoking associations.

Implicit memory may be defined as memory for previous events impacting on current behaviours without conscious awareness (Graf and Schacter 1985).

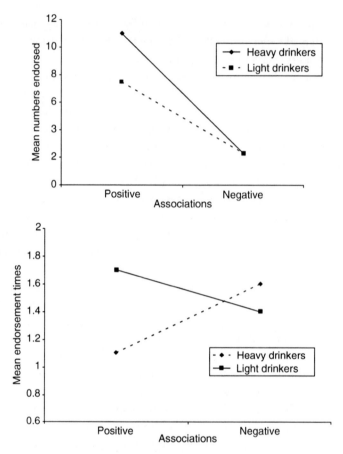

Fig. 5.3 Mean number of positive versus negative endorsements and response times to make those endorsements in heavy and light drinkers. Both groups endorse more positive than negative propositions. However, whilst light drinkers are faster in their response times to negative associations, the reverse is true for heavy drinkers (Armstrong 1997).

The efficiency of implicit memory processes for newly presented information is assumed to reveal the extent to which such information is *automatically* encoded and retrieved, putatively without conscious awareness (Schacter 1992). Furthermore, the degree to which such automatic processing occurs is likely to reflect the extent or strength of pre-existing cognitive architecture consistent with the information processed (Graf and Mandler 1984). Perform-ance on word-fragment completion tasks, where the individual is asked to complete word stems with the first word which comes to mind, has been shown to be influenced by previously primed, but not consciously recollected, words (Graf and Mandler 1984). McCusker and Gettings (1997) followed up their Stroop procedure with gamblers with such a word-stem task involving word stems from previously exposed gambling, drug and neutral Stroop tasks. They found that gamblers completed word fragments from the gambling Stroop task with more of these primed words than word fragments primed on control Stroop tasks involving drug and neutral stimuli. This was interpreted as indicative of a memory bias for gambling-related information operating at an automatic and implicit level of processing. Similar implicit memory biases, inferred from word-fragment completion tasks, have been demonstrated in heavy drinkers and smokers (Armstrong 1997; Leung 2000).

The studies reviewed in this section suggest that cognitive aspects of automaticity in addictive behaviours extend beyond attentional processes. Conceptual information, with motivational significance, also appears to be processed in an automatic manner. Moreover, this automatic processing appears biased in a way which differentiates excessive or addictive involvement with the behaviour from control samples. In the next section, a theoretical model is proposed which incorporates the findings discussed above of automaticity in autonomic, attentional and propositional response systems. It will be argued that such a model extends cognitive accounts of addictive behaviours and in a way which helps us better understand loss of control.

5.8 Theoretical integration: an automatic network theory of addictive behaviours

Although the research presented above has most to say about what will be described as a third, *stereotypy* phase of addictive behaviour, this is assumed to be preceded by two earlier phases, where automatic cognitive processes are also likely to be in operation.

At a *pre-exposure phase*, the child or young adult will have observed a varying balance of positive and negative propositions about drug use. For some drugs (e.g. alcohol), the balance sheet is likely to be overwhelmingly positive from both explicit instructions (e.g. 'alcohol is OK in moderation') and implicit communications in the family, peer group and popular culture (e.g. 'alcohol relieves stress . . . drinking alcohol means I have become adult'). Numbers who take up drinking behaviour will be consequently high (e.g. see the preponderance of positive expectancies explicitly endorsed by both heavy and light drinkers described in Section 5.7). For other drugs, which have explicit associations of a more negative nature (e.g. heroin or nicotine), the number of users may be fewer. However, many implicit communications from familial, peer or popular subcultures, which exert automatic influences on behaviour (Berkowitz 1984), may be positive and at odds with dominant social values (e.g. 'risk takers use heroin...coke is a sign of wealth and is cool...interesting people smoke'). These may very well have more immediate and affective significance for the individual, and thus motivate the onset of the drug-using behaviour.

During a second, *exposure phase*, drug–outcome associations are repeatedly paired and these associations are signalled by a stimulus complex of both external and internal cues (e.g. drug paraphernalia, pubs, mood states). Although early and salient inhibitory experiences may occur for some (e.g. being sick, losing repeated gambling bets), most of these experiences are likely to confer positive associations (reinforcement of a primary or secondary nature) and strengthen such behaviour–outcome propositions in long-term memory. This is represented in Fig. 5.4 by the larger circle of positive outcomes. Given that behaviour–outcome associations, and signals which predict their occurrence, have such affective significance, automaticity in the triggering of these associations in memory networks is likely to develop (Stacy 1997; Bargh and Chartrand 1999). As automatic processes dominate non-automatic processes, as noted above, these are likely to become salient mediators of behaviour—even in non-addicted samples—as indeed the evidence base suggests (Weingardt *et al.* 1996; Stacy 1997).

As the *exposure phase* continues and deepens, a third *stereotypy phase* may follow where a range of automatic responses to internal and external addiction-related stimuli converge to promote the familiar stereotyped pattern of addictive behaviour, relapse and loss of control described at the outset. Three such levels have been discussed in this chapter. These are summarized in Fig. 5.4 and include: (1) autonomic reactions; (2) selective and impairing attentional biases; and (3) activation of a biased propositional network related to the behaviour.

It is suggested that through the repeated and signalled associative experiences outlined in the exposure phase, automatic processes are triggered in three systems which independently and/or interactively compromise capacity for behavioural control. One pathway, suggested by Franken (2003), is that drug-related stimuli trigger a range of conditioned physiological responses (as reviewed in Section 5.4) the most important of which may be an increase in dopaminergic levels in the brain. This in turn serves to prompt an attentional bias for information which signals reward, and thus we see the automatic diversion of cognitive resources towards the processing of addiction-related information as outlined in Section 5.6. With attention locked on addiction-related information, a further stage in this automatic sequence will be the

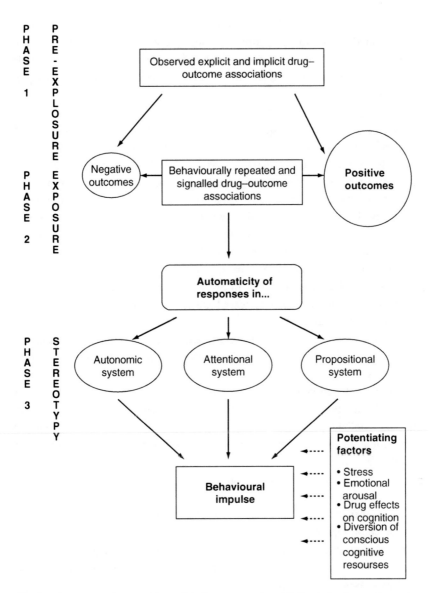

Fig. 5.4 An automatic network model of stereotypy in addictive behaviours. Through largely positive outcome associations with the drug/behaviour in early stages of the addiction career, a network of automatic reactions in autonomic, attentional and propositional systems becomes primed to internal and external signals of previous use. Automatic processes in the network precipitate the behavioural impulse related to the addiction, as a function of the strength of the network; contextual factors which deplete non-automatic cognitive resources potentiate activity in, and salience of, the automatic network for behavioural control.

triggering of those addiction-related propositions which are most accessible. As outlined in Section 5.7, although negative and inhibitory information may very well be more, or increasingly, available, automatically accessible associations are likely to be for positive information with a longer history of association.

Although elegant, such an automatic network model does not require that processes operate in this sequence. Each of the three systems may be triggered in parallel, rather than in sequence, and exert independent but additive effects on behavioural control. Indeed, it may be that the automatic triggering of positive or reward propositions increases autonomic arousal and stimulates biased attentional processing of target stimuli in a way which optimizes appetitive behaviour to secure the signalled reward. Alternatively, attentional bias may potentiate the appetitive behavioural impulse independently, by reducing cognitive resources for inhibitory information as described in Section 5.6. Indeed, autonomic arousal, which is subjectively experienced as anxiety provoking (as described in Section 5.4), may increase the incentive value of the automatically triggered positive propositions related to the addictive behaviours.

The degree to which such automatic autonomic and cognitive responses result in the automaticity of behaviour observed in the addictions will vary according to the strength of the network and converging stimulus conditions. It will be mediated, however, by personal, situational and other drug factors as also indicated in Fig. 5.4. High levels of stress and emotion can be expected to impair 'cool' hippocampal–cortical reasoning processes and potentiate the 'emotional memory system' of automatic and implicit cognitive processes (Metcalfe and Jacobs 1998). Thus the extent of personal and situational stress experienced by the individual is likely to mediate the extent to which activity in the automatic addiction network can be resisted and controlled, or follows on in a stereotyped manner towards behavioural loss of control. The relevance of 'high risk' situations has been a well-established feature of social learning accounts of relapse (e.g. Marlatt and Gordon 1985), but the automatic network model offers a different account of mediating mechanisms and processes.

Direct drug effects on cognitive processing are also likely to add impetus to the impact of automatic systems on subsequent behaviour. Alcohol, for example, impairs the organization and explicit retrieval of information by interfering with elaborative reasoning and recall processes (Sayette 1999). This is likely to render the most accessible propositions from memory most salient for behavioural regulation. As more and more alcohol is consumed, automatic cognitive processes are likely to assume greater and greater significance, resulting in the ultimate loss of control of continuing the appetitive drinking behaviour until physical incapacity or environmental blocks intervene.

Other, situational, factors may potentiate the impact of these automatic processes on addictive behaviours. An increased rate of drinking or smoking has, for example, been observed in arousing social situations. This social facilitation effect (Zajonc 1965) may partly be due to the fact that explicit cognitive resources are diverted elsewhere, i.e. in processing, evaluating and responding to a complexity of interpersonal information, leaving ongoing drinking or smoking behaviour to the acquisitive impulses of the automatic processes described above.

Although research has not yet addressed the question, there may be a final natural *recovery phase* in how automatic processes impinge on an addiction career. It is likely that as negative outcomes become increasingly, more immediately and pervasively experienced, or as therapeutic interventions increase awareness, and modify the nature, of automatic processes and potentiating factors, the cognitive network will become reconfigured. Autonomic responses may become extinguished, negative propositions may become more accessible, as well as more available, from memory, and strategies to counteract biases in attentional processing may be developed.

5.9 Future directions and clinical implications

Automatic processes in addictive behaviours have been emphasized in this chapter. Such processes have, however, been increasingly emphasized in the psychology of human behaviours more generally. Behavioural performance, especially early in the genesis of new behavioural preferences, may need to be controlled by non-automatic cognitive judgements and intentions. However, conscious cognitive resources are limited and evolutionary advantage would be gained from transferring as many behavioural operations as possible to an *automatic* control network (Posner and Snyder 1975). This proposition is perhaps most obviously illustrated by how the slow and effort-filled behaviours and reasoning style of infancy and childhood evolve into the more fluent and effortless execution of the same operations in later development and adulthood. Such a premise is extended to a wide range of social behaviours by Bargh and Chartrand (1999). They suggest that imperatives of cognitive efficiency mean that 'most of our day-to-day actions, motivations, judgements and emotions are not the products of conscious choice and guidance, but must be driven by mental processes put into operation directly by environmental features and events' (p. 465). In their paper they present a gamut of research from social psychology which appears to support such a thesis. From the implicit impact of perceptual exposure on subsequent behaviour, through to goal-seeking behaviours which, once rehearsed at crucial developmental periods become schematic in behavioural control, automaticity is proposed as a cardinal parameter of human behaviour. Such a thesis is consistent with the theory of addictive behaviours proffered here. The normal evolutionary advantages conferred by automaticity of appetitive responses, however, turns against the individual who crosses some threshold of appetitive behaviour rendering it an automatic addictive cycle of self-destructive proportions.

Future research in this area of automaticity in addictive behaviours should look at those individual characteristics and situational conditions which promote the biased encoding and accessibility of positive drug propositions at implicit levels of consciousness. A plausible heuristic has been outlined above (see Fig. 5.4), but the precise conditions which render negative and inhibitory information more accessible from the outset for some, and not others, requires clarification. So too do we need to examine more directly individual and situational features which connect the automaticity of cognitive and autonomic processes to behavioural operations. The model above suggests some useful parameters to explore. Research presented above suggests that automatically controlled processes and propositions may predict drug-using behaviours better than those of a non-automatic nature. Further experimental work is required, however, which tests the predictive utility of competing explicit and implicit propositions directly on behavioural per-formance relevant to the addictive behaviour. Again it is likely that situational, stage of drug use and person variables will mediate the relationship here, rather than a simple rule being derived. Nonetheless, such work will inform and certainly have important implications for preventive and clinical interventions.

A number of such implications may be discerned even at this stage of research into automaticity of addictive behaviours. Psychoeducation appears crucial. For all its problems, the disease model may be so seductive because it offers a metaphor, which has ecological validity, for the loss of control impulses experienced. As well as teaching self-control and coping strategies, cognitive–behavioural interventions need to inform the individual struggling with an addictive behaviour that automatic (probably 'hard-wired') reactions outside of their control will be triggered in certain situations. This will help provide some understanding and emotional distance from the autonomic reactions, which can be experienced with subjective anxiety and distress, from the ruminative thinking which becomes locked on the activity, and from the deluge of positive incentives, which may appear overwhelming and confusingly at odds with those which seemed to have been so solidly rehearsed in the therapy room.

The model supports the use of 'external' restraints, which will thwart the automaticity of internal processes on behavioural loss of control proposed above (e.g. drug antagonists, avoidance of high-risk situations, chaperoned exposure, etc.). However, these should be restricted to the early and more fragile stages of recovery. Ultimately, it is likely that the automatic network needs to be activated if it is to be modified. In much the same way that emotional processing of traumatic experiences requires exposure *in vivo* and imaginally (Foa and Kozak 1986), so too is guided exposure likely to be required in cognitive therapy in order to access the 'hot' associations putatively located in the amygdala system (Metcalfe and Jacobs 1998).

Prevention campaigns need to speak more to associations in the implicit memory system, rather than continuing to build on explicit negative associations which may already outweigh the positive in terms of their availability. Alternative ways to feel 'cool...good...chic', etc. need to be propagated in popular culture and amongst valued groups (family, peers, etc.) so that such behavioural operations will become as implicitly associated with these outcomes as drug use (Stacy *et al.* 2004).

Finally, assessment instruments which are used in clinical and research contexts—in an effort to indicate risk status or latency to relapse—need to be more informed by the principles of automatic cognition. The paradigms used in the research described above have been experimental in nature. Psychometric refinement and clinical standardization are required to enable routine attention to autonomic and attentional processes as well as speed of processing indices of propositional accessibility.

Bargh and Chartrand (1999) describe an 'automaticity of being' which they suggest serves us better than the deterministic stereotypes which humanistic and early cognitive theorists railed against. The challenge for those affected by addiction and their therapists, however, is somehow to switch off, or neutralize, such an automatic network and rebuild one of more controlled proportions.

Acknowledgements

Any correspondence concerning this chapter should be addressed to Dr Chris McCusker, Department of Psychology, The Queen's University Belfast, Belfast, BT7 1NN.

References

Ajzen, I. (1991). The theory of planned behaviour. *Organisational Behaviour and Human Decision Process,* **50,** 179–211.

American Psychiatric Association (1994). *Diagnostic and statistical manual of mental disorders,* 4th edn. Washington, DC: American Psychiatric Association.

Ames, S.L., Sussman, S., Dent, C.W. and Stacy, A.W. (2005). Implicit cognition and dissociative experiences as predictors of adolescent substance use. *American Journal of Drug and Alcohol Abuse,* 1, 129–162.

Armstrong, C. (1997). Automatic and non-automatic processing of alcohol-associations in heavy versus light drinkers. Paper presented at the London conference of the British Psychological Society, London.

Baker, T.B., Piper, M.E., Fiore, M.C., McCarthy, D.E. and Majeskie, M.R. (2004). Addiction motivation reformulated: an affective processing model of negative reinforcement. *Psychological Review,* 111, 33–51.

Balota, D.A. and Lorch, R.F. (1986). Depth of automatic spreading activation: mediated priming effects in pronunciation but not in lexicon decisions. *Journal of Experimental Psychology: Learning, Memory and Cognition,* 12, 336–345.

Bandura, A. (1969). *Principles of behaviour modification.* New York, NY: Holt, Rinehart and Winston.

Bandura, A. (1977). *Social learning theory.* Englewood Cliffs, NJ: Prentice Hall.

Bargh, J.A. and Chartrand, T.L. (1999). The unbearable automaticity of being. *American Psychologist,* 54, 462–477.

Bargh, J.A., Bond, R.N., Lombardi, N.J. and Tota, M.E. (1986). The additive nature of chronic and temporary sources of construct accessibility. *Journal of Personality and Social Psychology,* 59, 869–878.

Beck, A.T. (1976). *Cognitive therapy and the emotional disorders.* New York, NY: New American Library.

Beck, A.T., Wright, F.D., Newman, C.F. and Liese, B.S. (1993). *Cognitive therapy of substance abuse.* New York, NY: Guilford Press.

Berkowitz, L. (1984). Some effects of thoughts on anti- and prosocial influences of media events: a cognitive neoassociation analysis. *Psychological Bulletin,* 95, 410–427.

Biddle, B.J., Bank, B.J. and Martin, M.M. (1980). Social determinants of adolescent drinking: what they think, what they do and what I think they do. *Journal of Studies on Alcohol,* 41, 215–241.

Brown, S.A., Christiansen, B.A. and Goldman, M.S. (1987). The alcohol expectancy questionnaire: an instrument of the assessment of adolescent and adult expectancies. *Journal of Studies on Alcohol,* 48, 483–491.

Cappell, H. and Greely, J. (1987). Alcohol and tension reduction: an update on research and theory. In H.T. Blane and K.E. Leonard (ed.), *Psychological theories of drinking and alcoholism.* New York, NY: Guilford Press, pp. 15–54.

Carey, K.B. (1995). Alcohol-related expectancies predict quantity and frequency of heavy drinking among college students. *Psychology of Addictive Behaviors,* 9, 236–241.

Childress, R.F., McLellan, A.T. and O'Brien, C.P. (1986). Role of conditioning factors in the development of drug dependence. *Psychiatric Clinics of North America,* 9, 413–425.

Cohen, S. (1989). Debunking myths about self-quitting. *American Psychologist,* 44, 1355–1365.

Conner, M. and Sparks, P. (1996). The theory of planned behaviour and health behaviours. In M.Conner, P. Norman (ed.), *Predicting health behaviour: research and practice with social cognition models.* Buckingham, UK: Open University Press, pp. 122–162.

Connors, G.J., Tarbox A.R. and Faillace, L.A. (1993). Changes in alcohol expectancies and drinking behavior among treated problem drinkers. *Journal of Studies on Alcohol,* 53, 676–683.

Cooney, N.L., Gillepsie, R.A., Baker, L.H. and Kaplan, R.F. (1987). Cognitive changes after alcohol cue exposure. *Journal of Consulting and Clinical Psychology,* 55, 150–155.

Cooper, M.J., Anastasiades, P. and Fairburn, C.G. (1992) Selective processing of eating-, shape-, and weight-related words in persons with bulimia nervosa. *Journal of Abnormal Psychology,* 101, 352–355.

Cox, W.M., Yeates, G.N. and Regan, C.M. (1999). Effects of alcohol cues on cognitive processing in heavy and light drinkers. *Drug and Alcohol Dependence,* 55, 85–89.

Cox, W.M., Hogan, M.H., Kristian, M.R. and Race, J.H. (2002). Alcohol attentional bias as a predictor of alcohol abusers' treatment outcome. *Drug and Alcohol Dependence,* 68, 237–243.

Davis, J.R. and Glaros, A.G. (1986). Relapse prevention and smoking cessation. *Addictive Behaviors,* 11, 105–114.

Dawe, S., Powell, J., Richards, D., Gossop, M., Marks, I. and Gray, J.A. (1993). Does post-withdrawal cue-exposure improve outcome in opiate addiction? A controlled trial. *Addiction,* 88, 1233–1245.

Diamond, R. and Rozin, P. (1984). Activation of existing memory in amnesic syndrome. *Journal of Abnormal Psychology,* 93, 98–105.

Drummond, D.C., Tiffany, S.T., Glautier, S.P. and Remmington, B. (1995). *Addictive behaviour: cue exposure theory and practice.* Chichester, UK: Wiley.

Edwards, G. and Gross, M. (1976). Alcohol dependence: provisional description of a clinical syndrome. *British Medical Journal,* 1, 1058–1061.

Ehrman, R.N., Robbins, S.J., Bromwell, M.A., Lankford, M.E., Monterosso, J.R. and O'Brien, C.P. (2002). Comparing attentional bias to smoking cues in current smokers, former smokers, and non-smokers using a dot-probe task. *Drug and Alcohol Dependence,* 67, 185–191.

Falk, J.L. (1983). Drug dependence: myth or motive? *Pharmacology Biochemistry and Behavior,* 19, 385–391.

Festinger, L. (1962). Cognitive dissonance. *Scientific American,* 207, 93–106.

Foa, E.B. and Kozak, M.J. (1986). Emotional processing of fear: exposure to corrective information. *Psychological Bulletin,* 99, 20–35.

Franken, I.H.A. (2003). Drug craving and addiction: integrating psychological and neuropsychopharmacological approaches. *Progress in Neuro-Psychopharmacology and Biological Psychiatry,* 27, 563–579.

Franken, I.H.A., Kroon, L.Y., Wiers, R.W. and Jansen, A. (2000) Selective cognitive processing of drug cues in heroin dependence. *Journal of Psychopharmacology (Berlin),* 14, 395–400.

Graf, P. and Mandler, G. (1984). Activation makes words more accessible, but not necessarily more retrievable. *Journal of Verbal Learning and Verbal Behavior,* 23, 553–568.

Graf, P. and Schacter, D.L. (1985). Implicit and explicit memory for new associations in normal and amnesic subjects. *Journal of Experimental Psychology: Learning, Memory and Cognition,* 11, 501–518.

Gross, T.M., Jarvik, M.E. and Rosenblatt, M.R. (1993). Nicotine abstinence produces content-specific Stroop interference. *Psychopharmacology (Berlin)*, 110, 333–336.

Heather, N. and Robertson, I. (1997). *Problem drinking*. Oxford: Oxford University Press.

Hill, D.L. and Grey, N. (1983) Australian patterns of tobacco smoking and related health beliefs in 1983. *Community Health Studies*, 8, 307–316.

Hinson, R.E. and Siegel, S. (1982). Nonpharmacological bases of drug tolerance and dependence. *Journal of Psychosomatic Research*, 26, 495–503.

Ho, R. (1992). Cigarette health warnings: the effects of perceived severity, expectancy of occurrence and self-efficacy on intentions to give up smoking. *Australian Psychologist*, 27, 109–113.

Hughes, J.R. (1992). Smoking cessation among self-quitters. *Health Psychology*, 11, 331–334.

Ingram, R.E. (1990). Self-focused attention and clinical disorders: review and a conceptual model. *Psychological Bulletin*, 107, 156–176.

Jellinek, E.M. (1960). *The disease concept of alcoholism*. New York, NY: Hillhouse Press.

Johnsen, B.H., Laberg, J.C., Cox, W.M., Vaksdal, A. and Hugdahl, L.K. (1994). Alcoholic subjects' attentional bias in the processing of alcohol-related words. *Psychology of Addictive Behaviors*, 8, 111–115.

Jones, B.T. and McMahon, J. (1994). Negative alcohol expectancy predicts post-treatment abstinence survivorship: the whether when and why of relapse to a first drink, *Addiction*, 89, 1653–1665.

Jones, B.T., Corbin, W. and Fromme, K. (2001). A review of expectancy theory and alcohol consumption. *Addiction*, 96, 57–72.

Leigh, B.C. (1989). Confirmatory factor analysis of alcohol expectancy scales. *Journal of Studies on Alcohol*, 50, 268–277.

Leigh, B.C. and Stacy, A.W. (2004). Alcohol expectancies and drinking in different age groups. *Addiction*, 99, 215–227.

Leung, K.S. (2000) Automaticity of cognitive biases in smokers. Unpublished PhD thesis. The Queen's University of Belfast.

Leung, K.S. and McCusker, C.G. (1999). Accessibility and availability of smoking-related associations in smokers. *Addiction Research*, 7, 213–226.

Litz, B.T., Payne, T.J. and Colletti, G. (1987). Schematic processing of smoking information by smokers and never-smokers. *Cognitive Therapy and Research*, 11, 301–313.

Logan, G.D. (1988). Toward an instance theory of automatization. *Psychological Review*, 95, 492–527.

Lowman, C., Hunt, W.A., Litten, R.Z. and Drummond, D.C. (2000). Research perspectives on alcohol craving: an overview. *Addiction*, 95 (Suppl. 2), S45–S54.

Macleod, C.M. (1991). Half a century of research on the Stroop effect: an integrative review. *Psychological Bulletin*, 109, 163–203.

Mandler, G. (1980). Recognizing: the judgement of previous occurrence. *Psychological Review*, 87, 252–271.

Marlatt, G.A. (1983). The controlled drinking controversy—a commentary. *American Psychologist*, 38, 1097–1110.

Marlatt, G.A. and Gordon, J.R. (1985). *Relapse prevention: maintenance strategies in the treatment of addictive behaviors*. New York, NY: Guilford Press.

McAuliffe, W.E. (1982). A test of Wikler's theory of relapse: the frequency of relapse due to conditioned withdrawal sickness. *International Journal of the Addictions*, 17, 19–33.

McCusker, C.G. (2001). Cognitive biases and addiction: an evolution in theory and method. *Addiction*, 96, 47–56.

McCusker, C.G. and Bell, R. (1988). Conditioned ethanol preference in rats. *Alcohol and Alcoholism*, 23, 359–364.

McCusker, C.G. and Brown, K. (1991). The cue-responsivity phenomenon in dependent drinkers: 'personality' vulnerability and anxiety as intervening variables. *British Journal of Addiction*, 86, 905–912.

McCusker, C.G. and Brown, K. (1995). Cue-exposure to alcohol-associated stimuli reduces autonomic reactivity but not craving and anxiety in dependent drinkers. *Alcohol and Alcoholism*, 30, 319–327.

McCusker, C.G. and Gettings, B. (1997). Automaticity of cognitive biases in addictive behaviours: further evidence with gamblers. *British Journal of Clinical Psychology*, 36, 543–554.

McCusker, C.G., Roberts, G., Douthwaite, J. and Williams, E. (1995). Teenagers and illicit drug use—expanding the 'user vs. non-user' dichotomy. *Journal of Community and Applied Social Psychology*, 5, 221–241.

McMahon, J., Jones, B.T. and O'Donnell, P. (1994). Comparing positive and negative alcohol expectancies in male and female social drinkers. *Addiction Research*, 1, 349–365.

McNamara, T.P. (1992). Theories of priming: I Associative distance and lag. *Journal of Experimental Psychology: Learning, Memory and Cognition*, 18, 1173–1190.

Metcalfe, J. and Jacobs, W.J. (1998). Emotional memory: the effect of stress on 'cool' and 'hot' memory systems. *The Psychology of Learning and Motivation*, 38, 187–215.

Meyer, R.E. (1988). Conditioning phenomena and the problem of relapse in opiod addicts and alcoholics. *National Institute on Drug Abuse Research Monograph Series*, 88, 161–179.

Mischel, W. (2004). Toward an integrative science of the person, *Annual Review of Psychology*, 55, 1–22.

Miller, W.R. and Rollnick, S. (1984) Motivational interviewing: preparing people to change addictive behaviour. New York, NY: Guildford Press.

Monti, P.M., Abrams, D.B., Kadden, R.M. and Cooney, N.C. (1989). *Treating alcohol dependence*. New York, NY: Guilford Press.

Monti, P.M., Binkoff, J.A., Adrams, D.B., Zwick, W.R., Nirenberg, T.D. and Liepman, M.R. (1987). Reactivity of alcoholics and nonalcoholics to drinking cues. *Journal of Abnormal Psychology*, 96, 122–126.

Nelson, D.L., Bennett, D.J., Gee, N.R. and Schreiber, T.A. (1993). Implicit memory: effect of network size and interconnectivity on cued recall. *Journal of Experimental Psychology: Learning, Memory and Cognition*, 19, 747–764.

Niaura, R.S. (2000). Cognitive social learning and related perspectives on drug craving. *Addiction*, 95 (Suppl. 2), S155–S163.

Nisbett, R.E. and Wilson, T.D. (1977). Telling more than we can know: verbal reports on mental processes. *Psychological Review*, 84, 231–259.

Palfai, T. and Wood, M.D. (2001). Positive alcohol expectancies and drinking behaviour: the influence of expectancy strength and memory accessibility. *Psychology of Addictive Behaviour*, 15, 60–67.

Plant, M. and Plant, M. (1992). *Risk-takers*. London: Tavistock/Routledge.

Pomerleau, O.F., Fertig, J., Baker, L.H. and Cooney, N. (1983). Reactivity to alcohol cues in alcoholics and non-alcoholics: indications for a stimulus control analysis of drinking. *Addictive Behaviours*, **8**, 1–10.

Posner, M.I. and Snyder, C.R.R. (1975). Attention and cognitive control. I.R.L.Solo (ed.), *Information processing and cognition: the Loyola symposium*. pp. 55–85. Hillsdale, NJ: Erlbaum.

Prochaska, J.O. and DiClemente, C.C. (1982). Transtheoretical therapy: towards a more integrative model of change. *Psychotherapy: Theory, Research and Practice*, **19**, 174–183.

Rohsenow, D.J., Monti, P.M., Abrams, D.B., Rubonis, A.V., Niaura, R.S., Sirota, A.D. and Colby, S.M. (1992). Cue elicited urge to drink and salivation in alcoholics: relationship to individual differences. *Advances in Behavioural Research and Therapy*, **14**, 195–210.

Ryan, F. (2002). Attentional bias and alcohol dependence: a controlled study using the modified Stroop paradigm. *Addictive Behaviours*, **27**, 471–482.

Sayette, M.A. (1999). Cognitive theory and research. I.K. Leonard and H. Blane (ed.), *Psychological theories of drinking and alcoholism*. New York, NY: Guilford Press, pp. 247–291.

Sayette, M.A. and Hufford, M.R. (1994). Effects of cue-exposure and deprivation on cognitive resources in smokers. *Journal of Abnormal Psychology*, **103**, 812–818.

Sayette, M.A., Monti, P.M., Rohsenow, D.J., Bird-Gulliver, S., Colby, S., Sirota, A., Niaura, R. and Abrams, D. (1994). The effects of cue-exposure on attention in male alcoholics. *Journal of Studies on Alcohol*, **55**, 629–634.

Schacter, D.L. (1992). Understanding implicit memory: a cognitive neuroscience approach. *American Psychologist*, **47**, 559–569.

Schafer, J. and Brown, S.A. (1991). Marijuana and cocaine effect expectancies and drug use patterns. *Journal of Consulting and Clinical Psychology*, **59**, 558–565.

Schiffrin, R.M. and Sneider, W. (1977). Controlled and automatic human information processing: II. Perceptual learning, automatic attending, and a general theory. *Psychological Review*, **84**, 127–190.

Sher, K.J., Wood, M.D., Wood, P.K. and Raskin, G. (1996). Alcohol outcome expectancies and alcohol use: a latent variable cross-lagged panel study. *Journal of Abnormal Psychology*, **105**, 561–574.

Southwick, L.L., Steele, C.M., Marlatt, G.A. and Lindell, M.K. (1981). Alcohol-related expectancies: defined by phase of intoxication and drinking experience. *Journal of Consulting and Clinical Psychology*, **49**, 713–721.

Stacy, A.W. (1997). Memory activation and expectancy as prospective predictors of alcohol and marijuana use. *Journal of Abnormal Psychology*, **106**, 61–73.

Stacy, A.W., Newcomb, M.D. and Bentler, P.M. (1991). Cognitive motivation and problem drug use: a 9-year longitudinal study, *Journal of Abnormal Psychology*, **100**, 502–515.

Stacy, A.W., Ames, S.L., Sussman, S. and Dent, C.W. (1996). Implicit cognition in adolescent drug use. *Psychology of Addictive Behaviors*, **10**, 190–203.

Stacy, A.W., Ames, S.L. and Leigh, B. (2004). An implicit cognition assessment approach to relapse, secondary prevention, and media effects. *Cognitive and Behavioral Practice*, **11**, 139–149.

Stewart, J., de Wit, H. and Eikelboom, R. (1984). The role of unconditioned and conditioned drug effects in the self-administration of opiates and stimulants. *Psychological Review*, **91**, 251–268.

Stroop, J.R. (1935). Studies of interference in serial verbal reactions. *Journal of Experimental Psychology,* **18,** 643–661.

Sutton, S., Marsh, A. and Matheson, J. (1990). Microanalysis of smokers' beliefs about the consequences of quitting: results from a large population sample. *Journal of Applied Social Psychology,* **20,** 1847–1862.

Tiffany, S.T. (1990). A cognitive model of drug urges and drug-use behaviour: the role of automatic and non-automatic processes. *Psychological Review,* **97,** 251–268.

Tiffany, S.T. and Carter, B.L. (1998). Is craving the source of compulsive drug use. *Journal of Psychopharmacology,* **12,** 23–30.

Tulving, E. and Thomson, D.M. (1973). Encoding specificity and retrieval processes in episodic memory. *Psychological Review,* **80,** 359–380.

Weingardt, K.R., Stacy, A.W. and Leigh, B.C. (1996). Automatic activation of alcohol concepts in response to positive outcomes of alcohol use. *Alcoholism: Clinical and Experimental Research,* **20,** 25–30.

Wikler, A. (1965). Conditioning factors in opiate addiction and relapse. In D. I. Wiher and G. Kassebaum (ed.), *Narcotics.* New York, NY: McGraw-Hill, pp. 85–100.

Wikler, A. (1980). *Opioid dependence.* New York, NY: Plenum Press.

Zajonc, R.B. (1965). Social facilitation. *Science,* **149,** 269–274.

6

From DNA to conscious thought: the influence of anticipatory processes on human alcohol consumption

Mark S. Goldman, Jack Darkes,
Richard R. Reich and Karen O. Brandon

6.1 Introduction

It all comes together when we address alcohol use, abuse and alcoholism: genetically based individual differences in sensitivity, reactivity and metabolism; neurobiologically based learning and motivational pathways; neuropsychological processes of self-regulation; family/social/peer influences; environmental exposures; and culturally normative drinking practices. In addition, these domains are not independent; each interplays with the others. In fact, some domains that may be described and researched separately actually reflect the same overall processes addressed at different levels of explanation.

To complicate matters further, the knowledge base in each area has exploded in recent years. The resulting span of knowledge necessary to address alcohol use and associated disorders comprehensively is enormous. Few, if any, scientists can readily accommodate this span of information while also being expert in critical domain-specific details. For this reason, research most often is characterized by approaches that organize the domains of interest horizontally and vertically: horizontally as separate disciplines, and vertically into distinct levels of explanation. Within this compartmentalized conceptual structure, impressive scientific advances in genetics, neuroscience, cognitive science and affective/motivational science have been made rapidly and applied to the questions of why alcohol use is reinforcing and how it becomes excessive.

This 'local focus' also engenders, however, less than optimal byproducts. It can impair cross-talk among researchers in different domains, and even can promote competition among researchers for primacy of explanation (if not for funding and journal space). We assert that, contrary to the cliché, nature has no joints, and problems attendant to alcohol use cannot be understood effectively in the absence of a transdisciplinary approach.

This chapter describes and updates a framework for integrating the disparate domains while maintaining respect for local explanations. We offer this synthesis modestly. Our intent is not to supplant work in specific domains, but to offer a synthetic framework that we hope may facilitate cross-talk between domains and encourage interweaving of emerging findings. The synthesis offered is based on the concept of expectancy, which in psychology was applied originally to general learning theory by Tolman (1932). Applications of this psychological concept were subsequently, in the alcohol field, related to placebo effects (Marlatt *et al.* 1973), and used to predict consumption (Brown *et al.* 1980). Howwever, this concept has independently emerged in increasingly diverse scientific venues, including basic behavioural processes such as operant (Dragoi and Staddon 1999) and classical conditioning (Van Hamme and Wasserman 1994; Kirsch *et al.* 2004), comparative judgement (Ritov 2000; Heekeren *et al.* 2004), models of memory (Bower 2000), the neurobiology of animal and human reward and reinforcement (Schultz *et al.* 1997; Kupfermann *et al.* 2000; Breiter *et al.* 2001; McClure *et al.* 2004; Schultz 2004), perception of motion (Kerzel 2005), development of language (Colunga and Smith 2005), time perception (Correa *et al.* 2005), the neurobiology of interpersonal trust (King-Casas *et al.* 2005) and brain electrophysiology [event-related potential (ERP); Donchin and Coles 1988], among others.

In the clinical domain, it has been related to mood, fear, pain reduction, sexual dysfunction, asthma, drug abuse, alcohol abuse and alcoholism, smoking, placebo (Wager *et al.* 2004; Petrovic *et al.* 2005) and nocebo effects (psychologically induced illness or even death), psychotherapy, hypnosis (see Kirsch 1999) and medicinal effects of drugs (Kirsch and Scoboria 2001). Krumhansl and Toivaine (2000) even explained the psychological appreciation of music via expectation. That expectancy operation is fundamental to behaviour is evidenced by its having been connected with visual orienting behaviour in early infancy (Haith *et al.* 1988). Expectancy is also related to social functioning; Kilner *et al.* (2004) conclude that '... mere knowledge of an upcoming movement [in another person] is sufficient to excite one's own motor system, enabling people to anticipate, rather than react to, other's actions' (p. 1299). Social functioning is, of course, more than just the anticipation of movements. Baron-Cohen *et al.* (2005) describe empathy as '... the capacity to predict and to respond to the behaviour of agents (usually people) by inferring their mental states and responding to these with an appropriate emotion' (p. 819).

At more basic levels, Kupfermann *et al.* (2000) refer to similar processes when they say, 'Homeostatic regulation is often anticipatory' (p. 1007). This capacity to anticipate biological demands is not limited only to acute adjustments. Adjustment of the homeostatic set point to anticipate repeated stressors has been called allostasis, which has been implicated in alcohol and drug addiction (Koob 2000), and may be considered an extreme version of expectancy.

Perhaps one reason that expectancy/anticipatory processing has appeared so ubiquitously in biological and behavioural science is that it is consistent with a number of other concepts that serve as conceptual bridges across scientific domains (see Goldman 2002). Foremost among these concepts is consilience, recently advanced by E. O. Wilson (1998), who, quoting Whewell (1840), said that consilience '... takes place when an Induction, obtained from one class of facts, coincides with an Induction, obtained from a different class' (p. 8). Wilson argued for the underlying unity of knowledge; he felt, therefore, that the search for consilient concepts might confer scientific advantage. The spontaneous emergence of expectancy in so many domains of science renders it an example of a consilient concept with a natural capacity to amalgamate disparate findings. Equally important is the consistency of the expectancy concept with three other cross-cutting concepts in biological science: conservation (reflecting the repeated observation of similar function and structure), contingency [describing both the responses of biological elements (e.g. individual cells) and of the intact organism to varying external conditions] and emergence [referring to phenomena that arise from complex combinations and interactions (e.g. of genes and neurons)]. Expectancy may be seen as a function that is conserved across domains and levels of explanation, that underpins contingent responding and that explains the emergence of complex behaviour from the operation of simpler operations.

In its largest sense, expectancy even may be regarded as the functional outcome of epigenesis, an increasingly popular and ever-evolving term in this age of genomics. As described by Gottesman and Hanson (2005), the concept originated in embryology in reference to the emergence of complex organisms from undifferentiated cells, but has come to be understood by molecular biologists as encompassing the mechanisms by which gene expression is influenced to alter cell structure and function and then transferred to future cells in that cell line (Jablonka and Lamb 2002). Even broader definitions refer to gene expression profiles that include behavioural outcomes

(Gottesman *et al.* 1982). Although we cannot extensively review the divergent uses of the term in these pages, Gottesman and Hanson (2005) relate epigenetics to '... the complexities of how multiple genetic factors and multiple environmental factors become integrated over time through dynamic, often nonlinear, sometimes nonreversible, processes to produce behaviourally relevant endophenotypes and phenotypes' (p. 267).

From an expectancy perspective, 'behaviourally relevant' means essentially that the integration of genetic and environmental factors sets the organism up to anticipate upcoming circumstances in the best way. Put another way, this process may be construed as the transfer, via multilevel biological processes, of the organism's experiences with external circumstances into biological tissue (and neurophysiological processes sustained by this tissue) that represents those experiences, so as to prepare the organism for future encounters with similar circumstances. The multilevel biological processes at issue range from those occurring within the nucleus of cells, to those that occur within single cells, to more complex processes that regulate the function of proximal groups of cells, and ultimately to distributed cell groupings that constitute brain 'systems' (Breiter and Gasic 2004). Some of these processes are discussed below.

It is important to appreciate that the expectancy concept as described herein is not attached to a single domain of science, and nor to any specific theorist. The word expectancy is not critical, and nor are the varied conceptualizations developed in the fields in which it has been used. We have used the word expectancy because it is the word most connected with the domain of research in which we have engaged, and because we wish to relate this domain (centred upon overt human cognitive–behavioural output) to the larger research enterprise that addresses the same processes. Referring to these processes, some investigators have used the word 'anticipation', others 'prediction', and still others 'preparation'. Consider Calvin and Bickerton (2000): '... you won't go far wrong if you think of your brain as always preparing for action, trying to guess what happens next, and gathering sensory information in aid of tentative plans for action' (p. 58).

The essential character of the expectancy concept may be parsed into six interconnected aspects: (1) anticipation/prediction; (2) comparison of stored information patterns with incoming stimulus arrays to influence outputs; (3) constrained perception of incoming stimuli by stored information patterns; (4) representation of anticipatory templates via changes in the physical neural substrate; (5) the equivalence of expectancy/anticipation and 'memory'; and (6) inseparability of memory, cognition and affect/emotion. In earlier papers (Goldman 1999, 2002; Goldman *et al.* 2006), we reviewed literature that supported these ideas. The present chapter will selectively update this material. It does not attempt to be comprehensive; the totality of this literature is well beyond the scope of any single work. In some instances, the research that underpins these ideas uses the term 'expectancy' directly. In other cases, this research highlights one or more of the above aspects without using the word itself (or uses anticipation/prediction/preparation). In this chapter, we will minimize the use of the word expectancy to underscore the relative unimportance of the word itself to the synthesis we suggest.

6.2 Aspects of expectancy operation

6.2.1 Anticipation/prediction

This crucial aspect of expectancy defines its basic function, i.e. the nervous system has evolved to store information about experiences so as to anticipate (predict) and negotiate future circumstances. The evolutionary pressure that led to this fundamental characteristic of neurobehavioural functioning arises from an inexorable feature of existence; namely, that time always moves forward. No environment (context) is static. Each instantly changes into the next context, and that context into the subsequent one. Some changes may be quite noticeable and others barely perceptible. Nonetheless, the context shifts from moment to moment and survival depends on organismic adjustments that proactively anticipate and prepare for the next moment, both biologically and behaviourally (these domains are, of course, really the same). Diverse scientific authors have recognized this fundamental characteristic of the nervous system: Dennett (1991) described brains as 'in essence, just anticipatory machines' (p. 177) and Brembs (2003) further suggested that, 'Anticipating the future has a decided evolutionary advantage, and researchers have found many evolutionarily conserved mechanisms by which humans and animals learn to predict future events' (p. 218).

Although basing future actions on information about the past may seem inefficient, Calvin and Bickerton (2000) point out, 'We, like most other creatures, are built to make generalizations on inadequate evidence at very short notice, because that works better, in terms of evolutionary fitness, than making 100% correct generalization after a long period of cogitation' (p. 34). Neurobiological systems 'bet' on the optimal adaptive response in an upcoming moment (as used here, the word 'bet' is not just a metaphor; we refer to what are likely stochastic processes; see Glimcher 2005). Most mis-steps will fall within tolerance limits for survival. Lethal mistakes must stay below a critical threshold (from the perspective of species survival) or natural selection would favour a different working arrangement.

Anticipation is, of course, not unitary. There is anticipation of the next moment; will the organism be best served by maintaining an ongoing behavioural sequence, or by changing behavioural output to accommodate altered contextual conditions? Also, beyond the next moment, to what time span should behavioural output be accommodated? Is responding keyed to relatively immediate payoffs, or to longer term, but potentially higher value rewards? As research addresses these questions, it becomes clearer that there are many anticipatory mechanisms, each operating on different time scales and with different functions (see the quote of Brembs above). For example, brain ERPs signal the adjustment of attention over a span of a few hundred milliseconds to accommodate unexpected events, a process that has been called 'context updating' (Donchin and Coles 1988). Space and time can themselves serve as anticipatory cues. Posner *et al.* (1980) have showed faster and/or more accurate responses and amplification of visual evoked potentials to targets appearing at expected locations. A review by Nobre (2001) indicated that response times are shorter when the temporal expectancy for a target and the time interval that actually precedes its presentation coincide.

Very recent research has revealed different brain circuitry that is differentially sensitive in anticipation of the smaller of two rewards (as opposed to just responding to the magnitude of reward; Glimcher and Lau 2005; Minamimoto *et al.* 2005), of shorter and longer term payoffs (McClure *et al.* 2004) and of rewards that are more or less likely (Tobler *et al.* 2005). Recasting the most common understanding of dopaminergic pathways as the substrate for reward, Montague and Berns (2002) describe the function of this system as reward prediction. Specifically, they contend that the level of dopamine pathway activity signals, in advance of consummatory behaviours, that rewards available in a particular context differ from rewards anticipated. Such a mechanism obviously would serve to adjust ongoing behaviour toward the most beneficial outcomes, and would not be unrelated to the ERP notion of context updating (Donchin and Coles 1988). Cohen and Blum (2002) point out that such a model of dopamine activity links the literature on classical conditioning with the computational literature on machine learning.

Research with monkeys has shown that orbitofrontal neurons activate differentially in anticipation of the different phases of the reward sequence (Schultz *et al.* 2000). Three distinctive kinds of activity were found: (1) responses to reward-predicting instructions; (2) activation in the period immediately preceding reward; and (3) responses following reward. Schultz *et al.* (2000) concluded that, 'The processing of reward expectations suggests an access to central representations of rewards which may be used for the neuronal control of goal directed behaviour' (p. 272).

A newly developing discipline, neuroeconomics (see Lee 2005), has applied economic theories of utility and games to the operation of neurons, and has found that the output of neuronal circuits often corresponds to mathematical models of decision making and outcome prediction (Breiter *et al.* 2001; Glimcher and Rustichini 2004; Schultz 2004; Cohen and Ranganath 2005). Even interpersonal trust, a quintessential aspect of human social behaviour, is fundamentally anticipatory in nature, and can be modelled using economic exchange theory (King-Casas *et al.* 2005). Miller (2005) notes that King-Casas' '...results also suggest that trust isn't purely noble—it may stem from a cold calculation of expected rewards' (p. 36).

6.2.2 Comparison of stored information patterns with incoming stimulus arrays to influence outputs

The second key aspect of expectancy relates to its mode of operation. Grossberg (1995) underscores this mode of operation, and that it is built upon the nexus between behaviour and biology, noting that, 'Neural networks that match sensory input with learned expectations help explain how humans see, hear, learn and recognize information'. As Brembs (2003) suggested, however, all these operations cannot be supported by a single mechanism. Expectancy is best understood, therefore, as a principle—a functional approach to adaptation and survival that has been manifested over evolutionary time in multiple biological systems using different structures and processes. This principle is not just an abstraction, however. We also may theorize a common substrate, sometimes using common underpinnings in the nervous system and sometimes with distinctive underpinnings that accomplish the same function.

Regardless of the specific mechanism in question, the common function seems based upon the storage of information about previous circumstances, followed by comparison of the newly encountered circumstances with this stored information. Given sufficient overlap, behaviours are performed that were effective in these circumstances (in the original encounter, behavioural options may have developed through trial and error or algorithmic computation, i.e. deliberative thought). The stored information is not simply a static 'photograph'; it incorporates the time sequence of the original circumstance, as well as a wide variety of other characteristics specific to that encounter (Leutgeb *et al.* 2005). When reactivated, it can unfold over time to represent the temporal parameter of the original experience. Of course, the more experiences one has with a particular circumstance, the more indelible this record might be (Martin and Gotts 2005).

Consider the experience of driving home from work over a very familiar route; we often arrive home with little recollection of the drive. The 'map' that directed our movements unfolded automatically, even though we might not have registered (looked at) exactly the same external cues as in preceding trips. Also, the 'map' would not just include the space external to the automobile's windshield. It might include information on the size and shape of the car's interior, and even the type of transmission in the automobile one was driving (e.g. how one would shift gears; Leutgeb *et al.* 2005).

It is also not necessary for a precise match to occur for the production of linked behaviour. The new stimulus configuration must fall only within a certain confidence interval of the stored template; that is, a decision is made based on 'fuzzy' logic (i.e. decisions are built upon stochastic processes). Furthermore, recent research suggests that the most simple (i.e. first-order) version of pattern matching (i.e. straightforward comparison of one 'picture' with another) cannot fully account for the manner in which we are able to generalize to every new, and somewhat different, situation using previously recorded experience (Poggio and Bizzi 2004). Although the Poggio and Bizzi (2004) model for matching in the visual system is quite complex and technical, for the present purposes it is sufficient to report that they suggest multidimensional 'tuning' of a linear combination of hierarchically arranged neurons to an optimal stimulus. In fact, the tuning of inferotemporal cortex cells is accomplished using a hierarchical set of cortical stages that successively combines responses from neurons tuned to simpler features (i.e. simple features detected at the lowest stage are recognized in combination at the next stage up in the processing hierarchy as a particular pattern; these patterns are in turn recognized as even higher order patterns/combinations at a further step up in processing). Poggio and Bizzi (2004) point out that feedback loops in this hierarchical system increase the power of these detection devices considerably over simple first-order pattern detectors. We have described a simplified version of such a system in an earlier paper (see Goldman 2002, p. 740).

Because the emission of contextually appropriate behaviours is the essential function of these anticipatory mechanisms, representations of broad motor patterns, the beginnings of motor activity appropriate to the recognized stimulus inputs, must be completely integrated. These motor patterns are arranged in a manner that is essentially the inverse of sensory recognition patterns (Saper *et al.* 2000). At the highest level, stored information constitutes an abstract action pattern laid out over a temporal dimension. A useful way of conceptualizing this type of expectancy is as a script, a dynamic blueprint for an action sequence. As activation proceeds down the movement hierarchy, scripts are translated into more specific sequences of planned muscle activity that in turn control actual movements.

6.2.3 Constrained perception of incoming stimuli by stored information patterns

The amount of information impinging on a human's sensory system on an ongoing basis is enormous. Not only is detailed information received by all the senses in parallel, but this information is updated on a moment to moment basis. Neurophysiological studies of human information processing estimate that potentially billions of bits of information impinge on the human senses every second, and studies have also shown that after processing, no more than 50 bits of information per second are available to consciousness (Norretranders 1991). Even allowing for additional capacity to process some information outside of consciousness, organisms still must dramatically reduce the degrees of freedom in the incoming information. At the same time, they must remain responsive to the information that matters for survival and adaptation. It is the anticipatory systems described above that work to reduce incoming information to a manageable level. The information to which the organism is biased (selectively responsive) is governed by the nature of our sensory systems, selected over evolutionary time. As one example, the visual system has receptive elements configured to be sensitive to particular stimulus arrays (e.g. lines in various orientations, patterns, movement; Maunsell 1995). However, a good deal of the selectivity is based on past experience; complex adaptations require plasticity. Many receptive elements can be altered by experience (Grossberg 1995). Even basic elements of the visual system, once thought to be hard-wired, have been found to be adaptable or plastic (Singer 1995; Tsodyks and Gilbert 2004). Also, these filtering mechanisms do not operate passively. Grossberg (1995, p. 439; credited to Richard Warren) offered an interesting example:

Suppose you hear a noise followed immediately by the words 'eel is on the....' If that string of words is followed by the word 'orange,' you hear 'peel is on the orange.' If the word 'wagon' completes the sentence, you hear 'wheel is on the wagon.' If the final word is 'shoe,' you hear 'heel is on the shoe'.

In this example, Grossberg suggests that expectations activated by the final word in the sequence influence perception of the first word. One substrate for this type of active process that completes a pattern in the absence of full and precise information (but not necessarily in this specific verbal example) has been shown to be an area of the hippocampus with massively interconnected pyramidal cells (CA3; Nakazawa *et al.* 2002). It has been suggested (Miyashita 2004, p. 436), that because 'the configuration of environmental stimuli and their behavioural context in daily life are unique and rarely repeated exactly', it is necessary for the brain to be able to recover a complete memory from stimulus conditions that supply only partial cues.

Similarly, non-human primates categorize stimuli based on important features rather than from all available visible information (Hampson *et al.* 2004). Because only a small number of particular hippocampal neurons were found to fire in relation to the categorization task, the researchers concluded that it was the *pattern* of activating neurons that conveyed the information. Most interestingly, the features that were important differed by individual monkey. Hampson *et al.* (2004) concluded, '...firing of category neurons in hippocampus may herald the presence of items that are similar to past events and therefore *through prior experience*, have a high probability of being significant to the individual...the presence of such neurons may be critical for the ability to detect and/or encode critical features present in ambiguous stimuli' (p. 3189).

The tendency to use limited cues to anticipate future events is not restricted to completion of spatial or auditory patterns; primates have been shown to anticipate multiple action sequences based on slight changes in the context of an initial action (Fogassi *et al.* 2005). In this study, monkeys could predict the outcome of movement sequence when the object involved and the initial movements were identical, with only the ultimate repository of the object differing. Other recent work on 'mirror neurons' underscores that human and primate brains are wired to anticipate intentions and behavioural patterns of other living actors in their environments (Iacoboni *et al.* 2005). In related work, rodents were trained to distinguish circles and squares, and the corresponding neuronal activation patterns were mapped (Wills *et al.* 2005). When ambiguous blends of these stimuli were presented along a continuum, the firing patterns corresponded to either the circle or square, with abrupt shifts in firing occurring about the midpoint of the transformation. In humans, the phenomenon of 'false memory' (in which stimuli are 'remembered' that were never actually experienced) is analogous to these animal paradigms. Roediger and McDermott (2000) referred to false memory errors as 'intelligent', in that they expose the operation of processes that are critical for adaptive cognition. This paradigm will be covered in more detail later.

6.2.4 **Representation of anticipatory templates via changes in the physical/neural substrate**

How are these information patterns represented? The full answer to this question would require, of course, a review of an extensive array of neurophysiological mechanisms. For present purposes, however, a more limited point can suffice: each of our senses registers information in a manner that is consistent with the actual physical input of an external stimulus. The visual system registers patterns of light using sensors that are sensitive to these patterns (Frishman 2001), the auditory system registers where on a vibrating membrane sound waves produce the most displacement (Moore 2001), and so on. As a consequence, storage of this information may take the form of actual physical enhancement of the pathways that move sensory input through the nervous system (Chklovskii *et al.* 2004; Fitzpatrick 2005; Winocur *et al.* 2005). Referring to the human visual system, Tarr (2005) underscores this point, suggesting that object recognition depends on what he calls our astonishing memory capabilities, and the encoding of much of what we see as it originally appeared. Buzsaki (2005) reports that, 'ensembles of hippocampal "place cells" form a map-like representation of the environment' (p. 568). These representations are templates for anticipatory functioning.

Kandel and Squire (2000) further point out that cerebral '. . . cortices can be reshaped by experience. In one experiment, monkeys learned to discriminate between two vibrating stimuli applied to one finger. After several thousand trials, the cortical representation of the trained finger became more than twice as large as the corresponding areas of the other fingers' (p. 1119). More recently, evidence has emerged that regionally specific reshaping of the cortex based on increased myelination occurs in children who began piano lessons relatively early in life (before age 11), as compared with children who began later (Bengtsson *et al.* 2005). Also, Bundesen *et al.* (2005) provide evidence that receptive fields of (visual) cortical cells are routinely remapped so that more processing resources (cells) are devoted to behaviourally important objects than to those less important. In referring to the mechanism that underlies the physical changes, both Kandel and Squire (2000) and Miyashita (2004) refer to altered gene expression, increased protein synthesis, growth of new synaptic connections and reorganization of neural circuits. Hence, core cellular mechanisms are mustered in the nervous system for the building and organizing of neural tissue that stores information. Once again, such information may be conceived of as expectancy templates.

Very recent findings from the study of somatosensory receptive fields for whiskers in rodents have considerably advanced understanding of the mechanisms of cortical plasticity (Feldman and Brecht 2005). Each rodent whisker has a columnar receptive field for which a single whisker is the primary source of input, and these radial columns are arranged into a 'barrel map'. The functional arrangement of these maps is, however, quite plastic; they are continually restructuring in response to sensory input. By investigating these processes of plasticity, it has become evident that simple Hebbian principles (i.e. simple enhancement of pathways that are most used) are insufficient to explain the full range of changes. Instead, it has become clearer that there is interplay between the intrinsic organization of neural tissue and sensory experience that stimulates modification of this tissue. Among the many plastic changes observed: different cortical layers respond differently; unused tissue may recede in addition to utilized tissue elaborating, and different layers may be differentially responsive to changes at different points in development. The fundamental point remains the same, however. Neural tissue is constructed and modified to reflect experience in a manner that is in some way (perhaps loosely) isomorphic to the input it receives.

6.2.5 Equivalence of expectancy/anticipation and 'memory'

Because anticipatory adjustments can only be based upon information obtained in similar past circumstances (either long-term or immediate past), we conventionally refer to this stored information as memory. The word memory, and the concept that underlies it, probably evolved to capture our subjective experience of looking within ourselves to recall information, and, most typically, 'backward' in time to locate this information. However, given that memory operation is most often inferred from alterations in ongoing behaviour as a function of (past) experience, the wider function of memory probably supports preparatory or anticipatory processing. To serve this function(s), many memory storage mechanisms exist, and exhibit reciprocal influence, in an intact organism (Miyashita 2004). 'Memory' systems are ubiquitous in higher organisms; immunology researchers often consider the immune system to be a memory system, wherein templates are stored to guide later responses (e.g. Lanzavecchia and Sallusto 2002). Despite humans' subjective sense of memory as a device for looking backward, memory perhaps is better understood as a means of looking forward (Goldman 2002). As Holland and Gallagher noted (2004, p. 148), 'The utility of learning and memory lies not in reminiscence about the past, but in allowing us to act in anticipation of future events'.

It is beyond the scope of this chapter to review thoroughly the functional distinctions between 'types' of memory and the structures that support memory. For a review of one memory taxonomy, please consult Miyashita (2004).

A few points should be made in the present context, however. First, common taxonomies of memory divide long- and short-term stores. All long-term memory can be viewed as implementing the anticipatory function, i.e. information is stored that allows comparison of a current context with previous exposures to similar or related contexts. However, even short-term (working) memory can be seen as serving, in part, this function: because multiple mechanisms may operate in tandem to anticipate behavioural adaptations, and these mechanisms may compete (i.e. lead to different behavioural outputs), working memory (as part of what is often called 'executive functioning') may enter into adjudication between potential outputs. Recent research has indicated that prefrontal neurons accomplish just this memory comparison process (Latham and Dayan 2005; Machens et al. 2005). Once again, though, the function of this decision making is to optimize behavioural outputs for achieving payoffs in upcoming circumstances (although optimization is not always achieved; see the next section for discussion of these payoffs).

Secondly, although distinctions have been made between explicit (declarative) and implicit (non-declarative) memory, it is important to recognize that a number of discriminable memory processes can be subsumed under these umbrellas (see Miyashita 2004). These processes do not necessarily handle the same information, and they do not, therefore, represent a single repository of either explicit or implicit information. Tasks designed to access such information cannot be assumed to be interchangeable. At the same time, however, research in cognitive psychology has suggested that many of these processes and subprocesses inter-relate on a continuing basis, and may not be measurable at the behavioural level as truly distinct (Roediger et al. 1999). As we shall show later, anticipatory processes (in the extended sense we are describing) are not confined to either explicit or implicit domains, or to any of these subprocesses. In fact, early in this chapter we noted that some of the subprocesses within the implicit domain as described by Miyashita (2004; e.g. simple conditioning) have been explained using the expectancy concept.

Thirdly, information used as reference points (expectancy templates) in anticipatory processing is not restricted to that contained in specific memory pathways; such information also can come from brain-wide networks that are assembled deliberately in response to top-down signals from the frontal cortex (active retrieval), or automatically, using a retrieval signal from the temporal cortex. To some as yet unknown extent, these assemblies integrate information that resides in the memory pathways described above. These integrations of information seem to arise automatically in particular contexts based on the principles of associationism (see Miyashita 2004), and also can be understood as supporting the anticipatory function described above. Associational relationships represent the 'glue' that binds together otherwise disparate objects or events, and renders them useful for anticipating upcoming circumstances.

6.2.6 Inseparability of memory, cognition and affect/emotion: logic of the linkage

Associationist principles (e.g. Nelson *et al.* 2003) provide an effective framework for understanding how initially disparate information may be 'bonded' into a larger whole in the appropriate context. Such bonding ensures that relevant information can be assembled efficiently and effectively when needed. Quickly and effectively searching the information in the brain, and assembling such information into coherent wholes that are situationally appropriate, is a formidable task. In some circumstances faced by modern humans (e.g. many modern information service jobs, taking academic tests, completing many laboratory-based cognitive tasks, and others), searches that assemble purely abstract information even may be sufficient to produce adequate behavioural output.

A full explanation of how behaviour is directed in the real world requires, however, additional theoretical elements that encourage pure information to be translated into overt behaviour. After all, human memory systems evolved before the existence of many of the abstract tasks modern humans encounter. It must be the case, therefore, that evolutionary (selection) pressures (e.g. adaptation, survival) influenced the development of these memory systems. We can identify mechanisms built into the basic biological fabric of each individual that serve as proxies for monitoring evolutionary success. These mechanisms are what we typically mean when we refer to the constructs motivation, reward, incentive, reinforcement (positive and negative) and punishment. Hence, to govern real world behaviour, information that is accessed by associational search at some point must be associated with information about reward and punishment (Schultz 2004). It is this final, critical information that specifies what searches will be undertaken, and what behaviours will be carried out to consummate the reward (or the avoidance of aversive stimuli).

In higher organisms, at least two kinds of motivational pathways may be distinguished. In the first, basic needs are regulated by physiological signals that indicate that the internal system has moved, or will soon move, outside an acceptable biological range (e.g. hunger, thirst, etc.). Such signals activate an associational search for information about the behavioural steps that will bring the system back into balance (some adjustments do not require overt behavioural outputs and occur purely at the physiological level, e.g. homeostasis).

The second kind of pathway accommodates rewards that go beyond the adjustment of basic biological parameters. In this pathway, contextual stimuli signal (via a number of neural pathways) that a rewarding condition may be achieved. Information that in some way represents the behavioural steps that may achieve reward is then accessed via an associational pathway. Because many rewarding circumstances are not of the variety that would call for activation of the first kind of pathway (they signal an advantage rather than a current biological necessity), organisms have neural systems that appear designed to signal the availability of reward, i.e. systems that use contextual signals to anticipate that reward is imminent (Berridge and Robinson 1998). In humans, rewards even may be anticipated into the distant future; such long-term rewards are most often called goals (Bargh *et al.* 2001). It must also be noted that these pathways are not independent; the second system may provide payoffs for behaving in a manner that anticipates (and wards off) strong activation of the first pathway. Although a thorough examination of motivational systems is beyond the present scope, it is essential to note that motivation must be linked to cognition. Consider Holland and Gallagher (2004, p. 148):

Recently, conventional associative learning paradigms have been adapted to allow systematic study of expectancy and action in a range of species, including humans … 'expectancy' refers to the associative activation of such reinforcer representations by the events that predict them, before the delivery of the reinforcer itself.

It is essential to note that reinforcing circumstances go far beyond the traditional basic biological needs. It is becoming clearer that our nervous systems are set up to register (and anticipate) consequences that are related to evolutionary 'fit' (Glimcher 2002). For example, successful jockeying for status in the social hierarchy may be noticed by motivational systems because such status makes reproductive success more likely. Glimcher refers to genetic survival as the ultimate cause of behaviour, and notes the correspondence between work on this notion and formal economic theory. He also notes, however, that behaviour of individuals often fails to optimize long-term (evolutionary fit), or sometimes even short-term, outcomes, perhaps because people are 'tuned' to evaluate longer term payoffs inaccurately, or because emotion over-rides logical thinking. As he suggests, decision making by individuals may be stochastic but, when outcomes of these stochastic processes are distributed across many individuals, they may result in collective outcomes that optimize survival (passing on of genes) at the population level.

In sum, to produce behaviour, associational pathways at some point must lead to the anticipation/prediction of payoffs, be they of basic needs or of the evolutionary fit variety. Contextual signals of the availability of payoffs are made salient within the total pattern of neural activation by brain mechanisms that highlight the presence of those signals (see below). Complex associational pathways then link the highlighted activation patterns to the internal representations of behaviours (i.e. scripts, templates) that have some probability of achieving these outcomes. The essence of the expectancy model is that context engenders *anticipation/prediction* of payoffs, and the emission of reward-related behaviours. In this way, expectancy serves as the theoretical amalgam of cognition (association) and motivation/emotion.

6.3 Selection of biologically meaningful inputs

Even if one accepts that cognition and emotion must be linked, the question remains of how biologically meaningful inputs are selected from the vast array of stimulus material encountered by the organism. At the current time, neuroscientists identify a number of brain pathways for registering biological significance. In one of these pathways, dopaminergic neurons running from the striatum to the nucleus accumbens and on to the frontal lobes are thought to convert '...an event or stimulus from a neutral "cold" representation (mere information) into an attractive and "wanted" incentive that can "grab" attention' (i.e. incentive salience; Berridge and Robinson 1998, p. 313). Referring to these same pathways, Kupfermann *et al.* (2000, p. 1010) note, '...dopaminergic neurons encode *expectations* [italics added] about external rewards'. More recent work by Matsumoto and Tanaka (2004, p. 178) indicates that the prefrontal cortex (anterior cingulate cortex) links these signals of biologically important inputs with actions '...based on goal expectation and memory of action–outcome contingency'. Using the conceptualization of expectancy presented in these pages, memory of action–outcome contingency would be, of course, the essence of the expectancy. It is this 'memory' that is re-enacted in anticipation that the action will again lead to reward.

A second means of registering biological significance centres on the amygdaloid complex, which has been identified as important for the expression of emotion (Holland and Gallagher 2004; Iversen *et al.* 2000). Obviously, a neural system supporting the experience of pleasure and displeasure would be instrumental in encouraging certain behaviours and discouraging others.

Once again, however, the close linkage between this source of emotional expression and systems that subserve information processing makes certain sensed patterns of information more salient, and therefore more likely to be stored and acted upon (Holland and Gallagher 2004; Phelps 2004; Schultz 2004). This linkage is further demonstrated by research showing connections between motivational areas such as the nucleus accumbens and amygdala, and information-processing areas such as the hippocampus and frontal cortex (Cardinal and Everitt 2004; Phelps 2004). Corticosteroids released by the hypothalamic–pituitary–adrenal (HPA) axis in response to threatening circumstances also influence hippocampal memory storage (Heinrichs and Koob 2004). This linkage between neurally processed information and motiv-ation/emotion is well captured by the expectancy concept, and researchers in this domain routinely do just that (Holland and Gallagher 2004; Phelps 2004).

Bechara *et al.* (2000) have presented a 'somatic marker hypothesis' that widens the spectrum of signals that may arise from bioregulatory processes to influence decision making. They aver that final oversight of decisions (competition between anticipatory pathways) is not just a function of the orbitofrontal cortex ('executive cortex'), but arises from large-scale systems that include both cortical and subcortical structures. Among these structures are the amygdala, the somatosensory/insular cortices and the peripheral nervous system.

6.4 **Complex behaviour as anticipatory**

Because the anticipatory mechanisms discussed in this chapter can be used conceptually to integrate biobehavioural functioning vertically (across levels of explanation) as well as horizontally, it must be understood that overarching human behavioural patterns (e.g. personality) also can be understood by viewing functional outputs through this same lens. Infants clearly have iden-tifiable behavioural propensities (temperament) even at birth (and perhaps even *in utero*; e.g. DiPietro *et al.* 1996). Once the newborn reacts to environ-mental contexts in accord with these propensities, they begin to accrue experiences with payoffs and failures as a function of their behavioural outputs in these contexts. Anticipatory patterns will result. Over time, successions of these experiences (representing interplay between innate behavioural propensities and ensuing experiences in the particular environ-ments they encounter) will result in apparently consistent behavioural patterns. (Because contexts are never identical, however, the performance of these patterns also will be variable and probabalistic.) These relatively consistent patterns are what we mean by personality. [It is perhaps not an accident that the term 'personality' has not typically been applied in relation to the behavioural consistencies of children (see Caspi *et al.* 2005). Anticipatory patterns may take a while to regularize into what we commonly mean by personality.]

Because complex behavioural patterns always reflect interweaving of tonic propensities (e.g. temperaments, mood, affective tone), more phasic affective reactions to triggering stimuli and previously learned behavioural templates, the neurophysiology of these tonic propensities and phasic affective reactions may influence behavioural outputs. The ultimate source of these propensities is, of course, gene expression and the activation of consequent neurophysiological processes. Research on the genetics of these processes has accelerated in recent years, and it is much better understood how the genetics of neurophysiological pathways for reward, behavioural control and stress response interplay with environments to produce, in the terms of this chapter, anticipatory patterns of behaviour (see Goldman *et al.* 2005).

For more extensive discussion of the overlap between expectancy and personality, please consult Goldman (1999). Recently, Smith and Anderson (2001) have shown that even after years of accrual of these reciprocal events (propensities leading to outcomes, leading in turn to storage of anticipatory information), humans acquire new expectancies as a function of pre-existing personality patterns. Anderson *et al.* (2005) have linked these 'personality' processes to brain functioning: greater inhibitory reactivity on functional magnetic resonance imaging (fMRI; an indicator of trait-like inhibition) predicted fewer expectancies of cognitive and motor improvement from alcohol use, but more expectancies of cognitive and motor impairment. Smith and Anderson (2001) have referred to this process as 'acquired preparedness'.

6.6 Application to alcohol use, abuse and dependence

We have been emphasizing how anticipatory function can serve as a conceptual bridge between cognitive/information-processing systems and emotional/motivational influences. Because recent research has shown that alcohol use can be characterized as arising from the integrated operation of these influences, the concept of anticipatory processing is very easily applied to alcohol use, abuse and dependence. The anticipatory processes described earlier place everyday stimuli that co-occur with the availability of alcohol into memory as information templates. Alcohol itself works on the system for tagging stimuli as biologically significant (via direct effects on the dopamine system, i.e. 'incentive salience'), and on emotional/motivational systems, to make these memories more indelible and salient, and thereby more influential within the overall decision-making process that leads to (or is avoidant of) alcohol use.

That the physical neural substrate reacts in anticipation of events related to alcohol and substance use has been shown by using neuroimaging techniques. Consider Sell *et al.* (1999, p. 1042; cited in Robinson and Berridge 2000, p. S101): '...PET has also been used...both heroin and heroin related cues activated the same structures...including the anterior cingulate, amygdala, and dorsolateral prefrontal cortex'. Although this anticipatory reaction might be seen as an example of a classically conditioned response, recall that some major theories of classical conditioning are expectancy based. Even more important, however, is the fact that the brain regions activated in response to cues are the very areas that have been suggested to be implicated in the learning and storage of expectancy (memory) templates that guide future behaviours. By using fMRI to address directly the brain region and neuro-transmitter system noted by Kupfermann *et al.* (2000) to encode expectancies, Robinson and Berridge (2000, p. S100; citing work by Breiter *et al.* 1997) note, 'One region to show bilateral activation during a saline retest was the nucleus accumbens, which the authors speculate could be related to expectancy for cocaine'.

Given that we all carry the substrate for storing memories to anticipate and prepare for future events, are all of us at risk for excessive alcohol use? In the right circumstances, we all share some level of risk; environments can either encourage or discourage particular levels and patterns of use. At the popula-tion level, after all, the most important predictor of drinking and drinking-related problems is availability of alcohol (Gruenewald *et al.* 1993).

After holding availability constant, however, differential risk largely reflects individual differences. The individual difference characteristics routinely invoked, such as emotional reactivity, personality and sensitivity to alcohol, are the very mechanisms discussed above as making memories more or less salient in anticipatory processing. For example, Katner *et al.* (1996) compared genetically bred alcohol-preferring (P) rats with Wistar rats, concluding that, 'the mere expectation of ethanol availability enhances the efflux of DA (dopamine) in the NAc (nucleus accumbens) of the P, but not the Wistar rat, which may play a role in the initiation or maintenance of ethanol seeking behaviour in the P line' (p. 669).

In humans, McCarthy *et al.* (2000) assessed self-report alcohol expectancies in individuals who differed genetically in the alcohol dehydrogenase gene (*ALDH2*; those with the variant allele metabolize alcohol less effectively and may find use more aversive). They reported that one mechanism by which the allelic variation may influence use is by lowering positive expectancies and reducing the expectancy–drinking relationship. A follow-up study (McCarthy *et al.* 2001) provided preliminary evidence that it is indeed the level of response to alcohol that mediates the relationship between *ALDH2* status expectancies. Among men, those with the *ALDH2* allele had more adverse reaction to alcohol and increased expectancies for cognitive impairment. Statistical modelling showed that the *ALDH2*–expectancy relationship was fully explained (mediated) by the level of response to alcohol. In females, having the *ALDH2* allele decreased the likelihood that women would view alcohol as tension reducing, but a mediational relationship was not found.

Underscoring the conclusion that positive expectancies partially mediate the relationship between level of response to alcohol and alcohol use is work by Schuckit *et al.* (2005) showing that 'the level of response to alcohol ... [a] ... well established ... genetically influenced phenotype related to alcoholism risk' operated '... both directly and through alcohol expectancies ...' (pp. 181–182) in the prediction of alcohol-related outcomes. Not only does level of response to alcohol influence the nature of expectancies held and decisions to drink by an individual, but self-reported anticipated effects of alcohol influence the actual response to alcohol once drinking begins (Park and Grant 2005). These researchers also concluded that positive expectancies over-ride the negative outcomes that occur when actual drinking takes place.

Furthermore, it is well established that differences in temperament/personality and psychopathology place people at differential risk for problematic alcohol use (Sher *et al.* 2005), and that these differences may influence drinking through the operation of expectancies (Darkes *et al.* 2004; Finn *et al.* 2005; Ham *et al.* 2005). Because it is also well established that such individual differences may have heritable components, it would not be surprising to find that some piece of the expectancy process might also be heritable. A small body of studies using genetically informative designs suggests that genetic differences contribute a small, but significant, amount of variance in explaining differences in alcohol expectancies in individuals who drink regularly (Vernon *et al.* 1996; Merrill *et al.* 1999). Apparently in contrast, Slutske *et al.* (2002) found that shared experiences, rather than genetic variation, explained most of the variation in positive alcohol expectancies in a large twin sample of young women. However, when only twin pairs in which both members drank were included in the analyses, a significant genetic effect was found for performance enhancement expectancies. Consistent with the idea that expectancies reflect the impact of experience on genetically influenced mechanisms for selecting and retaining information, all genetically informative studies found that non-shared environmental influences predominated in determining expectancy variation. In sum, it is the interplay between basic mechanisms of incentive salience, emotional reactivity and alcohol reactivity, some aspects of which may be inherited, and contextual influences from the environment, that sculpt expectancies and, at least in part, determine subsequent usage patterns.

6.7 Language-based access to expectancies

6.7.1 Explicit 'questionnaire' approaches

Since 1984, over 1122 studies have investigated these processes as they relate to human alcohol use (based on a PsycINFO search using 'alcohol' and 'expectancies' as key words). The vast majority have related responses of children, adolescents and adults to verbal items aggregated in questionnaires to patterns of alcohol use. The relationship between expectancies measured in this fashion and alcohol use has proven robust: measured expectancies correlated with drinking in all groups, accounting for up to 50 per cent of the variance in drinking outcomes when used with statistical methods that attenuate error (e.g. Leigh and Stacy 1993; Darkes *et al.* 2004); measured well before drinking began, expectancies predicted drinking as children matured into adolescence (Christiansen *et al.* 1989) and even many years later, into adulthood (Stacy *et al.* 1991); as drinking increased, so did expectancies (Smith *et al.* 1995); as drinking showed its typical decrease in young adulthood, so did expectancies (Sher *et al.* 1996). The relationship between psychometrically measured expectancies and drinking was also found in longitudinal designs to be reciprocal: higher expectancies led to higher drinking, which, in turn, led to higher expectancies (Smith *et al.* 1995). Expectancies measured by questionnaires could also be inferred to be causal: in the context of appropriate designs (Baron and Kenny 1986), expectancies were shown to mediate up to 50 per cent of the variance of other known antecedents of consumption levels (e.g. Henderson *et al.* 1994; Scheier and Botvin 1997; Finn *et al.* 2000; Darkes *et al.* 2004).

6.7.2 Recent implicit approaches

These conclusions using questionnaire-based measures of alcohol expectancies, which are considered explicit in nature, have been recently supplemented using 'implicit/indirect' cognitive measures for accessing anticipatory information. Implicit approaches may be of great importance, given that much of the decisional process that leads to alcohol use is theorized to take place outside of awareness. To what extent probing implicit (or, indeed, explicit) language-based processes might reveal the output of neurally based decisional activity described earlier remains an open question, of course. We shall revisit this question later.

A foreshadowing of one of the critical questions regarding distinctions between explicit and implicit language-based measures (not to mention distinctions among these measures and neurophysiological measures of anticipation), arises from close consideration of the processes by which individuals select responses to such measures. Expectancy questionnaires are typically considered explicit in nature, meaning that responses are thought to be largely influenced by conscious retrieval and decision making. However, although agree/disagree or Likert-type response formats may be completed via conscious deliberation, it is not possible to say with complete certainty what determines such responses. For example, just by changing item order in such questionnaires, one can change observed factor structure (Weinberger *et al.* 2006). To respond to expectancy instruments, respondents may deliberately recollect specific experiences from 'autobiographical memory', or may simply 'go with their subjective feeling or impression'. The latter strategy is somewhat difficult to distinguish from implicit processing, and experts in this area have noted inherent difficulty in parsing one type of processing from the other, regardless of the techniques used (see Roediger 2003).

In our discussion of implicit assessment of expectancies, we will emphasize our own work, although we acknowledge substantial work by others in connection with implicit memory processes and alcohol use (e.g. Stacy 1997; Palfai and Wood 2001; Wall *et al.* 2001; Wiers *et al.* 2002). Each of these studies has demonstrated that the association between expected effects of alcohol and alcohol use can be measured implicitly. In particular, they have shown that, when primed with alcohol cues, heavy drinkers have strong associations for positive and/or arousing effects of alcohol.

To begin our own investigations of memory processes in alcohol expectancy operation (and concomitantly, implicit anticipatory processes), we mapped expectancy associational space in accord with Estes's (1991) suggestion that memory 'traces can be viewed as vectors or lists, as nodes in a network, or as points in multidimensional space' (p. 12). To this end, we collected individual associations to the prompt, 'Alcohol makes one...' subsequently placing scaled responses to this prompt into a hypothetical memory network using multidimensional scaling (MDS). In adults (Rather *et al.* 1992; Rather and Goldman 1994) and children (both prior to, and subsequent to drinking experience; Dunn and Goldman 1996, 1998), the resulting network was well described using two orthogonal dimensions, valence (positive–negative) and arousal (sedation–excitation). Each associate could be located in space by its coordinates on these two dimensions. Although the data collection methods required attention and deliberation, participants did not explicitly compare pairs of items and did not directly create networks generated based on the relationships among all expectancy items. These network maps served as models of implicit memory storage that could later be tested empirically in experimental studies. (An instrument based on conversion of this MDS solution into confirmatory factor structure that can be used in prediction equations has recently been made available; Goldman and Darkes 2004.)

In these models, proximal words were viewed as more likely to co-activate than words more distant. Ancillary analyses showed that heavier drinkers were more likely to associate positive and arousing outcomes with drinking and to have 'tighter' relationships (closer associations) among these outcomes than did lighter drinkers. If viewed as models for causal influence on drinking, the increased likelihood of co-activation of positive and arousing outcomes would increase heavier drinkers' likelihood of drinking. Interestingly, pre-adolescent children's maps suggested that they were more likely to activate negative expectancies. As age increased into adolescence (these were cross-sectional studies), however, gradual shifts toward an increased likelihood of activating positive/arousing expectancies, whether drinking had been initiated or not, occurred, i.e. changes in the expectancy network anticipated later drinking, moving toward patterns seen in drinkers.

We recently confirmed these maps of the expectancy association network using free association (Nelson et al. 2000) to quantify directly the strength of association between memory concepts as the probability that given one word or concept (e.g. 'salt') another will be produced (e.g. 'pepper'). These probabilities can be used to derive process models of memory operation (e.g. Nelson et al. 2003). Free associations were first obtained from 1465 children in the USA in the 2nd to 12th grades, and then from 4585 college students. Children's free associations essentially replicated the previous work using the MDS approach, validating the earlier MDS maps (Dunn and Goldman 2000). Obtaining free associates from a large number of students allowed us to accomplish a task that not only clarified the incentive matrix for drinking, but also was informative for future research in general cognitive psychology, which generally has not emphasized individual differences in memory processes. Previous free association research (Nelson et al. 2000) had established general population norms for common words by obtaining first associates to a limited list of prompts from groups of 100–200 participants.

By iterating this procedure many times to many prompts, norms were derived that would characterize the average response for all English language speakers (Nelson *et al.* 1998). In contrast, we used our sizeable sample of responses to a single prompt to establish norms for subgroups of drinkers (i.e. individual differences in alcohol cognitions; Reich and Goldman 2005). As drinker level increased, our free associate norms showed a steady shift from negative and sedating expectancies, toward arousing and positive expectancies. This finding provided direct confirmation that lighter and heavier drinkers activated different concepts in response to an alcohol prompt, and showed that humans differ in their associational network as a function of individual differences.

These models of alcohol expectancy network structure provided the initial framework for investigating a multistep process in which context activates anticipatory cognitions/affect that, in turn, influences behavioural output. Next, we tested whether the activation process suggested by these memory network models could be demonstrated via the experimental manipulation of alcohol cues, i.e. whether the expectancy network in fact activated as predicted in response to stimuli that might signal a drinking opportunity. To this end, we conducted a series of studies using tasks developed by cognitive psychologists to test memory activation following implicit primes. The first of these studies used the Stroop technique (Kramer and Goldman 2003); following an alcohol (e.g. 'vodka') or alcohol-neutral (e.g. 'milk') prime, expectancy target words were ink-named. Activation was indexed as the relative difference in latency to ink-name alcohol and neutral primed trials (interference; the greater the activation, the slower the ink naming). Expectancy activation differed as a function of prime (alcohol/neutral), level of customary consumption and location of the expectancies in a network defined by the valence/arousal dimensions. Following an alcohol cue, the heaviest drinkers had slower latencies (greater activation) to ink-name arousing expectancy words compared with other target words, whereas lighter drinkers had greater activation (interference) to sedating expectancy words; context differentially activated particular expectancies as a function of individuals' experience with alcohol.

Reich *et al.* (2005) demonstrated expectancy activation consistent with the network models via priming in a word list memory study in which the experimental manipulation was simply a change in the first word on the study list. The study list was comprised of alcohol expectancy words and grocery words (50 per cent each). The first word on the list studied was either 'MILK' or 'BEER'. Although either word can be considered grocery words, participants presented with BEER as the first word recalled more alcohol expectancy words than those who saw MILK. Further analysis revealed that this effect resulted from interaction between drinker type and type of expectancy word recalled. Heavy drinkers tended to remember proportionately more alcohol expectancy words, predominantly positive expectancies, than lighter or non-drinkers. In other words, a slight change in the initial study stimuli seemingly activated different memory patterns, and these patterns reflected differences that would be expected by different drinker types.

Another study used the false memory paradigm (Deese 1959; Roediger and McDermott 1995) to test the structure of the modelled networks. In this paradigm, several 12-word lists (e.g. hot, chilly, frigid, wet) that were each associated to with a non-studied word (e.g. cold) were studied. In previous studies, about 80 per cent of participants remembered studying the non-presented word. To extend this paradigm to expectancies, we developed a study list of alcohol expectancy words and excluded three high-frequency positive and arousing expectancy words that, according to our network maps, should have been part of that list. Participants studied this list in either a simulated bar or an alcohol-neutral room, so that we could evaluate how an alcohol context might activate alcohol expectancy elements that intentionally were not presented. In addition, participants were split into heavier, lighter and non-drinkers, to explore how context affected individuals with different patterns of consumption. After controlling for the tendency falsely to remember any type of word, heavier drinkers' false recognition for the three non-presented positive/arousing words was significantly higher in the bar context than in the neutral room, whereas false memory rates were similar across context for the other drinker groups. In other words, for heavier drinkers, the alcohol context created sufficiently strong memory activation to make participants 'remember' positive and arousing expectancies that were never presented.

After showing that elements within the network models could be activated as predicted by the associational structure, the next step was to show that cognitions activated by alcohol cues would translate into actual drinking. To this end, we used priming techniques and a disguised beer 'taste test' in two studies to demonstrate increased alcohol consumption following the presentation of alcohol-consistent cues. In Roehrich and Goldman (1995), participants in a 'memory' study were primed with either alcohol expectancy words or alcohol-neutral words and viewed one of two video clips from network television comedies. The clips were similar in content, with the central difference being the contextual location of the show. One show took place in a bar (*Cheers*; alcohol prime), the other in an inn at a breakfast table

(*Newhart*; alcohol-neutral prime). Following the manipulation, participants exposed to alcohol primes consumed more alcohol during what they believed was an entirely separate 'taste test' study; increased alcohol priming led to more alcohol consumption. The group exposed to both alcohol-consistent primes drank most, the group that had both alcohol-neutral primes drank least, and the other two groups with discrepant primes fell in between. These results supported the inference of a causal relationship between expectancy activation and alcohol consumption.

Because the above study did not assess actual activation of the memory networks following priming, a further study (Stein *et al.* 2000) used a similar design, but included a recognition task to determine whether priming had activated alcohol expectancies. Again, the level of consumption increased as the level of alcohol cues presented increased, and greater priming (activation) effects occurred for heavier drinkers. Other similar studies have also shown increases in consumption following exposure to alcohol expectancy-like cues (Carter *et al.* 1998; Palfai *et al.* 2000).

In sum, this line of research, coupled with findings from other laboratories, supports language-based memory models of alcohol expectancies, shows that these hypothesized networks can be activated implicitly, and shows that, as theorized, these models accurately reflect individual differences in previous drinking, and that both alcohol contexts and these individual differences can be engaged to influence actual drinking prospectively.

6.8 A fundamental question raised by expectancy challenge findings

How do language-based measures of expectancies correspond to the anticipatory processes discussed earlier in this chapter? To begin this discussion, we will briefly divert into what has become an increasingly controversial area of expectancy research, expectancy challenge. The expectancy challenge originally was designed to test experimentally alcohol expectancies' causal role in drinking. In the two original studies conducted in our laboratory (Darkes and Goldman 1993, 1998), groups of 10–15 moderate to heavy drinking males consumed, in a bar environment, beverages that they were told might or might not contain alcohol (i.e. alcohol or placebo beverages, randomly distributed) and then were asked to identify which group members had consumed actual alcohol. It appeared that the experience of being unable reliably to identify drinkers based on observed behaviours, coupled with information about alcohol expectancies, resulted in expectancy and drinking decreases as much as 6 weeks following the procedure.

Following these initial studies, researchers have tested challenge protocols modified from the original procedure in various ways. Exclusively female groups have been used (Dunn *et al.* 2000). Mixed gender, rather than same sex, groups have been used (e.g. Wiers *et al.* 2005), challenges that do not administer alcohol-containing beverages have been used (Corbin *et al.* 2001), the number of sessions has been modified (e.g. Corbin *et al.* 2001; Wiers *et al.* 2005), and videotaped (Keillor *et al.* 1999) and computer-based (Hunt *et al.* 2005) challenges have been tested. As might be anticipated given the variability in procedures, challenge studies have varied in the degree to which findings have replicated the original results. In some cases, findings have not replicated the original research. Some reviewers (e.g. Jones *et al.* 2001*a,b*) have suggested that these latter findings raise questions about the value or applicability of expectancy as an explanatory process.

Given the extensive body of research from widely divergent fields of study that have converged on some version of an expectancy (anticipation/prediction/preparedness) process as an explanatory device, it hardly seems that inconsistent findings in a single expectancy-based paradigm should raise questions about this entire approach to understanding behavioural outputs (Del Boca and Darkes 2001). Even within the alcohol field, the numerous survey studies and true experiments that have shown expectancies to predict drinking and to mediate the influence of other variables on drinking should support continuing research in this vein. If we restrict discussion to variations on the expectancy challenge itself, we may of course eventually conclude that this particular approach to disrupting anticipatory pathways is ineffective at changing expectancies, or does not reliably influence drinking. Alternatively, it remains possible that investigation of variables related to this paradigm, including treatment dose and duration, the need for a true drinking experience, the role of other individual difference characteristics, the need for expectancy specificity in challenge protocols and the temporal relationship between explicitly or implicitly measured expectancy changes and drinking changes, will lead to a more reliable procedure. Full tests of these parameters of the challenge remain to be fully evaluated.

These considerations do lead, however, to a more fundamental question that attaches to the multidimensional and multilevel characterization of anticipatory processes as presented in these pages. If, as asserted herein, the expectancy principle is manifested via a number of different mechanisms that may, or may not, process similar information, to what extent are these varied anticipatory pathways consistent in their outputs? A related question is to what extent can assessment of higher order (i.e. language-based) pathways provide information about more basic neurophysiological pathways, or even about each other (i.e. explicit to implicit, or vice versa)? Given the range of stimulus inputs handled by these kinds of systems, there is no reason that they should be entirely redundant and consistent in the outputs at which they might arrive. In fact, evolutionary advantage might accrue from somewhat divergent outputs, because variations in 'recommendations' for final outputs would offer more flexibility and adaptability than a single system that always offered the same choice. [As noted by Glimcher (2002), some degree of randomness might also be advantageous.] Of course, some means of adjudicating among varied output recommendations would have to be available and, as noted earlier, seems to be the function of specific nuclei in the prefrontal cortex (Latham and Dayan 2005; Machens *et al.* 2005). As a consequence of the multiple pathways for this kind of decision making, there is no reason to believe that assessing one automatically will provide information about what is going on in the other.

Within the explicit domain alone, expectancy instruments have been found to show overlap (Darkes *et al.* 1996). In contrast, early data indicate that comparisons of common measures of explicit and implicit expectancies show little overlap in variance accounted for (e.g. Stacy 1997; Wiers *et al.* 2002; Kramer and Goldman 2003). However, within the implicit domain alone, we do not yet know the extent to which measures overlap. For a variety of methodological reasons, such studies are difficult to carry out. In addition, while implicit processes need not be verbal in nature, much of the research in this domain uses language-based stimuli. Therefore, some empirical findings may not reflect the implicit/explicit distinction *per se*, but instead reflect the nature of the human language system. Even within the human language system, the general cognitive literature indicates the activity of at least two types of implicit processes; Blaxton (1989) has shown both perceptually based and conceptually based implicit processing for text read from a page or computer screen.

Given these unknowns about the overlap of expectancy processes, there is no reason to assume that specific interventions designed to influence non-specific expectancy pathways will necessarily influence all of the pathways involved in influencing particular decisions (assuming, of course, that they effectively influence *any* decision pathway), any more than we might assume that any single measure of such processes taps *all* relevant information. For example, should a verbal intervention inexorably influence a classically conditioned response, even if both were construed as based on anticipatory processes? It is possible, therefore, that a particular pathway could be influenced without impacting the decisional endpoint. If expectancy challenge procedures turn out to be erratic in their influence, this could be one reason why.

6.9 Conclusion

We offer this transdisciplinary synthesis in the spirit suggested by Reber's (1997) cautionary words: 'Psychologists just can't seem to resist dichotomies...We like to be able to decide that one theory is right and the other wrong...We seem ineluctably drawn to setting up poles rather than recognizing continua. Alas, this tendency often functions as a hindrance to doing good science' (p. 52). Even if the expectancy concept turns out not to be the optimal integrating device, the diverse literature presented in these pages begs us to decipher common themes that should be capitalized upon to advance the understanding of human decision making in general, and decision making related to alcohol use, abuse and dependence in particular.

Acknowledgements

We would like to acknowledge the contributions of a long list of colleague, graduate students and research assistants, whose hard work over the years has made this programme of research possible. Preparation of this chapter was supported by NIAAA Grants 2R01 AA008333 and R01 AA011925. Correspondence concerning this chapter should be addressed to Mark S. Goldman, Department of Psychology, PCD 4118G, University of South Florida, 4202 E. Fowler Ave., Tampa FL 33620–8200, USA or via E-mail to goldman@cas.usf.edu

References

Anderson, K., Schweinsburg, A. and Paulus, M. (2005). Examining personality and alcohol expectancies using functional magnetic resonance imaging (fMRI) with adolescents. *Journal of Studies on Alcohol*, 66, 323–331.

Bargh, J.A., Gollwitzer, P.M., Lee-Chai, A., Barndollar, K. and Trotschel, R. (2001). The automated will: nonconscious activation and pursuit of behavioral goals. *Journal of Personality and Social Psychology*, 81, 1014–1027.

Baron, R.M. and Kenny, D.A. (1986). The moderator–mediator variable distinction in social psychological research: conceptual, strategic, and statistical considerations. *Journal of Personality and Social Psychology*, 51, 1173–1182.

Baron-Cohen, S., Knickmeyer, R.C. and Belmonte, M.K. (2005). Sex differences in the brain: implications for explaining autism. *Science*, 4, 819–823.

Bechara, A., Damasio, H. and Damasio, A. (2000). Emotion, decision making and the orbitofrontal cortex. *Cerebral Cortex*, 10, 295–307.

Bengtsson, S.L., Nagy, Z., Skare, S., Forsman, L., Forssberg, H. and Ullén, F. (2005). Extensive piano practicing has regionally specific effects on white matter development. *Nature Neuroscience*, 8, 1148–1150.

Berridge, K.C. and Robinson, T.E. (1998). What is the role of dopamine in reward: hedonic impact, reward learning, or incentive salience? *Brain Research and Brain Research Reviews*, 28, 309–369.

Blaxton, T.A. (1989). Investigating dissociations among memory measures: support for a transfer appropriate processing framework. *Journal of Experimental Psychology: Learning Memory, and Cognition*, 15, 657–668.

Bower, G. (2000). A brief history of memory research. I.E. Tulving and F.I.M. Craik (ed.), *The Oxford handbook of memory*. New York, NY: Oxford University Press, 3–32.

Breiter, H. and Gasic, G.P. (2004). A general circuitry processing reward/aversion information and its implications for neuropsychiatric illness. In M. Gazzaniga (ed.). *The cognitive neurosciences*, 3rd edn. Boston, MA: MIT Press, 1043–1065.

Breiter, H.C., Gollub, R.L., Weisskoff, R.M., Kennedy, D.N., Makris, N., Berke, J.D., Goodman, J.M., Kantor, H.L., Gastfriend, D.R., Riorden, J.P., Mathew, R.T., Rosen, B.R. and Hyman, S.E. (1997). Acute effects of cocaine on human brain activity and emotion. *Neuron*, 19, 591–611.

Breiter, H.C., Aharon, I., Kahneman, D., Dale, A. and Shizgal, P. (2001). Functional imaging of neural responses to expectancy and experience of monetary gains and losses. *Neuron*, 30, 619–639.

Brembs, B. (2003). Operant reward learning in aplysia. *Current Directions in Psychological Science*, 12, 218–221.

Brown, S.A., Goldman, M.S. and Inn, A. (1980). Expectations of reinforcement from alcohol: their domain and relation to drinking patterns. *Journal of Consulting and Clinical Psychology*, 48, 419–426.

Bundesen, C., Habekost, T. and Kyllingsbæk, S. (2005). A neural theory of visual attention: bridging cognition and neurophysiology. *Psychological Review*, 112, 291–328.

Buszaki, G. (2005). Similar is different in hippocamal networks. *Science*, 309, 568–569.

Calvin, W.H. and Bickerton, D. (2000). *Lingua ex machina: reconciling Darwin and Chomsky with the human brain*. Boston, MA: MIT Press.

Cardinal, R.N. and Everitt, B.J. (2004). Neural and psychological mechanisms underlying appetitive learning: links to drug addiction. *Current Opinion in Neurobiology*, 14, 156–162.

Carter, J.A., McNair, L.D., Corbin, W.R. and Black, D.H. (1998). Effects of priming positive and negative outcomes on drinking responses. *Experimental and Clinical Psychopharmacology*, 6, 399–405.

Caspi, A., Roberts, B.W. and Shiner, R.L. (2005). Personality development: stability and change. *Annual Review of Psychology*, 56, 453–484.

Chklovskii, D.B., Mel, B.W. and Svoboda, K. (2004). Cortical rewiring and information storage. *Nature*, 431, 782–788.

Christiansen, B.A., Smith, G.T., Roehling, P.V. and Goldman, M.S. (1989). Using alcohol expectancies to predict adolescent drinking behavior at one year. *Journal of Consulting and Clinical Psychology*, 57, 93–99.

Cohen, J. and Blum, K. (2002). Reward and decision. *Neuron*, 36, 193–198.

Cohen, M.X. and Ranganath, C. (2005). Behavioral and neural predictors of upcoming decisions. *Cognitive, Affective and Behavioral Neuroscience*, 5, 117–126.

Colunga, E. and Smith, L.B. (2005). From the lexicon to expectations about kinds: a role for associative learning. *Psychological Review*, 112, 347–382.

Corbin, W.R., McNair, L.D. and Carter, J.A. (2001). Evaluation of a treatment appropriate cognitive intervention for challenging alcohol outcome expectancies. *Addictive Behaviors*, 26, 475–488.

Correa, A., Lupiáñez, J. and Tudela, P. (2005). Attentional preparation based on temporal expectancy modulates processing at the perceptual level. *Psychonomic Bulletin and Review*, 12, 328–334.

Darkes, J. and Goldman, M.S. (1993). Expectancy challenge and drinking reduction: experimental evidence for a mediational process. *Journal of Consulting and Clinical Psychology*, 61, 344–353.

Darkes, J. and Goldman, M.S. (1998). Expectancy challenge and drinking reduction: process and structure in the alcohol expectancy network. *Experimental and Clinical Psychopharmacology*, 6, 1–13.

Darkes, J., Greenbaum, P.E. and Goldman, M.S. (1996). Positive/arousal and social facilitation alcohol expectancies and concurrent alcohol use. Paper presented at the Annual Scientific Meeting of the Research Society on Alcoholism, Washington DC, USA, 1996.

Darkes, J., Greenbaum, P.E. and Goldman, M.S. (2004). Alcohol expectancy mediation of biopsychosocial risk: complex patterns of mediation. *Experimental and Clinical Psychopharmacology*, 12, 27–38.

Deese, J. (1959). On the prediction of occurrence of particular verbal intrusions in immediate recall. *Journal of Experimental Psychology*, 58, 17–22.

Del Boca, F.K. and Darkes, J. (2001). Is the glass half full or half empty? An evaluation of the status of expectancies as causal agents. *Addiction*, 96, 1670–1672.

Dennett, D.C. (1991). *Consciousness explained*. New York, NY: Little, Brown.

DiPietro, J.A., Hodgson, D.M., Costigan, K.A. and Johnson, T.R.B. (1996). Fetal antecedents of infant temperament. *Child Development*, 67, 2568–2583.

Donchin, E. and Coles, M.G. (1988). Is the P300 component a manifestation of context updating? *Behavioral and Brain Sciences*, 11, 357–427.

Dragoi, V. and Staddon, J.E.R. (1999). The dynamics of operant conditioning. *Psychological Review*, 106, 20–61.

Dunn, M.E. and Goldman, M.S. (1996). Empirical modeling of an alcohol expectancy network in elementary school children as a function of grade. *Experimental and Clinical Psychopharmacology*, 4, 209–217.

Dunn, M.E. and Goldman, M.S. (1998). Age and drinking-related differences in the memory organization of alcohol expectancies in 3rd-, 6th-, 9th-, and 12th-grade children. *Journal of Consulting and Clinical Psychology*, 66, 579–585.

Dunn, M.E. and Goldman, M.S. (2000). Validation of multidimensional scaling-based modeling of alcohol expectancies in memory: age and drinking-related differences in expectancies of children assessed as first associates. *Alcoholism: Clinical and Experimental Research*, 24, 1639–1646.

Dunn, M.E., Lau, H.C. and Cruz, I.Y. (2000). Changes in activation of alcohol expectancies in memory in relation to changes in alcohol use after participation in an expectancy challenge program. *Experimental and Clinical Psychopharmacology*, 8, 566–575.

Estes, W.K. (1991). Cognitive architectures from the standpoint of an experimental psychologist. *Annual Review of Psychology: Neuropsychology*, 8, 464–475.

Feldman, D.E. and Brecht, M. (2005). Map plasticity in somatosensory cortex. *Science*, 310, 810–815.

Finn, P.R., Sharkansky, E.J., Brandt, K.M. and Turcotte, N. (2000). The effects of familial risk, personality, and expectancies on alcohol use and abuse. *Journal of Abnormal Psychology*, 109, 122–133.

Finn, P.R., Bobova, L., Wehner, E., Fargo, S. and Rickert, M.E. (2005). Alcohol expectancies, conduct disorder and early-onset alcoholism: negative alcohol expectancies are associated with less drinking in non-impulsive versus impulsive subjects. *Addiction*, 100, 953–962.

Fitzpatrick, D. (2005). Zooming in on cortical maps. *Nature Neuroscience*, 8, 264.

Fogassi, L., Ferrari, P.F. and Gesierich, B. (2005). Parietal lobe: from action organization to intention understanding. *Science*, 308, 662–667.

Frishman, L.J. (2001). Basic visual processes. In E.B. Goldstein (ed.). *Blackwell handbook of perception*. Malden, MA: Blackwell, 53–91.

Glimcher, P.W. (2002). Decisions, decisions, decisions: choosing a biological science of choice. *Neuron*, 36, 323–332.

Glimcher, P.W. (2005). Indeterminacy in brain and behavior. *Annual Review of Psychology*, 56, 25–56.

Glimcher, P.W. and Lau, B. (2005). Rethinking the thalamus. *Nature Neuroscience*, 8, 983–984.

Glimcher, P.W. and Rustichini, A. (2004). Neuroeconomics: the consilience of brain and decision. *Science*, 306, 447–452.

Goldman, D., Oroszi, G. and Ducci, F. (2005). The genetics of addictions: uncovering the genes. *Nature Reviews: Genetics*, 6, 521–532.

Goldman, M.S. (1999). Expectancy operation: cognitive and neural models and architectures. In I. Kirsch (ed.). *How expectancies shape experience*. Washington, DC: American Psychological Association, 41–63.

Goldman, M.S. (2002). Expectancy and risk for alcoholism: the unfortunate exploitation of a fundamental characteristic of neurobehavioral adaptation. *Alcoholism: Clinical and Experimental Research*, 26, 737–746.

Goldman, M.S. and Darkes, J. (2004). Alcohol expectancy multi-axial assessment (A.E.Max): a memory network-based approach. *Psychological Assessment*, 16, 4–15.

Goldman, M.S., Darkes, J. and Del Boca, F.K. (1999). Expectancy mediation of biopsychosocial risk for alcohol use and alcoholism. In I. Kirsch (ed.). *How expectancies shape experience*. Washington, DC: American Psychological Association, 233–262.

Goldman, M.S., Reich, R.R. and Darkes, J. (2006). Expectancy as a unifying construct in alcohol-related cognition. In R.W. Wiers and A.W. Stacy (ed.), *Handbook of implicit cognition and addiction*, Thousand Oaks, CA: Sage, pp. 105–121.

Gottesman, I.I. and Hanson, D.R. (2005). Human development: biological and genetic processes. *Annual Review of Psychology*, 56, 263–286.

Gottesman, I.I., Shields, J. and Hanson, D.R. (1982). *Schizophrenia: the epigenetic puzzle.* London: Cambridge University Press.

Grossberg, S. (1995). The attentive brain. *American Scientist*, 83, 438–449.

Gruenewald, P., Millar, A. and Treno, A. (1993). Alcohol availability and the ecology of drinking behavior. *Alcohol Health and Research World*, 17, 39–45.

Haith, M.M., Hazan, C. and Goodman, G.S. (1988). Expectation and anticipation of dynamic visual events by 3.5-month old babies. *Childhood Development*, 59, 467–479.

Ham, L.S., Carrigan, M.H., Moak, D.H. and Randle, C.L. (2005). Social anxiety and specificity of positive alcohol expectancies: preliminary findings. *Journal of Psychopathology and Behavioral Assessment*, 27, 115–121.

Hampson, R.E., Pons, T.P., Stanford, T.R. and Deadwyler, S.A. (2004). Categorization in the monkey hippocampus: a possible mechanism for encoding information into memory. *Proceedings of the National Academy of Sciences of the USA*, 101, 3184–3189.

Heekeren, H.R., Marrett, S., Bandettini, P.A. and Ungerleider, L.G. (2004). A general mechanism for perceptual decision-making in the human brain. *Nature*, 431, 859–861.

Heinrichs, S.C. and Koob, G.F. (2004). Corticotropin-releasing factor in brain: a role of activation, arousal and affect regulation. *Journal of Pharmacology and Experimental Therapeutics*, 311, 427–440.

Henderson, M.J., Goldman, M.S., Coovert, M.D. and Carnevalla, N. (1994). Covariance structure models of expectancy. *Journal of Studies on Alcohol*, 55, 315–326.

Holland, P.C. and Gallagher, M. (2004). Amygdala–frontal interactions and reward expectancy. *Current Opinion in Neurobiology*, 14, 148–155.

Hunt, W.M., Darkes, J. and Goldman, M.S. (2005). Effects of participant engagement on alcohol expectancies and drinking outcomes for a computerized expectancy challenge intervention with college student drinkers. *Alcoholism: Clinical and Experimental Research*, 29, 73A.

Iacoboni, M., Molnar-Szakacs, I., Gallese, V., Buccino, G., Mazziotta, J.C. and Rizzolatti, G. (2005). Grasping the intentions of others with one's own mirror neuron system. *Public Library of Science Biology*, 3, 529–535.

Iversen, S., Kupfermann, I. and Kandel, E.R. (2000). Emotional states and feelings. In E.R. Kandel, J.H. Schwartz and T.M. Jessell (ed.), *Principles of neural science*, 4th edn. New York, NY: McGraw-Hill, 982–997.

Jablonka, E. and Lamb, M.J. (2002). The changing concept of epigenetics. *Annual Proceedings of the New York Academy of Sciences*, 981, 82.

Jones, B.T., Corbin, W. and Fromme, K. (2001*a*). A review of expectancy theory and alcohol consumption *Addiction*, 91, 57–72.

Jones, B.T., Corbin, W. and Fromme, K. (2001*b*). Half full or half empty, the glass still does not satisfactorily quench the thirst for knowledge on alcohol expectancies as a mechanism of change. *Addiction*, 96, 1672–1674.

Kandel, E.R. and Squire, L.R. (2000) Neuroscience: breaking down scientific barriers to the study of brain and mind. *Science*, 290, 1113–1120.

Katner, S.N., Kerr, T.M. and Weiss, F. (1996). Ethanol anticipation enhances dopamine efflux in the nucleus accumbens of alcohol-preferring (P) but not Wistar rats. *Behavioral Pharmacology*, 8, 669–674.

Keillor, R.M., Perkins, W.B. and Horan, J.J. (1999). Effects of videotaped expectancy challenges on alcohol consumption of adjudicated students. *Journal of Cognitive Psychotherapy: An International Quarterly*, 13, 179–187.

Kerzel, D. (2005). Representational momentum beyond internalized physics. Embodied mechanisms of anticipation cause errors in visual short-term memory. *Current Directions in Psychological Science*, 14, 180–184.

Kilner, J.M., Vargas, C., Duval, S., Blakemore, S.-J. and Sirigu, A. (2004). Motor activation prior to observation of a predicted movement. *Nature Neuroscience*, 7, 1299–1301.

King-Casas, B., Tomlin, D. and Anen, C. (2005). Getting to know you: reputation and trust in a two-person economic exchange. *Science*, 308, 78–83.

Kirsch, I. (ed.) (1999). *How expectancies shape experience*. Washington, DC: American Psychological Association.

Kirsch, I. and Scoboria, A. (2001). Apples, oranges, and placeboes: heterogeneity in a meta-analysis of placebo effects. *Advances in Mind-Body Medicine*, 17, 307–309.

Kirsch, I., Lynn, S.J., Vigorito, M. and Miller, R.M. (2004). The role of cognition in classical and operant conditioning. *Journal of Clinical Psychology*, 60, 369–392.

Koob, G. (2000). Addictions. Paper presented to the National Institutes of Health STEP Program, Washington, DC, USA, 2000.

Kramer, D.A. and Goldman, M.S. (2003). Using a modified Stroop task to implicitly discern the cognitive organization of alcohol expectancies. *Journal of Abnormal Psychology*, 112, 171–175.

Krumhansl, C.L. and Toivane, P. (2000). Melodic expectations: a link between perception and emotion. *Psychological Science Agenda*, 13, 8.

Kupfermann, I., Kandel, E.R. and Iversen, S. (2000). Motivational and addictive states. In E.R. Kandel, J.H. Schwartz and T.M. Jessell (ed.), *Principles of Neural Science*, 4th edn. New York, NY: McGraw-Hill, 998–1018.

Lanzavecchia, A. and Sallusto, F. (2002). Progressive differentiation and selection of the fittest in the immune reponse. *Naure Reviews: Immunology*, 2, 982–987.

Latham, P.E. and Dayan, P. (2005). Touché: the feeling of choice. *Nature Neuroscience*, 8, 408–409.

Lee, D. (2005). Neuroeconomics: making risky choices in the brain. *Nature Neuroscience*, 8, 1129–1130.

Leigh, B.C. and Stacy, A.W. (1993). Alcohol outcome expectancies: scale construction and predictive utility in higher order confirmatory models. *Psychological Assessment*, 5, 216–229.

Leutgeb, S., Leutgeb, J.K., Barnes, C.A., Moser, E.I., McNaughton, B.L. and Moser, M.-B. (2005). Independent codes for spatial and episodic memory in hippocampal neuronal ensembles. *Science*, 309, 619–623.

Machens, C.K., Romo, R. and Brody, C.D. (2005). Flexible control of mutual inhibition: a neural model of two-interval discrimination. *Science*, 307, 1121–1124.

Marlatt, G.A., Demming, B. and Reid, J.B. (1973). Loss of control drinking in alcoholics: an experimental analogue. *Journal of Abnormal Psychology*, 81, 233–241.

Martin, A. and Gotts, S.J. (2005). Making the causal link: frontal cortex activity and repetition priming. *Nature Neuroscience*, 8, 1134–1135.

Matsumoto, K. and Tanaka, K. (2004). The role of the medial prefrontal cortex in achieving goals. *Current Opinion in Neurobiology,* 14, 178–185.

Maunsell, J.H.R. (1995). The brain's visual world: representation of visual targets in cerebral cortex. *Science,* 270, 764–768.

McCarthy, D.M., Wall, T.L., Brown, S.A. and Carr, L.G. (2000). Integrating biological and behavioral factors in alcohol use risk: the role of ALDH2 status and alcohol expectancies in a sample of Asian Americans. *Experimental and Clinical Psychopharmacology,* 8, 168–175.

McCarthy, D.M., Brown, S.A., Carr, L.G. and Wall, T.L. (2001). ALDH2 status, alcohol expectancies, and alcohol response: preliminary evidence for a mediation model. *Alcoholism: Clinical and Experimental Research,* 25, 1558–1563.

McClure, S.M., Laibson, D.I. and Loewenstein, G. (2004). Separate neural systems value immediate and delayed monetary rewards. *Science,* 306, 503–507.

Merrill, K., Steinmetz, J.E., Viken, R.J. and Rose, R.J.(1999). Genetic influences on human conditionability: a twin study of the conditioned eyeblink response. *Behavior Genetics,* 29, 95–102.

Miller, G. (2005). Economic game shows how the brain builds trust. *Science,* 308, 36.

Minamimoto, T., Hori, Y. and Kimura, M. (2005) Complementary process to response bias in the centromedian nucleus of the thalamus. *Science,* 308, 1798–1801.

Miyashita, Y. (2004). Cognitive memory: cellular and network machineries and their top-down control. *Science,* 306, 435–440.

Montague, P.R. and Berns, G.S. (2002). Neural economics and the biological substrates of valuation. *Neuron,* 36, 265–284.

Moore, B.C.J. (2001). Basic auditory processes. I.E.B. Goldstein (ed.). *Blackwell handbook of perception.* Malden, MA: Blackwell, 379–407.

Nakazawa, K., Quirk, M.C. and Chitwood, R. (2002). A requirement for hippocampal CA3 NMDA receptors in associative memory recall. *Science,* 297, 211–218.

Nelson, D.L., McEvoy, C.L. and Schreiber, T.A. (1998). *The University of South Florida word association, rhyme, and word fragment norms.* Retrieved November 21, 2000, from http://w3.usf.edu/~fan/

Nelson, D.L., McEvoy, C.L. and Dennis, S. (2000). What is free association and what does it measure? *Memory and Cognition,* 28, 887–899.

Nelson, D.L., McEvoy, C.L. and Pointer, L. (2003). Spreading activation or spooky action at a distance. *Journal of Experimental Psychology: Learning, Memory, and Cognition,* 29, 42–51.

Nobre, A.C. (2001). Orienting attention to instants in time. *Neuropsychologia,* 39, 1317–1328.

Norretranders, T. (1999). *The user illusion: cutting consciousness down to size.* New York, NY: Penguin Group.

Palfai, T. and Wood, M.D. (2001). Positive alcohol expectancies and drinking behavior: the influence of expectancy strength and memory accessibility. *Psychology of Addictive Behaviors,* 15, 60–67.

Palfai, T.P., Monti, P.M. and Ostafin, B. (2000). Effects of nicotine deprivation on alcohol-related information processing and drinking behavior. *Journal of Abnormal Psychology,* 109, 96–105.

Park, C.L. and Grant, C. (2005). Determinants of positive and negative consequences of alcohol consumption in college students: alcohol use, gender, and psychological characteristics. *Addictive Behaviors*, **30**, 755–765.

Petrovic, P., Dietrich, T., Fransson, P., Andersson, J., Carlsson, K. and Ingvar, M. (2005). Placebo in emotional processing—induced expectations of anxiety relief activate a generalized modulatory network, *Neuron*, **46**, 957–969.

Phelps, E.A. (2004). Human emotion and memory: interactions of the amygdala and hippocampal complex. *Current Opinion in Neurobiology*, **14**, 198–202.

Poggio, T. and Bizzi, E. (2004). Generalization in vision and motor control. *Nature*, **431**, 768–774.

Posner, M.I., Snyder, C.R.R. and Davidson, B.J. (1980). Attention and the detection of signals. *Journal of Experimental Psychology*, **109**, 160-174.

Rather, B.C. and Goldman, M.S. (1994). Drinking-related differences in the memory organization of alcohol expectancies. *Experimental and Clinical Psychopharmacology*, **2**, 167–183.

Rather, B.C., Goldman, M.S., Roehrich, L. and Brannick, M. (1992). Empirical modeling of an alcohol expectancy memory network using multidimensional scaling. *Journal of Abnormal Psychology*, **101**, 174–183.

Reber, A.S. (1997). How to differentiate implicit and explicit modes of acquisition. In J.D. Cohen and J.W. Schooler (ed.), *Scientific approaches to consciousness*. Hillsdale, NJ: Lawrence Erlbaum Associates, Inc., 137–160.

Reich, R.R. and Goldman, M.S. (2005). Exploring the alcohol expectancy memory network: the utility of free associates. *Psychology of Addictive Behaviors*, **19**, 317–325.

Reich, R.R., Goldman, M.S. and Noll, J.A. (2004). Using the false memory paradigm to test two key elements of alcohol expectancy theory. *Experimental and Clinical Psychopharmacology*, **12**, 102–110.

Reich, R.R., Noll, J.A. and Goldman, M.S. (2005). Cue patterns and alcohol expectancies: how slight differences in stimuli can measurably change cognition. *Experimental and Clinical Psychopharmacology*, **13**, 65–71.

Ritov, I. (2000). The role of expectations in comparisons. *Psychological Review*, **107**, 345–357.

Robinson, T.E. and Berridge, K.C. (2000). The psychology and neurobiology of addiction: an incentive-salience view. *Addiction*, **95** (Suppl. 2), S91–S117.

Roediger, H.L. (2003). Reconsidering implicit memory. In J.S. Bowers and C.J. Marsolek (ed.), *Rethinking implicit memory*. New York, NY: Oxford University Press, pp. 3–18.

Roediger, H.L. and McDermott, K.B. (1995). Creating false memories—remembering words not presented in lists. *Journal of Experimental Psychology: Learning, Memory, and Cognition*, **21**, 803–814.

Roediger, H.L. and McDermott, K.B. (2000). Tricks of memory. *Current Directions in Psyuchological Science*, **9**, 123–127.

Roediger, H.L., Buckner, R.L. and McDermott, K.B. (1999). Components of processing. I.J.K. Foster and M. Jelicic (ed.), *Memory: systems, process, or function*. New York, NY: Oxford University Press, 31–65.

Roehrich, L. and Goldman, M.S. (1995). Implicit priming of alcohol expectancy memory processes and subsequent drinking behavior. *Experimental and Clinical Psychopharmacology*, **3**, 402–410.

Saper, C.B., Iversen, S. and Frackowiak, R. (2000). Integration of sensory and motor function: the association areas of the cerebral cortex and the cognitive capabilties of the brain. In E.R. Kandel, J.H. Schwartz and T.M. Jessell (ed.), *Principles of Neural Science*, 4th edn. New York, NY: McGraw-Hill, 349–380.

Scheier, L.M. and Botvin, G.J. (1997). Expectancies as mediators of the effects of social influences and alcohol knowledge on adolescent alcohol use: a prospective analysis. *Psychology of Addictive Behaviors*, 11, 48–64.

Schuckit, M.A., Smith, T.L., Danko, G.P., Anderson, K.G., Brown, S.A., Kuperman, S., Kramer, J., Hesselbrock, V. and Bucholz, K. (2005). Evaluation of a level of response to alcohol-based structural equation model in adolescents. *Journal of Studies on Alcohol*, 66, 174–184.

Schultz, W. (2004). Neural coding of basic reward terms of animal learning theory, game theory, microeconomics and behavioural ecology. *Current Opinion in Neurobiology*, 14, 139–147.

Schultz, W., Dayan, P. and Montague, P.R. (1997). A neural substrate of prediction and reward. *Science*, 275, 1593–1599.

Schultz, W., Tremblay, L. and Hollerman, J.R. (2000). Reward processing in primate orbitofrontal cortex and basal ganglia. *Cerebral Cortex*, 10, 272–283.

Sell, L.A., Morris, J., Bearn, J., Frackowiak, R.S., Friston, K.J. and Dolan, R.J. (1999). Activation of reward circuitry in human opiate addicts. *European Journal of Neuroscience*, 11, 1042–1048.

Sher, K.J., Wood, M.D., Wood, P.K. and Raskin, G. (1996). Alcohol outcome expectancies and alcohol use: a latent variable cross-lagged panel study. *Journal of Abnormal Psychology*, 103, 561–574.

Sher, K.J., Grekin, E.R. and Williams, N.A. (2005). The development of alcohol use disorders. *Annual Review of Clinical Psychology*, 1, 493–523.

Singer, W. (1995). Development and plasticity of cortical processing architectures. *Sciences*, 270, 785–764.

Slutske, W.S., Heath, A.C. and Madden, P. (2002). Personality and the genetic risk for alcohol dependence. *Journal of Abnormal Psychology*, 111, 124–133.

Smith, G.T. and Anderson, K.G. (2001). Personality and learning factors combine to create risk for adolescent problem drinking: a model and suggestions for intervention. In P.M. Monti, S.M. Colby and T.A. O'Leary (ed.), *Adolescents, alcohol, and substance abuse: reaching teens through brief interventions*. New York, NY: Guilford Press, 109–144.

Smith, G.T., Goldman, M.S., Greenbaum, P.E. and Christiansen, B.A. (1995). Expectancy for social facilitation from drinking: the divergent paths of high-expectancy and low-expectancy adolescents. *Journal of Abnormal Psychology*, 104, 32–40.

Stacy, A.W. (1997). Memory activation and expectancy as prospective predictors of alcohol and marijuana use. *Journal of Abnormal Psychology*, 106, 61–73.

Stacy, A.W., Newcomb, M.D. and Bentler, P.M. (1991). Cognitive motivation and problem drug use: a 9-year longitudinal study. *Journal of Abnormal Psychology*, 100, 502–515.

Stein, K.D., Goldman, M.S. and Del Boca, F.K. (2000). The influence of alcohol expectancy priming and mood manipulation on subsequent alcohol consumption. *Journal of Abnormal Psychology*, 109, 106–115.

Tarr, M. (2005). How experience shapes vision. Psychological Science Agenda Website. Available at: http://www.apa.org/science/psa/july05.pdf

Tobler, P.N., Fiorillo, C.D. and Schultz, W. (2005). Adaptive coding of reward value by dopamine neurons. *Science*, **307**, 1642–1645.

Tolman, E.C. (1932). *Purposive behavior in animals and man.* New York, NY: Appleton-Century-Crofts.

Tsodyks, M. and Gilbert, C. (2004). Neural networks and perceptual learning. *Nature*, **431**, 775–781.

Van Hamme, L.J. and Wasserman, E.A. (1994) Cue competition in causality judgments: the role of nonpresentation of compound stimulus elements. *Learning Motivation*, **25**, 127–151.

Vernon, P.A., Lee, D., Harris, J.A. and Jang, K.L. (1996). Genetic and environmental contributions to individual differences in alcohol expectancies. *Personality and Individual Differences*, **21**, 183–187.

Wager, T.D., Rilling, J.K. and Smith, E.E. (2004). Placebo-induced changes in fMRI in the anticipation and experience of pain. *Science*, **303**, 1162–1167.

Wall, A.M., Hinson, R.E., McKee, S.A. and Goldstein, A. (2001). Examining alcohol outcome expectancies in laboratory and naturalistic bar settings: a within-subject experimental analysis. *Psychology of Addictive Behaviors*, **15**, 219–226.

Weinberger, A.H., Darkes, J., Del Boca, F.K., Greenbaum, P.E. and Goldman, M.S. (2006). Items as context: the effect of item order on factor structure and predictive validity. *Basic and Applied Social Psychology*, in press.

Whewell, W. (1840, 1967). *The philosophy of the inductive sciences.* London: Cass Publishing.

Wiers, R.W., van Woerden, N., Smulders, F.T.Y. and de Jong, P.J. (2002). Implicit and explicit alcohol-related cognitions in heavy and light drinkers. *Journal of Abnormal Psychology*, **111**, 648–658.

Wiers, R.W., Van de Luitgaarden, J., Van den Wildenberg, E. and Smulders, F.T.Y. (2005). Challenging implicit and explicit alcohol-related cognitions in young heavy drinkers. *Addiction*, **100**, 806–819.

Wills, T.J., Lever, C. and Cacucci, F. (2005). Attractor dynamics in the hippocampal representation of the local environment. *Science*, **308**, 873–876.

Wilson, E.O. (1998) *Consilience: the unity of knowledge.* New York, NY: Knopf.

Winocur, G., Moscovitch, M. and Fogel, S. (2005). Preserved spatial memory after hippocampal lesions: effects of extensive experience in a complex environment. *Nature Neuroscience*, **8**, 273–275.

Opiate cognitions

Ross McD. Young, Barry T. Jones,
Carey Walmsley and Anthony Nutting

7.1 Introduction

The subjective effects of opiates have been described since antiquity:

> What an upheaving, from the lowest depths, of the inner spirit! What an apocalypse of the world within me! That my pains had vanished, was now a trifle in my eyes... here was the secret of happiness, about which philosophers had disputed for so many ages, at once discovered: happiness could be bought for a penny, and carried in the waistcoat pocket: portable ecstasies might be had corked up in a pint bottle: and peace of mind could be sent down in gallons by the mail coach.
>
> Thomas De Quincey (1821). Confessions of an opium eater.

De Quincey's expectation of joy and relief from using opium aligns with the reports of many who have documented their own subjective experience with opiates before and since. Equally, his subsequent difficulty coping with the barrage and complexity of opium-induced fantasies, and opium withdrawal, became burdensome over time and attest to the problems of chronic use (Plant 1999). Prior to De Quniceys introspection, the most famous example of the influence of opium on the literary imagination was Samuel Taylor Coleridge's *Kubla Khan* in 1797 (Day and Smith 2002). In Homer's *The Odyssey*, opium is described as having the 'power to rob grief and anger of their sting'. More contemporary works that illuminate the subjective effects of heroin include William S. Burrow's *Junky* in 1953 and Irvine Welsh's 1993 work *Trainspotting*. Focal research regarding individual differences in the subjective consequences of opiate use, both reinforcing and aversive, is much less well established than the literary tradition. This contrasts with the large body of work examining the subjective outcomes of other drugs, such as alcohol, where our attention has embraced both the writer's desk and the behavioural scientist's laboratory.

Due to a paucity of research examining any one class of cognitive construct *per se*, this review will draw on research examining a set of constructs that reflect the subjective experience of drug reinforcement. Despite this broad definition, there is only a small body of research to examine and this typically involves explicit cognition as opposed to implicit cognition. For example, implicit memory associations have been examined with alcohol (Stacy 1997; Gadon *et al.* 2004) but not with the opiates. The cognitive constructs include justifications (Del Pozo *et al.* 1998), motivations (Murphy *et al.* 2003), confidence (Sklar *et al.* 1997), drug desire (Franken *et al.* 2003), medication beliefs (Schieffer *et al.* 2005) expectancy (Walmsley 2004) and self-efficacy (Sklar and Turner 1999). Some studies are not included as they have measured cognitive constructs that are not related to opiates *per se*, although these constructs have been studied in opiate users. This includes research involving generalized attributions of responsibility for treatment outcome in methadone clinic attendees (Bradley *et al.* 1992), or reference to the use of other drugs, such as the effect of methadone on nicotine (Stein and Anderson 2003).

There is a range of pragmatic and attitudinal factors that are likely to have contributed to the lack of research advancement in opiate cognition. The more common use of drugs such as alcohol makes the opiates a less viable drug class to study. The common stereotype held about those who use opiates is negative and may have historically impaired research progress (Cirakoglu and Isin 2005). Typically, the popular media have portrayed those addicted to opiates as deviates that suffer from a moral sickness and are responsible for their own problems (Elliott and Chapman 2000). These negative attitudes extend to clinicians who treat opiate dependence (Gerlach and Capelhorn 1999), and to politicians who influence the scientific agenda through policy that is purported to be morally grounded (Lawrence *et al.* 2000). Furthermore, the view that the consequences of opiate use are more 'pharmacologically driven' than those of other drugs may have inhibited the examination of individual differences related to opiate outcomes.

The collective lack of interest in opiate cognitions is surprising, as the use of opiates has been part of human society for millennia. Opium is the Greek word for 'juice', referring to the juice of the opium poppy, *Papaver somniferium*: opium is typically the dried exudate of this flower. Opiates, such as morphine and codeine, are derived from opium, although opium itself is approximately 10 per cent morphine (Edwards 2005). Pharmacologically, the opiates are a class of substances with morphine-like effects that can be reversed by the specific opiate antagonist, naloxone.

Opiates have been described in Sumerian writings of around 4000 BC. Opium poppy heads have been found in Neolithic burial caves in Southern Spain that date to 4200 BC and in Swiss dwellings of the same era. This may indicate both routine use as well as their anaesthetic use in those with terminal illness, or the use of the drug as part of the burial ritual. Opium has been a vital part of the pharmacopaeia of many of the great physicians of antiquity including Hippocrates and Galen, and in more recent history Sir William Osler described morphine as 'God's own medicine'. Regarding the more recreational uses of the opiates, an opium-smoking epidemic in eighteenth century China was encouraged by the British and the Dutch who created a significant export trade from India (Lang 1998). Prior to this period in China, and in neighbouring India, opium amongst other psychotropic substances had been used for at least a thousand years. Opium was incorporated into religious festivals, social gatherings and was used for medicinal value. The hunger of the British for profit fuelled the creation of the opium epidemic as a social phenomenon in China. This resulted in the Chinese government seizing all opium in 1839 as a control measure and prompted the first of the Opium Wars between the Chinese and the British (Edwards 2005). It is estimated that at the peak of opium use in the nineteenth century about one-quarter of all Chinese adults were opium dependent. The tension between the economic forces of drug supply and government attempts to control opiates is no less relevant today.

Heroin (diacetylmorphine) is a semisynthetic chemical derivative of morphine, first commercially produced in 1898 by Bayer Pharmaceuticals. This represents a crucial historical turning point regarding the opiates. While heroin was recognized as a less effective analgesic than morphine, it was thought to hold great promise as a cough suppressant. However, in the twentieth century—the 'happiness and peace of mind' associated with opiates is strongly evident with heroin—heroin progressed to be a widely misused drug in many countries. In parallel with the worsening heroin misuse at a community level, the cost of the drug has generally lowered, and the purity generally increased (Office of National Drug Control Policy 1999). The reasons underlying the increased use of heroin are complex and are not simply related to the greater availability of the drug, consistent with the socio-cultural examples noted in the history of heroin use. For example, it has been suggested that contemporary Western social and cultural forces related to materialism have produced both an increased anxiety and a diminished sense of personal control amongst youth. This is proposed to result in self-esteem being viewed as contingent on external factors, thus increasing personal discontent (Eckersley 2005). These broader forces of social unease may contribute to the attractiveness of drugs such as opiates that 'increase pleasure and diminish pain'. In addition, a common sociological view of the contemporary response to established heroin use is that the prohibitionist 'war on drugs' is a means of distracting the voting populace away from the social and economic forces that create this unease (Elliott and Chapman 2000).

These historical glimpses indicate that our beliefs about the relative benefits of opiate use, and of our views of those who use opiates, are a consequence of multiple factors. These factors include shared cultural meanings, drug use environments, individual expectations of drug effects and the effects of the pharmacological agent on the brain. The understanding of the effects of opiates on central nervous system (CNS) function has advanced considerably since the discovery of the endogenous opiates in 1975. While it is logical that these alterations in brain function contribute to drug cognitions, the research evidence integrating opiate pharmacology and cognitions is sparse. A more extensive summary of clinical pharmacology can be found in Young *et al.* (2002), and Jaffe and Jaffe (2004).

7.2 **Psychopharmacological mechanisms underpin the subjective effects of the opiates**

Although there is a distinction that the opiates are naturally occurring and opioids are synthetic equivalents, both terms will be used interchangeably. In a brief overview of Young *et al.* (2002), the endogenous opioid system has three main types of opioid receptor (δ, κ and μ), and less well described types that show more limited distributions (e.g. ε). The σ receptor is no longer considered an opiate receptor. There are also subtypes of each. Four groups of endogenous opioid peptides (enkephalins, β-endorphins, dynorphins, endomorphins and nociceptin) are produced as required by peptidases that cleave inactive precursor peptides (e.g. pro-opiomelanocortin to β-endorphin). This allows for precise and efficient control of these compounds. The opioid peptides and their receptors are widely distributed in the brain and spinal cord, and each peptide acts at a specific receptor. This further contributes to the finely grained control of their actions (e.g. endorphins act at the μ receptor). Receptor activation typically results in an inhibitory response, inhibiting adenylate cyclase, opening potassium channels and blocking voltage-gated calcium channels. This leads to alterations in the release of multiple neurotransmitters including dopamine, serotonin, acetylcholine, glutamate and γ-aminobutyric acid (GABA). There is also release of Substance P, a pain-modulating neuropeptide. Collectively this forms the underlying basis of alterations to subjective experience (Waldhoer *et al.* 2004).

At the core of the subjective experience of opiate use is altered pain perception. There is strong evidence regarding the biologically adaptive advantages of pain control mediated by endogenous opiates. The endogenous opioid system is activated by stress—typically inescapable stress and particularly stress that is protracted and uncontrollable (Maier 1984). This activation may be associated with self-efficacy expectations given that differences in stress-induced analgesia have been demonstrated in the laboratory related to confidence in coping with pain (Bandura *et al.* 1988). Such opiate activation assists in the modulation of both physical and emotional pain. Social and physical pain appear to be subsumed by similar underlying physiological mechanisms in the mammalian brain (Panskepp 2003), and this may provide an adaptive advantage for the development of socially cohesive groups given the aversive nature of social exclusion (MacDonald and Leary 2005).

Experimental evidence of the interface between social behaviour and pain control can be found in humans and other mammalian species. For example, in an online 'ball tossing' game, participants were informed that they would be playing with another person. Those 'excluded' from playing, who were also sensitive to rejection, showed a stronger analgesic response than those who were sensitive to rejection but 'included' (MacDonald *et al.* 2005, cited in MacDonald and Leary 2005). Similarly, rat pups removed from the nest, their mother and littermates experience analgesia that is induced by isolation (Kehoe and Blass 1986). Being raised in an impoverished social environment can result in repeated activation of endogenous opioid systems that may eventually result in an aversive response akin to opiate withdrawal if social contact is not forthcoming (Panskepp 2003). Phylogenetically it is also likely that endogenous opiates are more generally involved in reward, for example songbird social vocalization is mediated by opioid neuropeptides (Riters *et al.* 2005). Mutant mice selectively bred to have a complete absence of μ opioid receptors show deficits in reward learning (Contarino *et al.* 2002).

This evolution of endogenous systems that reduce pain and assist in compensating for environmental impoverishment forms a powerful basis for the abuse liability of opiate drugs. The opiate drugs act on the same class of receptors as the endogenous opiates; however, considerable variability in the individual response to these agents is evident (Galer *et al.* 1992). Morphine itself is named after the Greek god of dreams, Morpheus. The principal effects of morphine are analgesia, drowsiness and mood change. This analgesic response, rather than removing pain, typically alters the perception of pain, making it less aversive. The mood change is usually experienced as relaxed contentment, or less commonly euphoria. Higher problem-solving functions and episodic memory are impaired, but consciousness and coordination are generally intact at low doses (Cleeland *et al.* 1996). In the CNS, morphine activates μ opiate receptors (μ1 and μ2) in an analogous manner to the enkephalins. This activation is particularly important in neurons that interface with the mesolimbic dopamine reward pathway and projections to the forebrain. These regions are rich in D2 dopamine receptors (one of five dopamine receptor subtypes) and this combined opioidergic and D2 dopamine receptor stimulation contributes significantly to the primary reinforcing effects of the drug (Di Chiara and Imperato 1988). In addition, animal research has confirmed the importance of a number of other neurotransmitter receptors related to opiate reinforcement, such as the CB1 anadamide (endogenous cannabinoid) receptor (Cossu *et al.* 2001; Le Foll and Goldberg 2005).

Heroin is more lipophilic than morphine and therefore enters the CNS more rapidly. It also results in greater dopamine release and stimulation of D2 dopamine receptors (Di Chiara and Imperato 1988). This leads to more intense feelings of well-being and, as such, is much more likely to be described as euphorigenic by users than is morphine. Both the distribution of neurotransmitter receptors and receptor structure are genetically mediated. Furthermore, the impact of self-administration itself and the associated learning process may also influence gene expression which could result in long-lasting changes in brain function, for example in the nucleus accumbens shell (Jacobs *et al.* 2005). Genetic variation related to opioid receptor genes is related to risk of dependence and with the nature of individual differences in response to opiates (Han *et al.* 2004). A review of the genes identified to date that are most likely to influence opiate addiction is given in Kreek *et al.* (2004, 2005) and will not be discussed in detail. For example, the μ opioid receptor gene (*OPRM1*) is not surprisingly related to variation in analgesic response to morphine. The A118G variant in particular appears to be associated with both pain perception and stress response. Differences in the clinical response of cancer patients to morphine have also been found to be related to beta arrestin 2 gene and stat 6 gene status but not to *OPRM1* variants (Ross *et al.* 2005).

Genes related to dopamine function may additionally influence subjective experience. For example, those who have the region A1+ allele (A1/A1 and A1/A2 genotypes) of the D2 dopamine receptor gene typically have a lesser density of D2 dopamine receptors than those with the more common A1− status (Noble *et al.* 1991). A1+ individuals also have a more severe substance misuse trajectory which has been hypothesized to be related to more potent reinforcement associated with drugs that stimulate dopamine release, such as the opiates (Young *et al.* 2004a). Those with A1+ allelic status are more likely to use high dose heroin (Shamoradgoli *et al.* 2005) and show a poor response to methadone treatment—as methadone results in a less intense dopaminergic response than heroin (Lawford *et al.* 2000). A1+ individuals also engage in more impulsive heroin use, with needle sharing (Lawford *et al.* 1999), than their counterparts with A1– status. Impulsivity in humans is associated with opiate misuse (Forman *et al.* 2004)—in rodents, opiates have been shown to exacerbate impulsivity and opiate antagonists to diminish impulsivity (Kieres *et al.* 2004). This may indicate that a genetic temperamental risk of impulsivity may be further worsened by chronic drug administration.

The primary reinforcing outcome of heroin administration, euphoria, is experienced about 30 min after heroin is 'snorted', 15 min after subcutaneous injection and within seconds after intravenous injection or the inhalation of smoke. This euphoric effect typically lasts for 3 or 4 h and is similar in subjective experience regardless of the mode of administration, other than the 'rush' obtained by intravenous use (Hendricks *et al.* 2001). In addition to euphoria, the subjective effects are described as spreading warmth, tranquillity and dreamy contentment. Heroin is deacetylated in the liver to 6-monocetyl morphine and then further deacylated to morphine, both of which are pharmacologically active metabolites and potentially continue, in a more protracted manner, to alter subjective experience. Codeine is also converted to morphine, its active form, via demethylation by the enzyme CYP2D6. Pethideine is converted into a metabolite, norpethidine, which is a CNS stimulant and thus increases the abuse potential of this agent given the mixed euphoric and energizing profile.

Some of the most aversive effects of opiates, including sedation, are related to the activation of κ opioid receptors. Significant aversive consequences relating to withdrawal are also reported by opiate misusers. Tolerance starts to develop after the first dose of opiate is administered and involves both downregulation (a decrease in the number of receptors) and desensitization (a diminished response to receptor activation), indicating that receptor function is under both genetic and homeostatic control related to drug exposure. Tolerance to opiates is characterized by a shortened duration and reduced intensity of the analgesic, euphorigenic and sedative effects. There is also marked interindividual variation in the development of tolerance, and this is likely to be due to individual differences in genetically determined neurotransmitter receptor density and the concentration of second messengers. However, prior learning, the impact of other substance use and the environment also exert an influence. Classically conditioned drug tolerance effects are well documented in animal research and are responsible for a preference for environments in which the drug has been previously administered

(McDonald and Siegel 2004). There is an increased likelihood of overdose when drugs are administered in a novel environment (Siegel 1982*a,b*). These conditioned effects may form a strong basis for the development of cognitions regarding drug effects in specific environments as even human analgesic effects show situational specificity (Siegel and Kim 2000). Cognitions may also influence experiences interpreted as physiological effects, including opiate withdrawal Phillips, Gossop, and Bradley (1986). The provision of accurate information about the withdrawal process leads to decreased withdrawal distress compared with providing no information at all (Green and Gossop 1988). Despite these learning effects, which appear to influence some aspects of tolerance and withdrawal, there are consequences of use that are independent of learning. Tolerance does not develop equally, or at the same rate, to all effects of opiates, and even chronic users get respiratory depression and miosis (unresponsive 'pin prick' pupils).

Regular heroin users experience withdrawal symptoms more than once a day if an opiate is not administered. Opioid binding to μ receptors causes increased dopamine concentrations in midbrain reward structures mediated by the inhibition of GABAergic interneurons. Opiate withdrawal involves a counter process of severe inhibition of dopamine release in the nucleus accumbens and the ventral tegmental area due to a rebound of GABA activity (Bailey and Connor 2005). Mood disturbance in withdrawal is related to increased GABAergic tone and dopamine depletion (Handelsman *et al.* 1992). In addition, activation of the locus coeruleus in withdrawal increases autonomic activation, leading to anxiety and mood deterioration (Jaffe and Jaffe 2004).

In summary, opiate drugs act at the same receptor sites as endogenous ligands. These CNS opiate systems have evolved as an adaptive response to psychological and physical pain. As a class of drugs, opiates alter mood, and heroin typically produces euphoria. Risk for misuse appears to be associated with genes that influence the subjective effects of the opiates themselves and the integrity of allied brain receptor systems. However, given that the endogenous opioid system evolved to facilitate an effective response to stress, it is likely that environmental and socio-cultural factors that contribute to stress also play a significant role in the experience of opioid effects. These appear to include acute reinforcement and withdrawal distress. The role of individual difference factors, including learning history, awaits clarification as the research agenda that examines response to opiates moves beyond a more exclusively biological frame of reference.

7.3 Socio-cultural influences on opiate use

Identifying the sources that influence our learning about drugs is an important research priority that has not been adequately addressed. The primary sources of information about opiates, including the electronic media, have typically promoted negative expectancies about opioids (Elliott and Chapman 2000) although 'heroin chic' is a relatively recent Western phenomenon. A broad array of influences shape commonly held alcohol expectancies, and they are largely positive. These influences include parental and peer modelling, multiple media influences and sophisticated marketing strategies. For example, in Australia, advertising on television has portrayed alcohol as enhancing personal identity, facilitating relationships and enabling relaxation (Young *et al.* 1991*b*). In contrast, the societal message that any heroin use is socially and personally deviant has been long documented (Zinberg *et al.* 1975). Lay theories of heroin addiction are more nuanced and vary according to social identity. Beliefs about the fundamentals of heroin addiction differ according to political affiliation, with more conservative voters being likely to offer individualistic and moralistic interpretations of use and less conservative voters being likely to view heroin use in the context of a psychological, personal and social problem (Furnham and Thomson 1996).

In societies where there is an absence of powerful messages outlining the reinforcing effects of alcohol, the balance between community heroin use and alcohol use may alter. In a comparison of psychiatric disorders reported between the UK and Iran, Hashemi and London (2003) reported that alcohol abuse disorders are rare in Iran, whereas the prevalence of opium addiction, while still relatively rare, is almost double that seen in the UK. As in the UK, opium is illegal in Iran but, as distinct from alcohol, it is not prohibited by Islam (Hashemi and London 2003). The role of subcultural influence is also likely to provide new opportunities to learn about opiates and potentially to develop different drug expectancies, although this has rarely been addressed. In a review of regional differences in India, Sharma (1996) describes a contrast in Punjab, where opiate use is high, from Rajasthan, where it is less common. In Punjab, opium was typically provided by landlords to labourers to facilitate hard physical labour without pain. In Rajasthan, use was more social or medicinal and was thus controlled through more functional socio-cultural structures and meanings. The largest increases in opium use in Rajasthan were in periods of drought, and this increased use was limited to these periods.

There is evidence that altering the socio-cultural meanings and rituals through externally imposed policies can have counterproductive results, as occurred in Laos in the 1960s where the policies to reduce local opium growing and use created a new market for imported heroin and a worsening of the opiate misuse (Westermeyer 1978). The context of opiate use within a single society can also change over time due to reasons that arise within a given cultural group. In Singaporean Chinese, in a similar manner to the effect described in Punjab, new Chinese workers brought to Singapore at the turn of the twentieth century were provided with free opium by their employers. They were encouraged to gamble and increase their indebtedness, thus giving employers greater potential for manipulation. This is likely to have created a greater number of opiate-dependent individuals given the less restrictive practices for use. However, by the 1950s, the most common beliefs amongst Singaporean Chinese regarding opium were related to medicating pain, or for the treatment of specific medical conditions, and suggest a very different use pattern. The medicinal use of opium at this time was typically to improve the vocal performance of actors, to treat respiratory complaints, to cure diabetes, to combat malaria, delay ejaculation, lower libido in widows, develop resilience in infants and to treat dysentery (Leong 1974). Only 13 per cent of those surveyed indicated that they associated opium use with pleasure (Leong 1959; cited in Leong 1974). Although cognition was not specifically measured in these studies, it appears as if the ritualized use of opiates to facilitate work, in all likelihood to reduce the pain associated with hard physical work, was associated with greater drug misuse than use for social or medicinal reasons.

There are a handful of studies that examine the potential influence of subculture and heroin cognition in a general sense. Beckerleg (2004) investigated heroin injection on the Kenya Coast and reported that considerable status was related to personal self-control in Swahili culture. Factors such as desperation, lack of control of intoxication and public displays of use or intoxication were undesirable qualities amongst heroin injectors. Those who could control their use of heroin and maintained financial obligations and housing were ascribed high status. This stratification within a group of drug users on the basis of drug outcomes that were valued to a greater or lesser extent could serve further to marginalize those with the most significant problems. Within subcultural groups, the expected benefits of heroin use can differ. A relationship between the prestige related to heroin and music subculture in the USA has been examined in a thesis which illustrates a social learning process related to drug attitudes. Those individuals who identified themselves as 'alternative music' fans were more likely to use high doses of heroin than those who were 'popular rock' fans (Kruse 1998). The relationship between heroin prestige and level of commitment to this musical identity varied between the two groups, with a positive relationship being evident only in those in the 'popular rock' category. No causality can be invoked from these findings; however, it does suggest that the broadly reinforcing outcomes of heroin use can vary according to social group or social identity. Zinberg had previously established in the 1970s that there were controlled users who balanced their heroin use with other aspects of their lives, just as controlled users of alcohol do. Zinberg *et al.* (1975) referred to both positive expectancies that controlled users held and negative 'risky' expectancies identified by them that characterized those with 'heroin problems'. These expectancies of risk primarily related to the capacity of heroin to relieve emotional discomfort in those who are distressed.

In addition to studying socially defined subgroups, other circumstantial evidence for opiate cognitions can be found in examining the developmental histories of heroin users themselves. Heroin misuse is associated with a history of mental health problems and early onset polysubstance misuse (Hopfer *et al.* 2002; Burns *et al.* 2004). Similar findings are evident with the adolescent misuse of prescription opioids (Sung *et al.* 2005). The prevalence of mental health problems for those with opiate abuse is as high as 87 per cent in clinic samples (Rousanville *et al.* 1982) and these co-morbidities have also been confirmed in those not seeking treatment (Fischer *et al.* 2005). The most commonly reported psychopathologies among opiate-using populations are depression and anxiety (Milby *et al.* 1996; Brooner *et al.* 1997; Mattick *et al.* 2000; Callaly *et al.* 2001; Teeson *et al.* 2005). Callaly *et al.* (2001) found that among a population of opiate misusers on methadone maintenance treatment, an anxiety disorder was evident in 68 per cent of the sample and an affective disorder in 53 per cent. Post-traumatic stress disorder (PTSD) is highly prevalent—in Mills *et al.* (2005) 41 per cent of heroin-dependent individuals received a diagnosis of PTSD, and 92 per cent had been exposed to traumatic events. Amongst this group, PTSD diagnoses were associated with more severe polydrug using histories, poorer mental and physical health overall and a greater demand on health services. The majority (84 per cent) of PTSD diagnoses were chronic, lasting for an average of 9.5 years.

The role of the increased trauma that may be due to drug use and associated lifestyle has not been typically accounted for. The majority of research is retrospective and focuses on trauma prior to substance misuse. Much of the research specifically examining PTSD and opiate misuse involved studies of US Vietnam War veterans. Heroin for smoking was widely available in Vietnam in the late 1960s and early 1970s. PTSD amongst US veterans was associated with an increased lifetime use of heroin (Saxon *et al.* 2001). Heroin was regularly used by 25 per cent of US troops in some units (Robins 1993) but was not documented in Australian or New Zealand troops where alcohol was a much more popular drug (Doyle *et al.* 2002). The majority of US veterans ceased use upon return home (Robins 1993). For those who continued opiate use upon return to the USA, the majority reported commencement of heroin while in the military, rather than prior to service, stating that opiates provided relief from fear related to war (Mintz *et al.* 1979). An increase in heroin use parallels the increase in PTSD symptoms, with PTSD symptom relief being associated with alcohol, marijuana, benzodiazepine and heroin use (Bremner *et al.* 1996). The retrospective nature of these studies renders it impossible to examine the pre-morbid features of those who developed heroin problems or psychological dysfunction. Thus issues of causality are not fully understood; however, self-medication for some individuals is strongly implied.

The notion of self-medication acknowledges that there are likely to be specific expectancies of use related to the presence of psychopathology. However, there is an absence of data examining this proposition. Psychological treatment for dual-diagnosis opiate-dependent clients, provided in addition to pharmacological substitution therapies (such as methadone), is associated with better outcomes when compared with pharmacological intervention and support alone (Nunes *et al.* 2004). Combined cognitive–behavioural therapy (or contingency management) with antidepressants reduces both depression and heroin use, again indicating the importance of co-morbid disorders (Stein *et al.* 2004). This may maintain addiction given the potential role that opiates have in providing relief from psychological pain. The addition of a psychological intervention to buprenorphine improves treatment retention significantly (Kakko *et al.* 2003). Recent Cochrane reviews have confirmed that adding any psychological treatment to methadone during methadone detoxification improves retention in treatment (Amato *et al.* 2004*a*) and decreases heroin use in methadone maintenance (Amato *et al.* 2004*b*). Dual diagnosis is not surprisingly associated with a more complex clinical presentation. This complexity includes increased severity of substance dependence, more psychosocial problems (Brooner *et al.* 1993), poorer treatment outcomes (Rousanville *et al.* 1982; McLellan 1986), higher human immunodeficiency virus (HIV) risk-taking behaviour and infection (Metzger *et al.* 1991; Brooner *et al.* 1993), higher rates of relapse (Safer 1987) and lower quality of life (Fassino *et al.* 2004). This forms a strong basis on which to imagine the greater psychological power invested in opiates in these individuals.

In summary, while there is an absence of cross-cultural or subcultural research examining opiate cognition, the genesis of opiate use does appear to be related to social and moral codes. In addition, exposure to traumatic events or chronic stress conveys risk. These risks are insufficient to result in heroin use, or indeed drug dependence generally. In addition, more subtle individual learning related to the specific functions of opiates may be reinforced on economic, social or medicinal grounds. These learning processes are likely to be influenced by a range of factors including personality and psychological adversity. Those with the most significant heroin dependence may hold expectations that heroin will be negatively reinforcing, i.e. that heroin will reduce distress.

7.4 **The measurement of cognitions that represent drug reinforcement**

Much of the laboratory placebo paradigm research examining drug cognitions assumes that expectancies exist but they are not directly measured (Wager 2005). For example, in a study where intense or mild electric shock was employed under the use of either placebo or anaesthetic cream, there was significant variability in placebo response, suggesting that there were individual differences worthy of more detailed exploration (Wager *et al.* 2004). Whether these responses relate to the strength of personal drug expectancies is not known. However, importantly in this study, where response was monitored using functional magnetic resonance imaging (fMRI), those with the greatest placebo response showed the greatest prefrontal activation during the shock phase. This is consistent with the presence of drug expectancies that are activated. There was also increased midbrain activity in anticipation of the shock, and this was higher in the placebo condition generally. Wager *et al.* noted that this effect could be mediated both by endogenous opioid activation as a result of learning or via a less direct path—the cognitive reappraisal of pain sensations. In a second study in Wager *et al.* (2004), which was restricted to placebo responders, the placebo reduced pain reports by 22 per cent and confirmed the involvement of key brain pain modulation areas. The role of μ opioid receptors appears to be central in the placebo response (in areas such as the nucleus accumbens where dopamine is also activated) (Zubieta *et al.* 2005). It has been estimated that informing the patient that a painkiller is being administered when a placebo is actually used is equivalent to 6–8 mg of morphine (Colloca and Benedetti 2005).

Drug expectancies represent 'if–then' contingencies such as '*if* I take heroin *then* my pain will diminish'. Essentially, 'if' the drug is taken 'then' certain anticipated drug effects will follow: these expectancies can be reciprocally (Smith *et al.* 1995) or causally related to the substance abuse (Oei *et al.* 1991).

Stronger replicated evidence is needed to clarify the precise relationship of cognitive expectancy and drug use over time (Jones *et al.* 2001) as the majority of this research is cross-sectional. With these caveats, social cognitive theory (Bandura 1977, 1999; Marlatt and Gordon 1985) defines two major expectancy constructs posited to be crucial in the acquisition and maintenance of behaviours; the first is *outcome expectancy*, which reflects a subjective appraisal of the specific outcomes that are likely to arise from a specific behaviour. The second is *self-efficacy*, or the subjective appraisal of one's ability to execute certain behaviours in specific situations. Although expectancies are typically based on previous experience, they may develop indirectly, via means such as observational learning (Millar and Dollard 1941; Rotter 1954). Expectancies may influence multiple consequences of drug ingestion including those that had been previously thought to be entirely related to the physiological actions of the drug. For example, alcohol expectancies are related to the considerable variation in the expression and acceptance of intoxication across cultures (MacAndrew and Edgerton 1969) in a similar way to that previously discussed in Kenyan subcultural groups of heroin users (Beckerleg 2004).

Drawing on Goldman *et al.* (1999), drug outcome expectancies are hypothesized to reflect the subjective reinforcement of the drug stored as memory templates. For example, 'if I use heroin I will feel more relaxed' is an outcome expectancy. Self-efficacy expectancies within the context of substance use have been variously defined but are typically related to confidence in resisting the drug in a given situation (Young *et al.* 1991*a*). For example, an opiate refusal self-efficacy expectancy could be 'I am likely to have difficulty resisting heroin when I am feeling depressed'.

The expectancy of having received a drug even though it is actually a placebo can produce some of the same effects that are associated with the actual consumption of the drug itself (Marlatt and Rohsenow 1980). Such studies have commonly employed a balanced placebo paradigm, in which either the drug (usually alcohol) or a placebo is administered. Within these two cells, half of the subjects are correctly informed of the drug being administered while the other half are misinformed of their allocated experimental condition. This creates a 2×2 matrix where the influences of expectancy and pharmacology can be separated. Research has demonstrated expectancy manipulation effects when examining opiates (Pollo *et al.* 2001), amphetamines (Mitchell *et al.* 1996), caffeine (Lotshaw *et al.* 1996), glucose (Green *et al.* 2001), insulin (Pohl *et al.* 1997) and high-fat (versus low-fat) ice cream (Bowen *et al.* 1992). In the study of Pollo *et al.* (2001), the participants, who had undergone thoracotomy, were all commenced on the mixed opiate agonist–antagonist buprenorphine and were randomly allocated to one of three expectancy groups. All groups were administered saline but provided with different expectancies regarding the additional infusion. Those who were told that the infusion was a potent additional analgesic showed a stronger analgesic effect (as shown by reduction in buprenorphine dose over 3 days) than those in the 'no instruction' control condition. Those who were told that they had been administered an additional substance that may be either a placebo or a powerful analgesic showed an intermediate effect. Importantly all three groups reported equivalent pain relief despite these differences in buprenorphine dose.

To enable the mapping of pre-held expectancies (as opposed to those that are manipulated experimentally), questionnaires have been developed using focus groups or structured interviews with samples of substance users. These resultant items subsequently undergo psychometric analysis (Jones *et al.* 2001). There are several long-established alcohol expectancy question-naires—for example, the Alcohol Expectancy Questionnaire (AEQ, Brown *et al.* 1980) the Drinking Expectancy Questionnaire (DEQ, Young and Knight 1989), the Negative Alcohol Expectancy Questionnaire (NAEQ, Jones and McMahon 1992)—and drug expectancy measures have also been developed for heroin (Young *et al.* 1999), nicotine (Brandon *et al.* 1999), cocaine (Jaffe and Kilbey 1994) and cannabis (Schafer and Brown 1991). These drug expectancies are statistically associated with drug use in a similar manner to the way that alcohol expectancies are associated with drinking parameters (Mitchell *et al.* 1996). Stronger endorsement of expectancies is associated with more problematic consumption. The variation in specific expectancies across different drugs is attributable, at least in part, to differences in pharmaco-logical profile (Aarons *et al.* 2001) and the context of drug administration.

7.5   The nature of opiate expectancy domains

Although some studies have evaluated either expectancies or refusal self-efficacy in individuals with opiate dependence (Gossop *et al.* 1990; Sklar *et al.* 1997; Sklar and Turner 1999), robust research that comprehensively evaluates measures designed specifically for opiate cognition representing drug-related reinforcement is uncommon. The studies of Gossop *et al.* (1990), Powell *et al.* (1993) and Walmsley (2004) are three key studies that have evaluated both opiate outcome expectancies (opiate expectancies) and refusal self-efficacy in opiate users. These three studies will be given the strongest emphasis in this review. An additional well designed study defined outcome expectancy in terms of the expectancies of abstinence (Saunders *et al.*

1995)—a neglected aspect of expectancy research. Examination of heroin desire using the Desires for Drug Questionnaire, which identified three broad domains of desire—Desire and Intention, Negative reinforcement and Control (Franken *et al.* 2002)—has been undertaken specifically in the context of relapse.

An early attempt to measure heroin expectancies related to aggression simply involved the modification of items on the already existing Buss–Durkee Hostility Inventory (Walter *et al.* 1990). A parallel modification was made to create a second measure of alcohol expectancies related to aggression. Only the alcohol aggression expectancies added to the prediction of diagnosis over measures of personality and psychopathology; there was no predictive effect for the heroin expectancies. Subsequently, Powell *et al.* (1992) developed the Motivational Checklist (MC), the earliest identified psychometric measure developed *de novo* to measure opiate expectancy. The MC was specifically employed to examine the relationship between opiate expectancies and craving. The questionnaire has two scales—one each for positive and negative outcomes—and provides a comprehensive coverage of potential heroin-related outcomes. The positive factors identified (along with sample items) were: 'emotional relief' (e.g. Helps to blot out problems/worries), 'lifestyle' (Provides a way of filling time), 'hedonism' (Pleasure of the rush), 'pain relief' (Relieves withdrawal symptoms) and 'social' (The importance of relationships with other users). The negative factors comprised 'aversiveness', 'poor quality of life' (Dislike effect on personality), 'criminal involvement' (Dislike of being involved in crime) 'personal failure' (Hurt caused to loved ones), 'pressure' (Pressure from other people to stop) and 'frustration' (Failure to get a rush any more). Cue-specific craving in response to drug-related stimuli was only associated with positive expectancies, the most highly correlated factors being 'emotional relief' and 'lifestyle'. The association between positive expectancy and craving was robust enough to remain after controlling for withdrawal-like symptoms, agonist-like effects and dysphoria. A relationship between positive expectancies and impulsivity was also evident, and again this association was due to the statistical influence of 'emotional relief' expectancies. This supports the previously discussed findings of high risk being associated with opiates and negative reinforcement. Drug-related cues may activate latent drug reinforcement expectancies in a similar way to that which occurs with alcohol (Gadon *et al.* 2004).

The 'justifications for cravings' in those who are abstinent or have relapsed to opiate use have been examined in research using structured interviews, as opposed to a questionnaire, such as the MC, but provide convergent validation (Del Pozo *et al.* 1998). The strongest difference between the relapse group and the abstainers related 'escaping from personal problems' and 'lack of satisfaction with abstinence'. Other less common but significant contrasts related to those who relapsed having more 'family reproaches with addictive behaviour' and having less 'physical uneasiness and insomnia'. Del Pozo *et al.* (1998) also documented motivation for use between the two groups of abstinent and relapsed heroin users, finding that the most common motivational factors for heroin use were dissatisfaction with life, psychological instability and loneliness. Similarly, although within a single environment, a prison community, the inmates who reported using heroin for hedonistic reasons (to get high, to be less inhibited, to feel more courage, to deal with boredom and to be more sexually responsive) were more likely to reject help from caseworkers that those who reported using for self-medicating reasons (to relax, to unwind, to feel happier, to feel better about yourself, to deal with negative feelings and to forget problems; Winfree *et al.* 1994). This collectively indicates that opiate cognitions representing more powerful heroin-related reinforcement are associated with underlying personality or psychopathology.

Principal component analysis of the Heroin Expectancy Questionnaire (HEQ; Young *et al.* 1999)—a measure that, like the MC, was developed using the definition of an expectancy as an 'if–then' contingency—revealed five factors. The first factor, 'Enhanced Self-perception and Cognition', represents mainly reinforcing consequences of heroin use that are specific to transformed cognitive and social abilities (Heroin sharpens my mind) and to the future (Heroin makes the future brighter). The second factor, 'Positive Affective Change', to a large part relates to negatively reinforcing consequences of heroin use that are specific to the reduction of aversive emotions (Using heroin helps me to forget painful memories) and the disinhibition (It is easier to open up and express my feelings on heroin). The third factor, 'Negative Affective Change', is entirely composed of aversive consequences of heroin use (Using heroin makes me tense). The fourth factor, 'Sexual Impairment', contains items relating to diminished sexual function (I tend to avoid sex on heroin). In addition, a broad factor of 'Dependence and Tension Reduction' represents negative consequences of heroin use that are specific to symptoms of opiate dependence (I am addicted to heroin). The tension reduction reported is likely to be partially due to withdrawal relief.

In a clinical sample of 160 heroin-dependent individuals, gender differences in heroin expectancy were evident on the HEQ. Average male total score was higher than the average female score despite no differences being evident in self-reported heroin use (Walmsley 2004). The factor to show the most significant difference was 'Enhanced Self-perception and Cognition'. For both males and females, symptoms of depression and anxiety were correlated significantly with 'Positive Affective Change' expectancies, indicating that psychopathological symptoms were associated with heroin expectancies of improved affect. This broadly supports a self-medication view, but of course gives no indication of causality. The possibility of self-medication was also evident in the relationship between self-reported neuropsychological symptoms and heroin expectancy. Self-reported deficits in planning and behavioural regulation correlated significantly with both the 'Positive Affective Change' and 'Dependence and Tension Reduction' factors. Similarly, reports of impaired executive memory and agitation correlated significantly with 'Enhanced Self-perception and Cognition' expectancies.

In Walmsley (2004), a self-reported early age of onset was associated with expectancies of 'Enhanced Self-perception and Cognition' and 'Positive Affective Change', perhaps indicating the chronicity of concomitant psychological difficulties. This might also reflect a less stable self-image and less developed coping skills observed in younger adults (Diehl *et al.* 1996). A similar effect has been established with cocaine users in expectancy research (Galen and Henderson 1999; Aarons *et al.* 2001) and with alcohol users in implicit expectancy research (Gadon *et al.* 2004). In terms of current use, in Walmsley (2004) typical heroin quantity used over the last month correlated significantly with expectancies of 'Dependence and Tension Reduction' and frequency of heroin use correlated significantly with both 'Dependence and Tension Reduction' and 'Sexual Impairment', consistent with a combination of positive and negative outcomes being associated with heavy, or more frequent, use.

Attempts to examine whether reinforcement expectancies are potential mechanisms for drug use, rather than constructs that simply co-vary with use, are not definitive. For example, in the use of a broad five-item unifactorial measure of 'pain medication beliefs', both state anxiety and beliefs regarding medication (e.g. beliefs regarding how much the individual perceives that pain can be better) statistically mediated the relationship between history of addiction and drug use (Schieffer *et al.* 2005). Indirect evidence that supports the relationship between endogenous opioid systems and drug expectancy can be found in a study examining the influence of the μ opioid receptor antagonist naltrexone on craving for alcohol (Palfai *et al.* 1999). Naltrexone significantly reduced the urge to drink only in those individuals who reported positive alcohol expectancies. However, there was no comparable effect on alcohol consumption, indicating a lack of correspondence between cognitive change and behaviour change. The dopamine antagonist haloperidol has also shown an effect on implicit cognition, as measured by improvements on the modified Stroop task, in heroin-dependent patients, but equally showed no effect on an allied cognitive measure—the desire to use heroin (Franken *et al.* 2004). The causal status of opiate expectancies is thus indeterminate.

In summary, the relationship between multifactorial measures of opiate cognition and drug use has only been examined in a handful of cross-sectional studies. These studies collectively indicate that emotional relief expectancies are associated with intensity of craving for opiates, and heroin expectancies of positive affective change are associated with the extent of depression and anxiety reported. In addition, impulsivity and impaired planning appear to be related to the relief of negative affect, or to positive affective change. Those reporting heavier heroin use report both stronger negative and positive expectancies than those with less heavy use, a similar situation to that evident with alcohol (Jones and McMahon 1992, 1994; Young 1994). There is, however, no evidence that these beliefs are causal agents in drug use.

7.6 **How might different cognitive subsets operate together to influence heroin use?**

Much of the literature reviewed has examined only a single cognitive construct with regard to opiate use, such as opiate expectancy, or has inferred the presence of cognitions that were not actually measured. As noted, social cognitive theory offers a means of examining the roles of both outcome expectancy and self-efficacy (Bandura 1989). Substance misuse has been hypothesized to occur as a consequence of an individual's perceived ability to cope in a given situation, combined with their expectations that the drug will assist them to cope because of anticipated positive outcomes from use (Marlatt and Gordon 1985). There is empirical evidence to support greater coping self-efficacy being associated with better treatment outcomes across a range of substances, including opiates (Gossop *et al.* 1990).

Powell *et al.* (1993) explored personality, cue-elicited craving, opiate expectancies (using the MC of Powell *et al.* 1992) and self-efficacy for resisting opiates in 43 opiate addicts undergoing inpatient detoxification. The majority of the participants also completed an adjunct relapse prevention programme involving cue exposure, or combined individual and group rehabilitation therapy. The main aim of the study was to predict prospective change. Participants who reported *lower* self-efficacy and *higher* positive outcome expectancies prior to detoxification were using opiates *less* often 6 months after detoxification than participants who reported higher self-efficacy and lower positive outcome expectancies. The authors speculated that heroin users who had greater awareness of their vulnerability developed more effective coping strategies. A contemporaneous study by the same research team, using the same measure of self-efficacy, found that self-efficacy increased significantly over time following treatment, but did not report the relationship between these changes and heroin use (Dawe *et al.* 1993).

Studies examining self-efficacy in heroin users have generally supported a positive relationship between self-efficacy and abstinence, or treatment adherence (Kleinman *et al.* 2002; Murphy *et al.* 2003). Perry and Hedges-Duroy (2004) found lower relapse self-efficacy was associated with the severity of presenting problem on admission. Over a 12 month follow-up, higher self-efficacy was associated with better outcomes. Other research has indicated that early predictive effects of self-efficacy evident during treatment do not predict drug use over the subsequent 9 months follow-up (Burleson and Kaminer 2005). Saunders *et al.* (1995), who did not find a relationship between self-efficacy and drug use *per se*, found a relationship between low self-efficacy and attrition. Several studies have used the Drug-taking Confidence Questionnaire (DTCQ) to capture self-efficacy, but this is primarily in cocaine users (Sherrilyn, Sklar and Turner (1997, 1999)). The psychometric characteristics of the full DTCQ (Sklar *et al.* 1997), and a brief eight-item version (Sklar and Turner 1999) are presented with regard to alcohol and cocaine misuse, but in both studies only 58 heroin misusers were recruited and these data were not analysed further. One study using the DTCQ (Reilly *et al.* 1995) tests a causal model where inference may be made of the role of self-efficacy in heroin misusers as a mechanism underlying response to treatment. While self-efficacy changes as phases of treatment change, for example where confidence increases as methadone dose is stabilized, the data did not support the role of self-efficacy as a causal agent. The study concludes that self-efficacy merely reflects the extent of previous behavioural attainments in abstaining from use. The majority of these studies have examined self-efficacy alone, and as such are an inadequate examination of the social cognitive model (Bandura 1989).

A similar principal components analysis to that already described with the HEQ was conducted on the Heroin Refusal Self Efficacy Questionnaire (Young *et al.* 1999), revealing three factors. The first factor of the HRSEQ, 'Emotional Relief', represents the ability to refrain from using heroin when experiencing a negative emotional state (e.g. 'When I feel upset'). The second factor, 'Social Enhancement', represents the ability to refrain from using heroin when exposed to social and/or personal contextual cues associated with the use of heroin (e.g. 'When I want to feel more accepted by friends'). The third factor,

'Opportunistic Use', reflects the ability to refrain from using heroin when the drug is simply available (e.g. 'When my spouse or partner is using'). Walmsley (2004) reported that females were more confident than males in resisting heroin under circumstances of social enhancement. Symptoms of depression and anxiety showed a significant negative correlation with 'Emotional Relief' expectancies, providing further confirmation of the role of negative reinforcement processes. Similarly, self-reported impairment in planning and behavioural regulation showed a significant negative relationship with 'Emotional Relief' self-efficacy and '*Opportunistic*' self-efficacy. Greater impairment in emotional perception and expression were also associated with lower 'Social Enhancement' self-efficacy. Both heroin quantity and frequency of use were negatively correlated with 'Emotional Relief' and 'Social Enhancement' self-efficacy.

Using multiple regression, Walmsley (2004) confirmed the importance of executive dysfunction, heroin expectancy and self-efficacy in predicting consumption. This analysis thus embodies both central constructs of social cognitive theory. Self-report of impaired executive memory, agitation, heroin expectancies reflecting positive affective change, and emotional relief self-efficacy collectively predicted heroin use. An analogous regression examining frequency of heroin use found that the only two significant predictors were weaker opiate expectancies of sexual impairment and lower emotional relief self-efficacy. The relationship between low self-efficacy for drug resistance when confronted by a need for emotional relief resonates strongly with the previous research reviewed. This research confirmed the difficult developmental trajectory frequently associated with heroin misuse, typically featuring depression, anxiety and exposure to traumatic events. These findings support the worth of a prospective examination of both of these key sets of opiate cognitions and developmental issues.

7.7   How similar are opiate expectancies to the expectancies associated with other drugs?

Comparing the domains of expected reinforcement that exist across different pharmacological classes allows a crude relative appraisal. Equally, the potentially reinforcing profile unique to each drug may be important in understanding their relative risk for addiction. Table 7.1 contains a summary of four expectancy scales that have been developed using identical factor analytic approaches by the same research group. These scales are the Heroin Expectancy Profile (HEP; Young *et al.* 1999), the Drinking Expectancy Profile (DEP; Young and Oei 1996), the Cannabis Expectancy Profile (CEP; Young *et al.* 2003) and the Nicotine Expectancy

Table 7.1 Comparison of the Heroin Expectancy Profile (HEP) with other substance use expectancy scales developed using similar methods

	Expectancy scales			
Drug expectancy domain/dependence	**HEP**	**DEP**	**CEP**	**NEP**
Positive perceptual and cognitive change[1]	✓	✓	✓	✓
Tension reduction[2]	✓	✓		✓
Dependence[3]	✓	✓	✓	
Negative affective change	✓	✓	✓	
Positive affective change	✓			
Sexual impairment	✓			
Social facilitation and assertion[4]		✓	✓	✓
Sexual enhancement		✓		
Negative consequences and sensations				✓
Structuring goals and activities				✓
Negative physical consequences				✓
Self-efficacy				
Social pressure self-efficacy[5]	✓	✓	✓	
Emotional relief self-efficacy	✓	✓	✓	✓
Opportunistic self-efficacy	✓	✓		✓

[1] Cognitive change (DEP), enhanced self-perception and cognition (HEP), positive effects (CEP).
[2] Tension reduction (DEP), dependence and tension reduction (HEP), and both distress reduction and relaxation (NEP); these are two separate factors.
[3] Dependence and tension reduction (HEP), habit (CEP).
[4] Assertion (DEP), social facilitation (CEP), social reward (NEP).
[5] Social enhancement self-efficacy (HEP), exposure to cannabis use self-efficacy (CEP).

Profile (NEP; Hausdorf *et al.* 2003). A similar methodology and psychometric evaluation has been used with each scale. It is striking that the expectancy domain common across heroin, alcohol, cannabis and nicotine relates to perceptual and cognitive change and reflects a primary reinforcement domain relating to an altered view of the world or our environment. An expectancy of tension reduction was also evident with three of these drug outcome expectancy scales, including heroin. The ubiquity of expectancies of tension reduction provides support for the 'tension reduction hypothesis' originally postulated by Conger (1956) regarding alcohol use. This states that individuals use alcohol because they believe that drinking reduces aversive sensations and cognitions that are tension related. It also provides support for self-medication. Drug consumption may conveniently serve as a means of self-medicating feelings of anxiety that is reflected in expectancies (Young *et al.* 1990). There is evidence that these beliefs have predictive utility in both treatment (Steindl *et al.* 2003) and non-treatment samples (Young and Oei 2000). In Table 7.1, social reinforcement was not evident with heroin but was with alcohol, cannabis and nicotine. The only scale to show a separate domain for positive affect was for heroin expectancy. The comparative picture of expectancy across several drugs gives some indication of which type of individual may select specific drugs and under what circumstances. While heroin clearly shares some similarity to other dugs of abuse in that it can transform perception, alter cognition and reduce tension, unlike the other agents reviewed, it is not typically reported to enhance social or sexual experience. Heroin is, however, described as enhancing positive affect. This paints a general picture of heroin as a drug with outcomes that are more consistent with an introspective perceptual–emotional experience rather than a collective or social one. This is a similar picture to the descriptions found in fiction about the opiates.

Despite the absence of outcomes related to social functioning, social cues were an important risk in encouraging the use of heroin reflected in the self-efficacy domains. The three factors for the HRSEQ Emotional Relief, Social Enhancement and Opportunistic self-efficacy are almost identical to those found for the Drinking Refusal Self-Efficacy Questionnaire (DRSEQ) (Young *et al.* 1991*a*) and show some overlap with parallel measures of cannabis and nicotine refusal self-efficacy (Young *et al.* 2003; Hausdorf *et al.* 2004). While some of this is an artefact of measure development, there is also a high degree of similarity between the HRSEQ factors and the domains of self-efficacy associated with triggers for drug use, or precipitants for relapse identified in other research (Marlatt and Gordon 1985). The three factors of the HRSEQ show overlap with the three second-order factors found in the Drug-taking Confidence Questionnaire (DTCQ; Sklar *et al.* 1997). These second-order factors (and their first-order component factors) include Negative Situations (Unpleasant Emotions, Physical Discomfort and Conflict with Others), Positive Situations (Pleasant Emotions, Pleasant Times with Others) and Temptation Situations (Testing Personal Control, Urges and Temptations, Social Pressure to Use). They are also similar to three of the four factors of negative affect, social, physical and other concerns, and cravings and urges identified using a version of the Alcohol Abstinence Self-Efficacy Scale modified for heroin (Hiller *et al.* 2000). Thus, it seems that while the reinforcing parameters of drug consequences may be quite different across drug classes, the cues associated with diminished confidence to resist use are much more similar relating to negative affective state, social pressure and availability or opportunity. Self-efficacy to resist heroin has been compared with that relating to the resistance of other drugs in several studies. For example, in a study of treatment-seeking adolescents, although the numbers were small ($n = 22$, heroin, $n = 34$ no heroin), those presenting with heroin misuse had lower self-efficacy than their non-heroin-using counterparts presenting with other drug problems (Perry and Hedges-Duroy 2004). Similar findings have been demonstrated in a comparison of adult heroin misuers and alcohol misusers (Salah, Gaily, and Bashir 2004). The low confidence that heroin users report regarding change underscores to the need to investigate the reinforcing parameters of heroin and the need to understand how learning processes may lead to addictive behaviour.

7.8 **Conclusions**

There is a simple and compelling conclusion that there is a paucity of research examining opiate cognitions compared with the bodies of research undertaken examining other drugs. This is likely to be due to the attitudes of society generally and a legacy of inaccurate perception that the mechanisms responsible for the outcomes of opiate consumption are of a pharmacological nature rather than representing pharmacological and psychological processes. However, there is no evidence to suggest that the opiates should be considered differently from other psychoactive agents; they need to be considered in terms of cognitive set and setting. Converging circumstantial evidence indicates that individual differences in response to the opiate drug class exist and there is evidence of a significant genetic influence. There is a lack of research examining the additional contribution of environment and learning processes to these individual differences. Cultural and subcultural beliefs regarding opiate use are likely to interact with environmental developmental disadvantage related to poor mental health and exposure to traumatic events to create risk. It is not known how the cognitions related to potential heroin reinforcement of altered perception, cognition and positive affect arise, but cognitions regarding the confidence to resist heroin endorse the importance of negative affect. Psychometric research to enable a more precise explication of the role of cognition has advanced at a slow pace. There is little solid research that examines the role of learning processes in opiate dependence, the risks for opiate use or the treatment of those addicted. Examination of the role of cognitions, such as expectancies, in this research process is needed. Until this research advances, our collective understanding of the 'joy and relief' of opiates will continue to owe more to literature than it does to behavioural science.

Acknowledgements

Any correspondence concerning this chapter should be addressed to Professor Ross Young, School of Psychology and Counselling, Queensland University of Technology, Australia.

References

Aarons, G.A., Brown, S.A., Stice, E. and Coe, M.T. (2001). Psychometric evaluation of the Marijuana and Stimulant Effect Expectancy Questionnaires for adolescents. *Addictive Behaviors*, 26, 219–236.

Amato, L., Minozzi, S., Davoli, M., Vecchi, S., Ferri, M. and Mayet, S. (2004*a*). Psychosocial combined with agonist maintenance treatments versus agonist maintenance treatments alone for treatment of opioid dependence. *The Cochrane Database of Systematic Reviews*, Issue 4, Art No.: CD004147. DOI: 10.1002/14651858. CD004147.

Amato, L., Minozzi, S., Davoli, M., Vecchi, S., Ferri, M. and Mayet, S. (2004*b*). Psychosocial and pharmacological treatments versus pharmacological treatments for opioid detoxification. *The Cochrane Database of Systematic Reviews*, Issue 4, Art No.: CD005031. DOI: 101002/14651858: CD005031.

Bailey C.P. and Connor, M. (2005). Opioids: cellular mechanisms of tolerance and physical dependence. *Current Opinion in Pharmacology*, 5, 60–68.

Bandura, A. (1977). *Social learning theory*. New York, NY: Prentice-Hall.

Bandura, A. (1989). Human agency in social cognitive theory. *American Psycholgist*, 44, 1175–1184.

Bandura A. (1999). A sociocognitive analysis of substance abuse: an agentic perspective. *Psychological Science*, 10, 214–217.

Bandura, A., Cioffi, D., Barr Taylor, C. and Brouillard, M.E. (1988). Perceived self-efficacy in coping with cognitive stressors and opioid activation. *Journal of Personality and Social Psychology*, 55, 479–488.

Beckerleg, S. (2004). How 'cool' is heroin injection at the Kenya Coast. *Drugs: Education, Prevention and Policy*, 11, 67–77.

Bowen, D.J., Tomoyasu, N., Anderson, M., Carney, M. and Kristal, A. (1992). Effects of expectancies and personalised feedback on fat consumption, taste and preference. *Journal of Applied Psychology*, 22, 1061–1079.

Bradley, B.P., Gossop, M., Brewin, C.R., Phillips, G. and Green, L. (1992). Attributions and relapse in opiate addicts. *Journal of Consulting and Clinical Psychology*, 60, 470–472.

Brandon, T.H., Juliano, L.M. and Copeland A.L. (1999). Expectancies for tobacco smoking. In I. Kirsch (ed.). *How expectancies shape experience*. Washington, DC: American Psychological Association, pp. 263–299.

Bremner, J.D., Southwick, S.M., Darnell, A. and Charney, D.S. (1996). Chronic PTSD in Vietnam combat veterans: course of illness and substance abuse. *American Journal of Psychiatry*, 153, 369–375.

Brooner, R., Greenfield, L., Schmidt, C. and Bigelow, G. (1993). Antisocial personality disorder and HIV infection among intravenous drug users. *American Journal of Psychiatry*, 150, 53–58.

Brooner, R., King, V., Kidorf, M., Schmidt, C. and Bigelow, G. (1997). Psychiatric and substance use co-morbidity among treatment-seeking opioid abusers. *Archives of General Psychiatry*, 54, 71–80.

Brown, S.A., Goldman, M.S., Inn, A. and Anderson, L.R. (1980). Expectations of reinforcement from alcohol: their domain and relation to drinking patterns. *Journal of Consulting and Clinical Psychology*, 48, 419–426.

Burleson, J.A. and Kaminer, Y. (2005). Self-efficacy as a predictor of treatment outcome in adolescent substance use disorders. *Addictive Behaviors*, 30, 1751–1764.

Burns, J.M., Martyres, R.F., Clode, D. and Boldero, J.M. (2004). Overdose in young people using heroin: associations with mental health, prescription drug use and personal circumstances. *Medical Journal of Australia*, 181 (Suppl. 7), S25–S28.

Callaly, T., Trauer, T., Munro, L. and Whelan, G. (2001). Prevalence of psychiatric disorder in a methadone maintenance population. *Australian and New Zealand Journal of Psychiatry*, 35, 601–605.

Cirakoglu, O.C. and Isin, G. (2005). Perception of drug addiction among Turkish university students: causes, cures and attitudes. *Addictive Behaviors*, 30, 1–8.

Cleeland, C.S., Nakamura, Y., Howland, E.W., Morgan, N.R., Edwards, K.R. and Backonja, M. (1996). Effects of oral morphine on cold pressor tolerance time and neuropsychological performance. *Neuropsychopharmacology*, 15, 252–262.

Colloca, L. and Benedetti, F. (2005). Placebos and painkillers: is mind as real as matter? *Nature Reviews: Neuroscience*, 6, 545.

Conger, J.J. (1956). Reinforcement theory and the dynamics of alcoholism. *Quarterly Journal of Studies on Alcohol*, 17, 296–305.

Contarino, A., Picetti, R., Matthes, H.W., Koob, G.F., Kieffer, B.L. and Gold, L.H. (2002). Lack of reward and locomotor stimulation induced by heroin mu-opioid receptor-deficient mice. *European Journal of Pharmacology*, 446, 103–109.

Cossu, G., Ledent, C., Fattore, L., Imperato, A., Bohme, G.A., Parmentier, M. and Fratta, W. (2001). Cannabinoid CB1 receptor knockout mice fail to self-administer morphine but not other drugs of abuse. *Behavioural Brain Research*, 118, 61–65.

Dawe, S., Powell, J., Richards, D., Gossop, M., Marks, I., Strang, J and Gray, J.A. (1993). Does post-withdrawal cue exposure improve outcome in opiate addiction? A controlled trial. *Addiction*, 88, 1233–1245.

Day, E. and Smith, I. (2002). Literary and biographical perspectives on substance use. *Advances in Psychiatric Treatment*, 9, 62–68.

De Quincey, T. (2003). *Confessions of an English opium eater: and other writings*. London: Penguin Classics.

Del Pozo J.M.L., Gomez C.F., Fraile, M.G. and Perez, I.V. (1998). Psychological and behavioural factors associated with relapse among heroin abusers treated in therapeutic communities. *Addictive Behaviors*, 23, 155–169.

Di Chiara, G. and Imperato, A. (1988). Drugs abused by humans preferentially increase synaptic dopamine concentrations in the mesolimbic system of freely moving rats. *Proceedings of the National Academy of Sciences of the USA*, 85, 5274–5278.

Diehl, M., Coyle, N. and Labouvie-Vief, G. (1996). Age and sex differences in strategies of coping and defense across the life span. *Psychology and Aging*, 11, 127–139.

Doyle, J., Grey, J. and Pierce, P. (2002). *Australia's Vietnam War*. College Station, TX: Texas A&M University Press.

Eckersley, R. (2005). 'Cultural fraud': the role of culture in drug abuse. *Drug and Alcohol Review*, **24**, 157–163.

Edwards, G. (2005). *Matters of substance drugs: is legalization the right answer—or the wrong question?* London: Penguin.

Elliott, A.J. and Chapman, S. (2000). 'Heroin hell their own making' construction of heroin users in the Australian press 1992–97. *Drug and Alcohol Review*, **19**, 191–201.

Fassino, S., Daga, G.A., Delsedime, N., Rogna, L. and Boggio, S. (2004). Quality of life and personality disorders in heroin abusers. *Drug and Alcohol Dependence*, **76**, 73–80.

Fischer, B., Rehm, J., Bristette, S., Brochu, S., Bruneau, J., El-Guebaly N., Noel L., Tyndall, M., Wild, C., Mun, P. and Baliunas, D. (2005). Illicit opioid use in Canada: comparing social, health and drug use characteristics of untreated users in five cities (OPICAN study). *Journal of Urban Health–Bulletin of the New York Academy of Medicine*, **82**, 250–266.

Forman, S.D., Dougherty, G.G., Casey, B.J., Siegle, G.J., Braver, T.S., Barch, D.M., Stenger, V.A., Wick-Hull, C., Pisarov, L.A. and Lorensen, E. (2004). Opiate addicts lack error-dependent activation of rostral anterior cingulate. *Biological Psychiatry*, **55**, 531–537.

Franken, I.H.A., Hendricks, V.M. and van den Brink, W. (2002). Initial validation of two opiate craving questionnaires: the Obsessive Compulsive Drug Use Scale and the Desires for Drug Questionnaire. *Addictive Behaviors*, **27**, 675–685.

Franken, I.H.A., Stam, C.J., Hendriks, V.M. and van den Brink, W. (2003). Neuropshysiological evidence for abnormal cognitive processing of drug cues in heroin dependence. *Psychopharmacology*, **170**, 205–212.

Franken, I.H.A., Hendriks, V.M., Stam, C.J. and Van den Brink, W. (2004). A role for dopamine in the processing of drug cues in heroin dependent patients. *European Neuropsychopharamcology*, **14**, 503–508.

Furnham, A. and Thomson, L. (1996). Lay theories of heroin addiction. *Social Science and Medicine*, **43**, 29–40.

Gadon, L., Bruce G., McConn ochie, F., and Jones, B.T. (2004). Negative alcohol consumption outcome associations in young and mature adult social drinkers: a route to drinking restraint. *Addictive Behaviours*, **29**, 1373–1384.

Galen, L.W. and Henderson, M.J. (1999). Validation of cocaine and marijuana effect expectancies in a treatment setting. *Addictive Behaviors*, **24**, 813–817.

Galer, B.S., Coyle, N., Pasternak G.W. and Portenoy, R.K. (1992). Individual variability in the response to different opioids: a report of five cases. *Pain*, **49**, 87–91.

Gerlach, R. and Caplehorn, J. R.M. (1999). Attitudes and beliefs of doctors prescribing methadone to addicts in the Westfalen-Lippe region of Germany. *Drug and Alcohol Review*, **18**, 163–170.

Goldman, M.S., Del Boca, F.K. and Darkes, J. (1999). Alcohol expectancy theory: the application of cognitive neuroscience. In K.E. Leonard and H.T. Blane (ed.), *Psychological theories of drinking and alcoholism*. New York, NY: Guilford Press, pp. 203–246.

Gossop, M., Green, L., Phillips, G. and Bradley, B. (1990). Factors predicting outcome among opiate addicts after treatment. *British Journal of Clinical Psychology*, **29**, 209–216.

Green, L. and Gossop, M. (1988). Effects of information on the opiate withdrawal syndrome. *British Journal of Addiction*, **83**, 305–309.

Green M.W., Taylor, M.A., Elliman, N.A. and Rhodes, O. (2001). Placebo expectancy effects in the relationship between glucose and cognition. *British Journal of Nutrition*, 86, 173–179.

Han, W., Ide, S., Sora, I., Yamamoto, H. and Ikeda, K. (2004). A possible genetic mechanism underlying individual and interstrain differences in opioid actions focus on the mu opioid receptor gene. *Annals of the New York Academy of Sciences*, 1025, 370–375.

Handelsman, L., Aronson, M.J., Ness R., Cochrane, K.J. and Kanof, P.D. (1992). The dysphoria of heroin addiction. *American Journal of Drug and Alcohol Abuse*, 18, 275–287.

Hashemi, N. and London, M. (2003). Psychiatric practice in Iran and UK. *Psychiatric Bulletin*, 27, 190–191.

Hausdorf, K., Young, R. McD., Connor, J.P., Saunders, J.B., Owen, N. and Hamilton, G. (2004). Targeted self-help materials for smoking cessation. *Proceedings of the 39th Conference of the Australian Psychological Society*, 138–142.

Hendricks, V.M., van der Brink, W., Blanken, P., Bosman, J.J. and van Ree, J.M. (2001). Heroin self-administration by means of 'chasing the dragon': pharmacodynamics and bioavailability of inhaled heroin. *European Neuropsychopharmacology*, 11, 241–252.

Hiller, M.L., Broome, K.M., Knight, K. and Simpson, D.D. (2000). Measuring self-efficacy among drug-involved probationers. *Psychological Reports*, 86, 529–538.

Hopfer, C.J., Khuri, E., Crowley, T.J. and Hooks, S. (2002). Adolescent heroin use: a review of the descriptive and treatment literature. *Journal of Substance Abuse Treatment*, 23, 231–237.

Jacobs, E.H., Smit, A.B., De Vries, T.J. and Schoffelmeer, A.N.M. (2005). Long-term gene expression in the nucleus of accumbens following heroin administration is subregion-specific and depends on the nature of drug administration. *Addiction Biology*, 10, 91–100.

Jaffe, A.J. and Kilby, M.M. (1994). The Cocaine Expectancy Questionnaire (CEQ): construction and predictive utility. *Psychological Assessment*, 6, 18–26.

Jaffe, J.H and Jaffe, A.B. (2004). Neurobiology of opioids. In M. Galanter and H.D. Kleber (ed.), *Textbook of substance abuse treatment*. Washington, DC, American Psychiatric Press.

Jones, B.T. and McMahon, J. (1992). Negative and positive expectancies in group and lone problem drinkers. *British Journal of Addiction*, 87, 929–930.

Jones, B.T. and McMahon, J. (1994). Negative and positive alcohol expectancies as predictors of abstinence after discharge from a residential treatment program: a one month and three month follow-up study in men. *Journal of Studies on Alcohol*, 55, 543–548.

Jones, B.T., Corbin, W. and Fromme, K. (2001). A review of expectancy theory and alcohol consumption. *Addiction*, 96, 57–72.

Kakko, J., Svanborg, K., Kreek, M. and Heilig, M. (2003). 1-year retention and social function after buprenorphine-assisted relapse prevention treatment for heroin dependence in Sweden: a randomised, placebo-controlled trial. *Lancet*, 361, 662–668.

Kehoe, P. and Blass, E.M. (1986). Behaviorally functional opioid system in infant rats: II. Evidence for pharmacological, physiological and psychological mediation of pain and stress. *Behavioral Neuroscience*, 100, 624–630.

Kieres, A.K., Hausknecht, K.A., Farrar, A.M., Acheson, A., de Wit, H. and Richards, J.B. (2004). Effects of morphine and naltrexone on impulsive decision making in rats. *Psychopharmacology*, 173, 167–174.

Kleinman, B.P., Millery, M., Scimeca, M. and Pollisar, N.L. (2002). Predicting long term treatment utilization among addicts entering detoxification: the contribution of help seeking models. *Journal of Drug Issues*, 32, 209–230.

Kreek, M.J., Neilsen, D.A. and LaForge, K.S. (2004). Genes associated with addiction—alcoholism, opiate, and cocaine addiction. *Neuromolecular Medicine*, 5, 85–108.

Kreek, M.J., Bart, G., Lilly, C., LaForge, K.S. and Nielsen, D.A. (2005). Pharmacogenetics and human molecular genetics of opiate and cocaine addictions and their treatments. *Pharmacological Review*, 57, 1–26.

Kruse, S. (1998). The importance of music in adolescent life: the relationship between music preference, values, attitudes, and behaviors. *Dissertation Abstracts International: Section B.The Sciences and Engineering*, 59, 0913.

Lang, E. (1998). Drugs in society: a social history. In M. Hamilton, A.Kellehear and G. Rumbold (ed.), *Drug use in Australia: a harm minimisation approach*. Melbourne: Oxford University Press, pp. 1–11.

Lawford, B.R., Young, R. McD., Crawford, D.H.G., Foley, P.F., Ritchie, T., Noble, E.P. and Cooksley, W.G.E. (1999). The DRD$_2$ A$_1$ allele: a behavioural genetic risk factor in hepatitis C infection of persistent drug abusers. *Addiction Biology*, 4, 61–66.

Lawford, B.R., Young, R. McD., Noble, E.P., Sargent, J., Powell, J., Shadforth, S., Zhang, X. and Ritchie, T. (2000). The D$_2$ dopamine receptor A$_1$ allele and opioid dependence. *American Journal of Medical Genetics (Neuropsychiatric Genetics)*, 96, 592–598.

Lawrence, G., Bammer, G. and Chapman, S. (2000). Sending the wrong signal: analysis of print media reportage of the ACT heroin prescription trial proposal. *Australian and New Zealand Journal of Public Health*, 24, 254–264.

LeFoll, B. and Goldberg, S.R. (2004). Cannabinoid CB1 receptor antagonists as promising new medications for drug dependence. *Journal of Pharmacology and Experimental Therapeutics*, 318, 875–883.

Leong, J. (1974). Cross-cultural influences on ideas about drugs. *Bulletin on Narcotics*, 26, 1–8.

Lotshaw, S.C., Bradley, J.R. and Brooks, L.R. (1996). Illustrating caffeine's pharmacological and expectancy effects utilizing a balanced placebo design. *Journal of Drug Education*, 26, 13–24.

MacAndew, L.R. and Edgerton, B. (1969). *Drunken comportment: a social explanation*. Chicago, IL: Aldine.

Macdonald, G. and Leary, M.R. (2005). Why does social exclusion hurt? The relationship between social and physical pain. *Psychological Bulletin*, 131, 224–230

Maier, S.F. (1984). Learned helplessness and animal models of depression. *Progress in Neuro-Psychopharmacology and Biological Psychiatry*, 8, 435–446.

Marlatt, G.A. and Gordon, J.R. (1985). *Relapse prevention: maintenance strategies in the treatment of addictive behaviour*. New York, NY: Guilford Press.

Marlatt, G.A. and Rohsenow, D.J. (1980). Cognitive processes in alcohol use: expectancy and the balanced placebo design. In N.K. Mello (ed.). *Advances in substance abuse: behavioral and biological research*. Greenwich: JAI Press, pp. 159–199.

Mattick, R., Oliphant, D., Bell, J. and Hall, W. (2000). *Psychiatric morbidity in methadone maintenance patients: prevalence, effect on drug use and detection*. Sydney, NSW: National Drug and Alcohol Research Centre, The University of New South Wales.

McDonald, R.V. and Siegel, S. (2004). The potential role of drug onset cues in drug dependence and withdrawal: reply to Bardo (2004), Bosset and Shaham (2004), Boulton (2004), and Stewart (2004). *Experimental and Clinical Psychopharmacology*, 12, 23–26.

McLellan, A. (1986). 'Psychiatric severity' as a predictor of outcome from substance abuse treatments. In R. Meyer (ed.). *Psychopathology and addictive disorders*. New York, NY: Guilford Press, pp. 97–139.

Metzger, D., Woody, G., de Philippis, D., McLellan, A., O'Brien, C. and Platt, J. (1991). Risk factors for needle sharing among methadone-treated patients. *American Journal of Psychiatry*, 148, 636–640.

Milby, J., Sims, M., Khuder, S., Schumacher, J., Huggins, N., McLellan, T., Woody, G. and Haas, N. (1996). Psychiatric co-morbidity: prevalence in methadone maintenance treatment. *American Journal of Drug and Alcohol Abuse*, 22, 95–100.

Miller, N.E. and Dollard, J. (1941). *Social learning and imitation*. New Haven, CT: Yale University Press.

Mills, K., Lynskey, M., Teesson, M., Ross, J. and Darke, S. (2005). Post-traumatic stress disorder among people with heroin dependence in the Australian treatment outcome study (ATOS): prevalence and correlates. *Drug and Alcohol Dependence*, 77, 243–249.

Mintz, J., O'Brien, C.P. and Pomerantz, B. (1979). The impact of Vietnam service on heroin-addicted veterans. *American Journal of Drug and Alcohol Abuse*, 6, 39–52.

Mitchell, S.H., Laurent, C.L. and de Wit, H. (1996). Interaction of expectancy and the pharmacological effects of d-amphetamine: subjective effects and self-administration. *Psychopharmacology*, 125, 371–378.

Murphy, P.N., Bentall, R.P., Ryley, L.D. and Ralley, R. (2003). Predicting postdischarge opiate abstinence from admission measures of motivation and confidence. *Psychology of Addictive Behaviors*, 17, 167–170.

Noble, E.P., Blum, K., Ritchie, T., Montgomery, A. and Sheridan, P. (1991). Allelic association of the D2 dopamine receptor gene with receptor binding characteristics in alcoholism. *Archives of General Psychiatry*, 48, 648–654.

Nunes, E., Sullivan, M. and Levin, F. (2004). Treatment of depression in patients with opiate dependence. *Biological Psychiatry*, 56, 793–802.

Oei, T.P.S., Lim, B. and Young, R.McD., (1991). Cognitive processes and cognitive behavior therapy in the treatment of problem drinking. *Journal of Addictive Diseases*, 10, 63–80.

Office of National Drug Control Policy (1999). *The National Drug Control Policy*. Washington, DC: Government Printing Office, Office of National Drug Control Policy Clearinghouse.

Panskepp, J. (2003). Neuroscience. Feeling the pain of social loss. *Science*, 320, 237–239.

Palfai, T., Davidson, D. and Swift, R. (1999). Influence of naltrexone on cue-elicited craving among hazardous drinkers: the moderational role of positive outcome expectancies. *Experimental and Clinical Psychopharmacology*, 7, 266–273.

Perry, P.D. and Hedges-Duroy, T.L. (2004). Adolescent and young adult heroin and non heroin users: a quantitative and qualitative study of experiences in a therapeutic community. *Journal of Psychoactive Drugs*, 36, 75.

Phillips, G.T., Gossop, M. and Bradley, B. (1986). The influence of psychological factors on the opiate withdrawal syndrome. *British Journal of Psychiatry*, 149, 235–238.

Plant, S. (1999). *Writing on drugs*. London: Faber and Faber.

Pohl, J., Frohnau, G., Kerner, W. and FehmWolfsdorf, G. (1997). Symptom awareness is affected by the subjects' expectations during insulin-induced hypoglycaemia. *Diabetes Care*, 20, 796–802.

Pollo, A., Amanzio, M., Arslanian, A., Casadio, C., Maggi, G. and Benedetti, F. (2001). Response expectancies in placebo analgesia and their clinical relevance. *Pain*, 93, 77–84.

Powell, J., Bradley, B. and Gray, J. (1992). Classical conditioning and cognitive determinants of subjective craving for opiates: an investigation of their relative contributions. *British Journal of Addiction*, **87**, 1133–1144.

Powell, J., Dawe, S., Richards, D., Gossop, M., Marks, I., Strang, J and Gray, J. (1993). Can opiate addicts tell us about their relapse risk? Subjective predictors of clinical prognosis. *Addictive Behaviors*, **18**, 473–490.

Reilly, P.M., Sees, K.L., Shopshire, M.S., Hall, S.M., Delucchi, K.L., Tusel, D.J., Banys, P., Clark, H.W. and Piotrowski, N.A. (1995). Self-efficacy and illicit opioid use in a 180-day methadone detoxification treatment. *Journal of Consulting and Clinical Psychology*, **63**, 158–162.

Riters L.V., Schroeder, M.B., Auger, C.J., Eens, M., Pinxten, R. and Ball G.F. (2005). Evidence for opioid involvement in the regulation of song production in male European starlings (*Sturnus vulgaris*). *Behavioral Neuroscience*, **119**, 242–255.

Robins, L.N. (1993). Vietnam Veterans' rapid recovery from heroin addiction—a fluke or normal expectation. *Addiction*, **88**, 1041–1054.

Ross, J.R., Rutter, D., Welsh, K., Joel, S.P., Goller, K., Wells, A.U., Du Bois, R. and Riley, J. (2005). Clinical response to morphine in cancer patients and genetic variation in candidate genes. *Pharmacogenomics Journal*, **5**, 324–336.

Rotter, J. B. (1954). *Social learning and clinical psychology*. New York, NY: Prentice-Hall.

Rousanville, B., Weissman, M., Crits-Christoph, K., Wilber, C. and Kleber, H. (1982). Diagnosis and symptoms of depression in opiate addicts. *Archives of General Psychiatry*, **39**, 151–166.

Safer, D. (1987). Substance abuse by young adult chronic patients. *Hospital and Community Psychiatry*, **38**, 511–514.

Salah, E., Gaily, E.S. and Bashir, T.Z. (2004). High-risk relapse situations and self-efficacy: comparison between alcoholics and heroin addicts. *Addictive Behaviors*, **29**, 753–758.

Saunders, B., Wilkinson, C. and Phillips, M. (1995). The impact of a brief motivational intervention with opiate users attending a methadone programme. *Addiction*, **90**, 415–424.

Saxon, A.J., Davis, T.M., Sloan, K.L., McKnight, K.M., McFall, M.E. and Kivlahan, D.R. (2001). Trauma, symptoms of posttraumatic stress disorder, and associated problems among incarcerated veterans. *Psychiatric Services*, **52**, 959–964.

Schafer, J. and Brown, S.A. (1991). Marijuana and cocaine effect expectancies and drug use patterns. *Journal of Consulting and Clinical Psychology*, **59**, 558–565.

Schieffer, B.M., Pham, Q., Labus, J., Baria, A., Van Vort, W., David, P., David, F. and Naliboff, B.D. (2005). Pain medication beliefs and medication misuse in chronic pain. *Journal of Pain*, **6**, 620–629.

Sell, L.A., Morris, J.S., Bearn, J., Frackowiak, R.S.J., Friston, K.J. and Dolan, R.J. (2000). Neural responses associated with cue evoked emotional states and heroin in opiate addicts. *Drug and Alcohol Dependence*, **60**, 207–216.

Shamoradgoli N.M., Ohadi, M., Joghataie, M.T., Valaie, F., Riazalhosseini, Y., Mostavi, H., Mogammadbeigi, F. and Najmabadi, H. (2005). Association between the DRD2 A1 allele and opium addiction in the Iranian population. *American Journal of Psychiatric Genetics*, **135**, 39–41.

Sharma, H. (1996). Socio-cultural perspective of substance use in India. *Substance Use and Misuse*, **31**, 1689–1714.

Sherrilyn, M., Sklar, H.M.A. and Turner, N.E. (1997). Development and validation of the drug-taking confidence questionnaire: a measure of coping self-efficacy. *Addictive Behaviors*, **22**, 655–670.

Sherrilyn, M., Sklar, H.M.A. and Turner, N.E. (1999). A brief measure for the assessment of coping self-efficacy among alcohol and other drug users. *Addiction*, **94**, 723–729.

Siegel, S. (1982a). Heroin 'overdose' death: contribution of drug-associated environmental cues. *Science*, **216**, 436–437.

Siegel, S. (1982b). Opioid expectation modifies opioid effects. *Federation Proceedings*, **41**, 2339–2343.

Siegel, S. (1986) Alcohol and opiate dependence: re-evaluation of the Victorian perspective. In H. Cappell, F. Glaser, Y. Israel, H. Kalant, W. Schmidt, E. Sellers and E. Smart (ed.), *Research advances in alcohol and drug problems*, Vol. 9. London: John Wiley and Sons, pp. 279–314.

Siegel, S. and Kim, J.A. (2000). Absence of cross tolerance and the situational specificity of tolerance. *Palliative Medicine*, **14**, 75–77.

Sklar, S.M. and Turner, N.E. (1999). A brief measure for the assessment of coping self-efficacy among alcohol and other drug users. *Addiction*, **94**, 723–729.

Sklar, S.M., Annis, H.M. and Turner, N.E. (1997). Development and validation of the Drug-taking Confidence Questionnaire: a measure of coping self-efficacy. *Addictive Behaviors*, **22**, 655–670.

Smith, G.T., Goldman, M.S., Greenbaum, P.E. and Christiansen, B.A. (1995). Expectancy for social facilitation from drinking: the divergent paths of high-expectancy and low-expectancy adolescents. *Journal of Abnormal Psychology*, **104**, 32–40.

Stacy, A.W. (1997). Memory activation and expectancy as prospective predictors of alcohol and marijuana use. *Journal of Abnormal Psychology*, **106**, 61–73.

Stein, M.D. and Anderson, B.J. (2003). Nicotine and drug interaction expectancies among methadone maintained cigarette smokers. *Journal of Substance Abuse Treatment*, **24**, 357–361.

Stein, M.D., Solomon, D.A., Herman, D.S., Anthony, J.L., Ramsey, S.E., Anderson, B.J. and Miller, I.W. (2004). Pharmacotherapy plus psychotherapy for treatment of depression in active injection drug users. *Archives of General Psychiatry*, **61**, 152–159.

Sung, H., Richter, L., Vaughan, V., Johnson, P.B. and Thom, M.S. (2005). Nonmedical use of prescription opioids among teenagers in the United States: trends and correlates. *Journal of Adolescent Health*, **37**, 44–51.

Teesson, M., Havard, A., Fairbairn, S., Ross, J., Lynskey, M. and Darke, S. (2005). Depression among entrants to treatment for heroin dependence in the Australian Treatment Outcome Study (ATOS): prevalence, correlates and treatment seeking. *Drug and Alcohol Dependence*, **78**, 309–315.

Wager, T.D. (2005). The neural bases of placebo effects in pain. *Current Directions in Psychological Science*, **14**, 175–179.

Wager, T.D., Rilling, J.K., Smith, E.E., Sokolik, A., Casey, K.L., Davidson, R.J., Kosslyn, S.M., Rose, R.M. and Cohen, J.D. (2004). Placebo-induced changes in fMRI in the anticipation and experience of pain. *Science*, **303**, 1162.

Waldhoer, M., Bartlett, S.E. and Whistler, J.L. (2004). Opioid receptors. *Annual Review of Biochemistry*, **73**, 953–990.

Walmsley, C.J. (2004). Factors associated with opiate dependence: an interaction of cognitive, genetic and psychosocial influences on acquisition and outcome. Unpublished PhD thesis. The University of Queensland, Australia.

Walter, D., Nagoshi, C., Muntaner, C. and Haertzen, C.A. (1990). The prediction of drug-dependence from expectancy for hostility while intoxicated. *International Journal of the Addictions*, **25**, 1151–1168.

Westermeyer, J. (1978). Social events and narcotic addiction: the influence of war and law on opium use in Laos. *Addictive Behaviors*, **3**, 57–61.

Winfree, L.T., Mays, G.L., Crowley, J.and Peat, B. (1994). Drug history and prisonization: towards understanding variations in inmate adaptations. *International Journal of Offender Therapy and Comparative Criminology*, **38**, 281–296.

Young, R. McD. (1994). Expectancies and drinking behaviour: the measurement of alcohol expectancies and drinking refusal self-efficacy. Unpublished PhD thesis. The University of Queensland, Australia.

Young, R.M.D., and Oci, T.P. (1996). Manual of the Drinking Expectancy Profile. Brisbane: Behaviour Research and Therapy Centre.

Young, R. McD. and Knight, R.G. (1989). The Drinking Expectancy Questionnaire: a revised measure of alcohol related beliefs. *Journal of Psychopathology and Behavioral Assessment*, **112**, 99–112.

Young, R. McD. and Oei, T.P. (2000). The predictive utility of drinking refusal self-efficacy and alcohol expectancy: a diary-based study of tension reduction. *Addictive Behaviors*, **25**, 415–421.

Young, R. McD., Oei, T.P. and Knight, R.G. (1990). The tension reduction hypothesis revisited: an alcohol expectancy perspective. *British Journal of Addiction*, **85**, 31–40.

Young, R. McD., Oei, T.P. and Crook, G.M. (1991a). Development of a drinking self-efficacy questionnaire. *Journal of Psychopathology and Behavioral Assessment*, **13**, 1–15.

Young, R. McD., Oei, T.P.S. and Crook, G.M. (1991b). Differences in the perception of alcoholic versus non-alcoholic beverage advertisements. *Psychologia*, **34**, 241–247.

Young, R. McD., Lawford, B.I.., Walmsley, C.J. and McFarland, K. (1999). Heroin expectancy profile. Unpublished questionnaire. The University of Queensland.

Young, R. McD., Connor, J.C. and Lawford, B.R. (2003). Cannabis expectancy profile. Unpublished questionnaire. Queensland University of Technology.

Young, R. McD., Lawford, B.R., Nutting, A. and Noble, E.P. (2004). Advances in molecular genetics and the prevention and treatment of substance misuse: implications of association studies of the A1 allele of the D2 dopamine receptor gene. *Addictive Behaviors*, **29**, 1275–1294.

Young, R. McD., Saunders, J.B., Hulse, G., McLean, S., Martin, J.and Robinson, G. (2002). Opioids. In G. Hulse, J.White and G. Cape (ed.), *Management of alcohol and drug problems*. Melbourne: Oxford University Press, pp. 79–99.

Zinberg, N., Jacobson, R. and Harding, W. (1975). Social sanctions and rituals as a basis for drug abuse prevention. *American Journal of Drug and Alcohol Abuse*, **2**, 165–182.

Zubieta, J.K., Bueller, J.A., Jackson, L.R., Scott, D.J., Xu, Y., Koeppe, R.A., Nichols, T.E. and Stohler, C.S. (2005). Placebo effects mediated by endogenous opioid activity on μ-opioid receptors. *Journal of Neuroscience*, **25**, 7754–7762.

8

Neurocircuitry of attentional processes in addictive behaviours

Marcus R. Munafò and Brian Hitsman

8.1 Introduction

There is increasing consensus regarding the core neurobehavioural mechanisms underlying drug craving, in particular implicating dopaminergic neurotransmission in mediating drug-taking behaviour, but also suggesting a role for other neurotransmitter pathways such as the serotonergic pathway. Historically, this evidence has relied heavily on animal models, but recently neuroimaging technologies, molecular genetic techniques, and safe and effective neurobiological challenges, such as tyrosine/phenylalanine and tryptophan depletion, have offered the potential to investigate these mechanisms in human participants. In this chapter, we briefly review the utility of these methodologies, and in particular their application in combination with sophisticated behavioural and cognitive measures of drug-taking behaviour and correlates such as craving.

Given the long-recognized role of dopamine in brain reward function, researchers have traditionally focused on the role of dopaminergic neurotransmission in addiction-related processes. The major concentration of dopamine-containing neurons is in the midbrain (substantia nigra, ventral tegmental regions), with primary pathways being the nigrostriatal, mesocorticolimbic (reward circuit) and tuberoinfundibular pathways. A large number of animal studies have demonstrated that acute administration of nicotine causes dopamine release, especially in the ventral tegmental and nucleus accumbens areas of the basal ganglia (George *et al.* 1998, 2000; Balfour *et al.* 2000; Koob and LeMoal 2001, Pontieri *et al.* 1996). While well documented among animal studies, there have been fewer studies of humans. Recent advances in neuroimaging, however, have enabled researchers to focus attention on the effect of nicotine and smoking status in humans. Using positron emission tomography (PET), Salokangas and colleagues observed greater presynaptic dopamine activity in the basal ganglia of male smokers than in male non-smokers (Salokangas *et al.* 2002).

In a study using functional magnetic resonance imaging (fMRI) to explore the brain regions modulated by smoking cues, Due and colleagues found significant activations throughout the mesocorticolimbic reward circuit, including the ventral tegmental, right posterior amygdala, posterior hippocampus and medial thalamus regions (Due *et al.* 2002).

In addition to the importance of dopamine neurotransmission in addictive behaviour, serotonin plays a role in the physiology of a range of neuropsychiatric disorders and normal behaviours, including addictive behaviours, and serotonergic agents are of central importance in neuropharmacology (Veenstra-VanderWeele *et al.* 2000). Serotonin is hypothesized to be involved in reward-related behaviour and cigarette smoking-related processes (Lee and Kornetsky 1998; Harrison *et al.* 2001; Olausson *et al.* 2002; Rogers *et al.* 2003). Most serotonergic neurons originate in the upper brainstem region, with the greatest concentration occurring in the raphe nucleus, and project throughout much of the brain. Specific sites of innervation include the limbic system (hippocampus and amygdala), basal ganglia (nucleus accumbens) and cerebral cortex (Cooper *et al.* 2003). Evidence of serotonergic involvement in nicotine addiction has emerged in the last decade from smoking cessation treatment studies of agents that enhance serotonergic neurotransmission. L-Tryptophan (Bowen *et al.* 1991), dexfenfluramine (Spring *et al.* 1991), paroxetine (Killen *et al.* 2001) and fluoxetine (Niaura *et al.* 2002) all have demonstrated modest, short-term efficacy in enhancing abstinence. Both sertraline (Covey *et al.* 2002) and paroxetine (Killen *et al.* 2001), which are selective serotonin re-uptake inhibitors, reduce craving after quitting smoking, though their ability to enhance even short-term cessation has been limited (Hitsman *et al.* 1999). Manipulations of serotonin function in animals influence the administration of nicotine, alcohol and cocaine in a manner consistent with these findings from the smoking treatment literature (Ciccocioppo 1999; Olausson *et al.* 2002).

Recent technological and methodological advances have, therefore, enabled the more detailed dissection of the neurobiological mechanisms which subserve components of nicotine addiction and cigarette smoking behaviour. The majority of research has focused on the dopamine pathway, although, as described above, the serotonin pathway is also of likely importance. One limitation to the study of nicotine addiction is that this is a complex construct, so that considerable research effort has been invested in the investigation of implicit measures of nicotine addiction and smoking behaviour which index various facets of these constructs. Some of these are described in detail elsewhere in this monograph, and are also discussed below.

The hypothetical advantage of multiple, implicit measures is that they do not suffer the potential limitations of self-report measures, while also offering the capacity to dissect the construct under investigation and index various facets independently.

Sections 8.2 and 8.3 briefly review implicit measures of nicotine addiction and smoking behaviour, with Section 8.2 focusing on cognitive measures, and in particular measures of attentional bias, while Section 8.3 reviews evidence from neuroimaging studies, including fMRI, PET and single photon emission computed tomography (SPECT). Sections 8.4 and 8.5 review the utility of these implicit measures in studies which are informative with respect to the neurobiological mechanisms underlying nicotine addiction and cigarette smoking, with Section 8.4 focusing on molecular genetic techniques, and Section 8.5 focusing on neurobiological challenge techniques.

8.2 **Attentional bias**

Selective processing bias refers to the tendency for information processing resources to be allocated disproportionately towards certain categories of stimuli. An example of this is the 'cocktail party effect', where an individual will attend to a single person's speech against a background of competing noise, but retain the ability to switch attention to salient information in that background, such as one's own name (Munafò *et al.* 2003*a*). Various methodologies exist for investigating processing bias in experimental contexts, the two most common of which are the modified Stroop and the attentional probe tasks.

The modified Stroop task requires participants to engage in a colour-naming task, while lexical stimuli are simultaneously presented. These words may be coloured, so that the task is to name the colour of the word, or may be superimposed on a coloured patch that is to be named. The words may belong to a neutral or relevant category, and evidence for a processing bias towards the relevant category consists of a relative slowing of colour-naming latencies in conditions when a relevant stimulus is present relative to when a neutral stimulus is present. This is predicated on the assumption that if a processing bias exists, this will be reflected in a decrement in performance on a primary task when relevant distractor stimuli are present, as these stimuli will 'capture' attention and thereby limit the information-processing resources available to the primary task.

The attentional probe task requires participants to engage in a reaction time task, with either lexical or pictorial stimuli being presented prior to the appearance of a visual probe to which participants are required to respond.

The stimuli may belong to a neutral or relevant category, and appear as pairs of relevant and neutral stimuli for a specified period, after which these stimuli are removed and replaced with a visual probe in the previous location of either the relevant or the neutral stimulus. Evidence for a processing bias towards the relevant category consists of a relative speeding of visual probe response latencies in conditions when the probe appears in the previous location of a relevant stimulus relative to when the probe appears in the previous location of a neutral stimulus.

The prediction that use of addictive substances may be associated with processing biases towards stimuli associated with those substances arises from the positive incentive model of addiction (Stewart *et al.* 1984) and, more recently, Robinson and Berridge's incentive sensitization model of addiction (Robinson and Berridge 1993, 2000). The Tiffany (1990) model of addiction also posits a role for selective processing of substance-associated stimuli. The incentive sensitization model extends the positive incentive model and suggests that the repeated use of substances that enhance meso-limbic dopamine neurotransmission results in permanent or semipermanent neuroadaptations in these neural mechanisms, which assign incentive salience to stimuli. This leads to stimuli associated with these substances being assigned high levels of incentive salience, which corresponds to the subjective 'wanting' of these substances and related stimuli. One hypothesized conse-quence of this is a processing bias towards these stimuli.

A number of studies have employed the modified Stroop and attentional probe paradigms to investigate selective processing of smoking-related cues in smokers compared with non-smokers (Johnsen *et al.* 1997; Ehrman *et al.* 2002; Bradley *et al.* 2003, 2004; Mogg *et al.* 2003; Munafò *et al.* 2003*a*), in heavy smokers compared with light smokers (Hogarth *et al.* 2003; Mogg *et al.* 2005), and in abstinent smokers compared with non-abstinent smokers (Gross *et al.* 1993; Waters and Feyerabend 2000; Mogg and Bradley 2002; Munafò *et al.* 2003*a*; Field *et al.* 2004). Measures of attentional bias have also been used to predict smoking cessation (Waters *et al.* 2003*a,b*), and investigate the impact of priming with alcohol (Field *et al.* 2005) and the perceived availability of cigarettes (Wertz and Sayette 2001). These studies are summarized in Table 8.1.

While there is growing evidence for some degree of attentional bias for smoking-related cues in smokers compared with non-smokers, there is still uncertainty regarding whether acute abstinence in smokers is associated with increased attentional bias, and whether light and heavy smokers differ in the degree of attentional bias which they demonstrate. These issues are described

Table 8.1 Studies of attentional bias to smoking-related cues among cigarette smokers

Study	Country	Measure	Participants	Results
Gross et al. (1993)	USA	Modified Stroop (card) with unmasked, blocked presentation	10 abstinent smokers and 10 non-abstinent smokers	Abstinent smokers demonstrated greater RT bias compared with non-abstinent smokers
Johnsen et al. (1997)	Norway	Modified Stroop (computer) with unmasked, blocked presentation	11 smokers, 11 ex-smokers and 11 non-smokers	Smokers demonstrated greater RT bias compared with ex-smokers
Waters and Feyerabend (2000)	UK	Modified Stroop (computer) with unmasked, blocked and unblocked presentation	48 abstinent smokers and 48 non-abstinent smokers	Abstinent smokers demonstrated greater RT bias compared with non-abstinent smokers in the blocked condition
Wertz and Sayette (2001)	USA	Modified Stroop (computer) with unmasked, unblocked presentation	92 abstinent smokers	Smokers told they would or would not be able to smoke demonstrated RT bias, but not those told they might be able to smoke
Ehrman et al. (2002)	USA	Visual probe (pictorial) with unmasked presentation	7 smokers and 23 non-smokers in Study 1; 67 smokers, 16 ex-smokers and 25 non-smokers in Study 2	Smokers demonstrated greater RT bias than non-smokers (Studies 1 and 2); ex-smokers demonstrated intermediate level bias (Study 2)
Mogg and Bradley (2002)	UK	Modified Stroop (computer) with masked and unmasked, blocked presentation; visual probe (pictorial)	27 smokers tested when abstinent and non-abstinent	No effect of abstinence manipulation; evidence of RT bias on both unmasked modified Stroop and visual probe tasks, but not masked modified Stroop task

(Continued)

Table 8.1 (*Continued*)

Study	Country	Measure	Participants	Results
Bradley *et al.* (2003)	UK	Visual probe (pictorial) with unmasked presentation (500 ms in Studies 1 and 2; and 2000 ms in Study 2)	20 smokers and 10 non-smokers in Study 1; 25 smokers and 20 non-smokers in Study 2	Smokers demonstrated greater 2000 ms presentation RT bias than non-smokers (Study 2); no evidence of 500 ms presentation RT bias (Studies 1 and 2)
Hogarth *et al.* (2003)	UK	Visual probe (pictorial) with unmasked presentation	10 smokers and 10 non-smokers in Study 1; 36 smokers and 24 non-smokers in Study 2	Light smokers demonstrated greater RT bias than heavy smokers and non-smokers, who did not differ (Studies 1 and 2)
Mogg *et al.* (2003)	UK	Visual probe (pictorial) with unmasked presentation and eye movement monitoring	20 smokers and 25 non-smokers	Smokers demonstrated greater duration of gaze and RT bias than non-smokers; no evidence of initial orienting bias
Munafò *et al.* (2003a)	UK	Modified Stroop (computer) with masked and unmasked, blocked presentation	43 smokers tested when abstinent and non-abstinent, 22 ex-smokers and 30 non-smokers	No effect of abstinence on RT bias; smokers demonstrated greater bias than ex-smokers and non-smokers, who did not differ
Waters *et al.* (2003a)	USA	Modified Stroop (computer) with unmasked, blocked presentation	81 smokers on nicotine replacement patch and 41 smokers on placebo patch	RT bias predicted smoking cessation success in participants on placebo patch only
Waters *et al.* (2003b)	USA	Modified Stroop (computer) with unmasked, blocked presentation	81 smokers on nicotine replacement patch and 41 smokers on placebo patch	RT bias on first half of the task predicted pre-task craving, but did not predict smoking cessation success

Bradley et al. (2004)	UK	Visual probe (pictorial) with masked and unmasked (200 and 2000 ms) presentation	20 smokers and 20 non-smokers	Smokers demonstrated greater RT bias than non-smokers with both 200 and 2000 ms presentation in the unmasked condition only
Field et al. (2004)	UK	Visual probe (pictorial) with unmasked presentation and eye movement monitoring	23 smokers tested when abstinent and non-abstinent	Effect of abstinence on duration of gaze bias; no effect of abstinence on either initial orienting or RT bias
Field et al. (2005)	UK	Visual probe (pictorial) with unmasked presentation and eye movement monitoring	19 abstinent smokers tested when given alcoholic drink and placebo drink	Effect of alcohol on duration of gaze and RT bias; RT bias evident in alcohol condition only; no evidence of initial orienting bias
Mogg et al. (2005)	UK	Visual probe (pictorial) with unmasked presentation and eye movement monitoring	41 smokers	Low dependence smokers demonstrated duration of gaze bias, while moderate dependence smokers did not; no evidence of initial orienting or RT bias

RT = reaction time.

in more detail elsewhere (see Chapter 3). What is notable for the purposes of the present chapter is that in a number of cases there is at best only a weak correlation between measures of attentional bias and subjective reports of craving. This suggests, as discussed above, that cognitive measures may index facets of nicotine addiction and smoking behaviour not necessarily entirely captured by self-report measures. In addition, the use of eye-tracking technology further allows the component processes (e.g. initial orienting, failure to disengage, etc.) involved in attentional bias to be dissected with considerable sophistication. The likelihood is that some of these facets of nicotine addiction and cigarette smoking, and their component processes, will be underpinned by somewhat distinct neurobiological mechanisms.

8.3 **Neuroimaging**

Functional brain imaging techniques, using fMRI, PET and SPECT, have begun to elucidate the neural correlates of nicotine administration and tobacco consumption. fMRI is used primarily to detect changes in brain blood flow in response to stimuli presented during scanning, while PET and SPECT use radiotracers to measure blood flow, metabolism or molecule densities (e.g. receptors). One common finding is that administration of nicotine, either directly or via cigarette smoking, results in decreased global brain activity (Yamamoto *et al.* 2003) and decreased cerebral blood flow (Kubota *et al.* 1987), although this latter finding has not been universally replicated (Terborg *et al.* 2002). With respect to regional activation, nicotine administration is typically associated with increased activation in the prefrontal cortex (Domino *et al.* 2000; Rose *et al.* 2003), thalamus (Stein *et al.* 1998; Domino *et al.* 2000) and visual system (Domino *et al.* 2000).

Functional neuroimaging has demonstrated that environmental cues associated with drug use activate an integrated network of brain regions involved in the motivational and appetitive processes of addiction to drugs of abuse (Breiter and Rosen 1999; Koob and Le Moal 2001; Volkow *et al.* 2003). To date, however, relatively few studies have investigated the neural correlates of these behavioural effects with respect to nicotine addiction. Studies using PET (Brody *et al.* 2002, 2004) and fMRI (Due *et al.* 2002) have demonstrated that smoking-related cues activate regions associated with dopamine-dependent incentive sensitization processes in multiple cortical and subcortical limbic regions.

Using fMRI, we recently demonstrated a difference between smokers and non-smokers in ventral striatum activation to smoking-related cues compared with neutral cues (see Fig. 8.1), which were presented while participants were

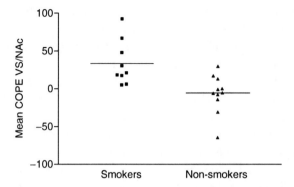

Fig. 8.1 Differences between smokers and non-smokers in activation to smoking versus neutral cues in ventral striatum. Scatter plot comparing smokers and non-smokers in contrast of parameter estimates (COPE) for smoking versus neutral pictorial cues. Region of interest analysis performed was within a bilateral mask of ventral striatum/ nucleus accumbens (VS/NAc).

engaged in the ostensibly neutral task of identifying the sex of target cues of males and females (David *et al.* 2005). The observation of activation in ventral striatum, oribitofrontal cortex, anterior cingulate cortex and fusiform gyrus in addicted smokers presented with smoking-related cues (versus neutral cues) is consistent with the findings of other studies examining drug-related cue reactivity. As in our study, Due and colleagues, using a similar design, observed activation in prefrontal gyrus and fusiform gyrus and a statistical trend toward activation in anterior cingulate cortex (Due *et al.* 2002). Despite differences in paradigm design, the convergence of findings amongst smokers in both studies reinforces the notion posed by Due and colleagues that mesocorticolimbic and visuospatial–attention circuits may work in concert to increase attention to stimuli of heightened salience, such as the sight of a burning cigarette, to an addicted smoker. The observation of ventral striatum activation to smoking-related cues is also consistent with the findings of Heinz *et al.* (2004) who demonstrated activation in ventral striatum (inclusive of nucleus accumbens and ventral caudate) to alcohol-related picture cues.

Another recent study has reported evidence for a difference in ventral anterior cingulate gyrus activation to smoking-related cues compared with neutral cues using fMRI (McClernon *et al.* 2005), which correlated with self-reported cigarette craving. However, this study failed to demonstrate a difference in activation between abstinent and non-abstinent smokers, as would be predicted by incentive sensitization models of dependence.

In our original study (David *et al.* 2005), we observed a main effect of smoking status on global reaction times, which is likely to be the result of nicotine deprivation in abstinent smokers (Trimmel and Wittberger 2004); we did not observe any differential effect of cue type on reaction time in smokers compared with non-smokers. In a more recent study, we investigated the effects of abstinence, using a within-subjects design where smokers were tested on separate days when they had been smoking normally or when they had been abstinent for 12 h. Using a paradigm similar to that used previously (David *et al.* 2005), we demonstrated greater ventral striatum/nucleus accumbens activation in the smoking condition than in the abstinent condition (unpublished data), although there was some evidence that this effect was lateralized (see Fig. 8.2); however, it would be premature to assert that there is clear lateralization of the effect of abstinence on ventral striatum/nucleus accumbens activation.

In particular, we observed a trend for increased reaction time interference when smoking-related cues were present compared with neutral cues, and a trend for this effect to be greater in the abstinent condition compared with the non-abstinent condition. This suggests that behavioural tasks similar to those

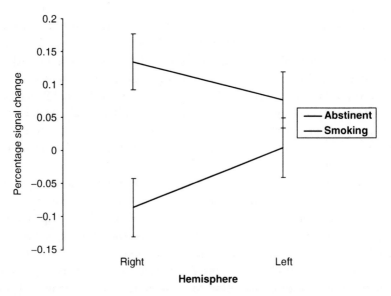

Fig. 8.2 Ventral striatum/nucleus accumbens (VS/NAc) mean contrast of parameter estimates (COPE) by hemisphere and condition. Mean COPE (smoking versus neutral stimulus contrast) within the ventral striatum including the nucleus accumbens for right and left hemispheres and both conditions (smoking and abstinent). Error bars represent the standard error of the relative activation (expressed as percentage signal change) for each hemisphere. Relative activation values were calculated within an anatomically defined VS/NAc mask.

employed in the cognitive psychology domain (described above and elsewhere) may be appropriate tasks to be employed in the context of fMRI studies, and validate the results of such neuroimaging studies. We also observed a correlation between craving and ventral striatum/nucleus accumbens activation in the non-abstinent condition, which may be explained, in part, in the context of prevailing nicotine dependence models (Robinson and Berridge 1993, 2000; Balfour 2004). Chronic nicotine use results in neuroadaptions which alter the smoking experience from hedonic reinforcement to aversive subjective experiences such as drug craving or 'wanting' and motivation to seek the drug at the expense of other activities (Balfour 2004). The measurement of craving has been a contentious issue in the nicotine research community as to whether craving is a subcomponent of nicotine withdrawal or an entirely different construct (Sayette *et al.* 2000), particularly because craving can exist in the absence of other nicotine withdrawal symptoms, and when withdrawal symptoms have been treated with nicotine replacement therapy (Waters *et al.* 2004). Thus, the measurement of craving during nicotine withdrawal and during nicotine satiation may, in fact, be capturing different neuropsychological processes. For this reason, it is interesting that there was a significantly positive correlation between tobacco craving and ventral striatum/nucleus accumbens activation in the non-abstinent condition but not in the abstinent condition. Although the craving scores were lower in the non-abstinent condition than in the abstinent condition, it could be argued that craving in this condition may be a more valid measure as the subjective experience is not confounded by the constellation of affective and somatic symptoms comprising the nicotine withdrawal syndrome such as anxiety, irritability, impatience, depression and sleep disturbances (Sayette *et al.* 2001).

8.4 **Molecular genetics**

Smoking behaviour and nicotine addiction have generated far less behavioural genetics research than other addictive behaviours, such as alcoholism. This is despite evidence from animal studies suggesting that key factors such as the number and distribution of nicotinic receptors and the development of nicotine tolerance are under strong genetic influence (Stitzel *et al.* 2000). The evidence that does exist from twin, adoption and separated twin studies, however, has consistently suggested a strong genetic component to smoking behaviour (Gilbert and Gilbert 1995). Behavioural genetics studies allow the relative contribution of genetic, shared environmental and unique environmental influences to be distinguished, with the heritability coefficient itself reflecting the former. This means, then, that between 28 and 85 per cent of the observed variation in current smoking behaviour in the population from which the data were drawn can be accounted for by inherited factors. Indeed, it has been suggested that the evidence for a genetic influence on smoking behaviour is stronger than the evidence for a genetic influence on alcoholism (Heath *et al.* 1995).

Consistent evidence for the heritability of smoking behaviours has led to molecular genetic studies designed to elucidate the specific genetic factors and biological mechanisms involved in nicotine addiction. Two general scientific approaches to address this question include genetic linkage analysis and genetic association or candidate gene studies. In linkage analysis, genetic variants or markers throughout the genome are tested within families (e.g. sibling pairs) and examined to identify markers that co-segregate with the trait of interest (e.g. nicotine dependence). This approach is hypothesis generating and does not require *a priori* knowledge about the biological pathways involved. In contrast, genetic association studies use a case–control methodology to compare the prevalence of candidate genetic variants in these two unrelated groups (e.g. nicotine-dependent and non-dependent individuals). Although case–control studies have greater statistical power and are less costly than linkage analysis, these designs may not be ideal for dissecting complex polygenic traits. There are likely to be many genes involved in complex behaviours such as cigarette smoking, including those that influence response to the rewarding effects of nicotine (Pomerleau 1995), along with determinants of susceptibility to addiction more generally (Nestler 2000). Interacting effects such as personality and environment must also play an important role (Heath *et al.* 1995). Issues such as population heterogeneity (e.g. ethnicity, gender and age) and possible stratification may also have a substantial impact on the outcome of association studies and may contribute to problems of non-replication, which seem to be endemic (Munafò and Flint 2004).

There have been 37 published candidate gene studies investigating genes involved in the dopamine pathway, of which the majority have investigated the dopamine receptor D2 (*DRD2*) gene. Although a majority of studies of the *DRD2* Taq1A polymorphism have reported an association with smoking behaviour (typically smoking status), a substantial number have failed to show an association. Moreover, the functional significance of the Taq1A polymorphism remains unclear, although some evidence for an association with D2 receptor density has been reported. One study (Yoshida *et al.* 2001) has investigated the functional *DRD2* –141C *Ins/Del* polymorphism and reported a significant association with smoking status. A modest number of studies have investigated other dopamine receptor genes (*DRD1*, *DRD4* and *DRD5*), the dopamine transporter (*DAT*) and genes involved in dopamine synthesis and metabolism, including tyrosine hydroxylase (*TH*), dopamine beta hydroxylase (*DBH*) and catechol-*O*-methyl transferase (*COMT*).

There have been 14 published candidate gene studies investigating genes involved in the serotonin pathway, of which eight have investigated the serotonin transporter (*5-HTT*) gene and three the tryptophan hydroxylase (*TPH*) gene, which is involved in serotonin synthesis. In all but one of the studies of the functional *5-HTTLPR* polymorphism, an association with smoking behaviour has been reported. A further three studies have investigated the monoamine oxidase A (*MAO-A*) gene, which is involved in the metabolism of both dopamine and serotonin. Two of these studies have reported an association with smoking behaviour, including both smoking status and cigarette consumption.

Meta-analysis is a potentially powerful tool for assessing population-wide effects of candidate genes on complex behavioural phenotypes such as smoking, and may also provide evidence for previously unrevealed diversity, for example by revealing heterogeneity in studies of apparently similar populations (Munafò and Flint 2004). Despite the large number of studies published reporting data on the association between specific candidate genes and smoking behaviour, however, a recent meta-analysis (Munafò *et al.* 2004) highlights the lack of depth of research to match the breadth that exists, and concludes that the '... evidence for a contribution of specific genes to smoking behaviour remains modest'. In this analysis, the only candidate genes for which there was evidence of association with smoking behaviour were the *5-HTT* and cytochrome P450 2A6 (*CYP2A6*, involved in nicotine metabolism) genes. In addition, the evidence for substantial between-study heterogeneity reported in this study suggests that extreme care is necessary in the design of case–control genetic association studies.

An alternative to elucidating the role of genetic variation in individual differences in addictive behaviours is to study intermediate smoking-related phenotypes, also known as endophenotypes. These are phenotypes that are biologically more proximal to their genetic antecedents than the complex behavioural phenotypes described above, on the assumption that biological proximity affords a more homogeneous phenotype and a stronger genetic signal. Biobehavioural endophenotypes can also be measured in the context of human behavioural pharmacology studies in order to gauge the potential effects of genotype on medication response before proceeding to a large-scale clinical trial. In the smoking arena, relevant measures include the acoustic startle response (including pre-pulse inhibition and affective modulation of the acoustic startle) (Hutchison *et al.* 2000), measures of the relative reinforcing value of nicotine in a behavioural choice paradigm (Blendy *et al.* 2005), various cue reactivity and cue-induced craving paradigms (Tiffany *et al.* 2000), and measures of attentional bias such as the modified Stroop task (Munafò *et al.* 2003*a*) and the dot-probe task (Waters *et al.* 2003*a*). The list of candidate endophenotypes is growing rapidly, and these may offer powerful measures for pharmacogenetic analysis. This suggests that behavioural tasks similar to those employed in the cognitive psychology domain (described above and elsewhere) may be appropriate tasks to be employed in the context of genetic association studies.

The functional polymorphism in the promoter region of the *5-HTT* gene (*5-HTTLPR*) has been investigated in relation to smoking behaviour, and our recent meta-analysis supported an association with smoking behaviour, in particular smoking cessation (Munafò *et al.* 2004). We recently demonstrated a possible association between *5-HTTLPR* genotype and attentional bias for smoking-related stimuli, suggesting a greater degree of attentional bias among those possessing at least one copy of the short allele of *5-HTTLPR*, although *post hoc* analyses suggested that this was only the case among ex-smokers (Munafò *et al.* 2005). Current models of addiction (Robinson and Berridge 1993, 2000) would suggest that drug-related cues retain their incentive salience for some time following cessation of drug use, resulting in continued selective processing of these cues among former drug users. However, we previously failed to show an attentional bias for smoking-related stimuli among ex-smokers using a modified Stroop task (Munafò *et al.* 2003*a*). Our present data suggest that this may be due to a moderating effect of the *5-HTTLPR* genotype on modified Stroop interference among ex-smokers. These data are presented in Fig. 8.3.

The mechanism by which this moderating effect might operate is an important question that future studies should address. One possibility is that individuals with the short allele of the *5-HTT* gene are more likely to experience the negative affective consequences of smoking cessation, possibly due to the putative association between the *5-HTT* gene and anxiety-related traits (Munafò *et al.* 2003*b*), although why this should be associated with selective processing of smoking-related stimuli is unclear. The lack of correlation between measures of anxiety and attentional bias in our data argues against this interpretation. Alternatively, the *5-HTT* gene may index some other aspect of smoking behaviour that is itself correlated with attentional bias, such as nicotine dependence (Munafò *et al.* 2005). The lack of correlation between measures of dependence and attentional bias in our data argues

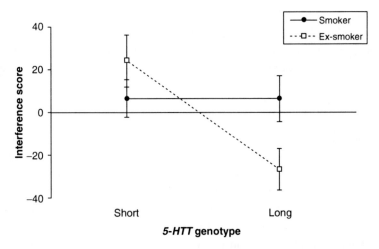

Fig. 8.3 Modified Stroop interference among smokers and ex-smokers, by *5-HTTLPR* genotype. Modified Stroop intereference, as indexed by the relative slowing of colour-naming latencies to smoking-related cues compared with neutral cues, is presented separately for smokers and ex-smokers, grouped by *5-HTTLPR* genotype. *5-HTTLPR* genotype is grouped as short or long.

against this interpretation. Another possibility is that individual differences in central serotonergic neurotransmission result in altered neuromodulation of brain regions related to associative learning, whereby stimuli acquire incentive motivational value (Rogers *et al.* 1999). Rogers and colleagues (1999) report that participants who undergo experimentally induced acute reduction in central serotonin levels (via tryptophan depletion, a neurobiological challenge technique reviewed in Section 5) exhibit impaired ability to learn changed stimulus–reward associations. Therefore, smokers with the short allele of the *5-HTT* gene may fail to learn the change from an association of smoking-related stimuli with a reward to one with no reward.

The influence of genetic variation may therefore operate via catechol-O methyl transferase mechanisms of the processes which underlie nicotine addiction. For example, the enzyme (COMT) is critical in the metabolic degradation of synaptic dopamine and norepinephrine (Lachman *et al.* 1996), which are key neurotransmitters hypothesized to influence human cognitive function, especially visuospatial attention and eye movements. The *COMT* gene contains a functional polymorphism (Val[158]Met) that determines high and low activity of this enzyme. Homozygosity for the low-activity (Met) allele is associated with a 3- to 4-fold reduction of COMT enzyme activity compared with homozygotes for the high-activity (Val) variant, resulting in reduced degradation of synaptic catecholamines in individuals with the Met allele. The *COMT* gene has also been reported to be associated with smoking behaviour (Redden *et al.* 2005).

In a pilot experiment, we have recently found evidence for an association between *COMT* genotype and attentional bias for smoking-related cues, using a modified Stroop task similar to that describe above, among 32 current smokers, $F(1,30) = 5.07$, $P = 0.03$, with greater attentional bias among participants with the Val/Val genotype compared with those with either the Val/Met or Met/Met genotype (unpublished data). This may account for inconsistent reports of association between the *COMT* genotype and various measures of smoking behaviour (Redden *et al.* 2005), given that the *COMT* genotype may influence facets of smoking behaviour and nicotine dependence such as visual capture by and attentional bias for smoking-related cues, which are not typically indexed by questionnaire measures of smoking status or dependence. This is also consistent with evidence that bupropion, which enhances catecholaminergic neurotransmission, attenuates cue-induced craving and corresponding anterior cingulate cortex activation in smokers (Brody *et al.* 2004), although clearly further work employing far larger sample sizes is required.

8.5 Neurobiological challenge

Neurobiological challenge techniques range from drug administration studies that typically involve repeated doses of neurotransmitter receptor-specific antagonists, to drug challenges that acutely block the synthesis of neurotransmitters, to dietary challenges that acutely restrict uptake of the necessary amino acid precursors. In contrast to drug challenges that can be used safely in the context of the human laboratory, dietary challenge paradigms broadly target specific monamine systems (serotonin or dopamine/norepinephrine) and do not have the significant side effect profile often found with acute drug challenges. Drug side effects are a significant concern in studies involving cognitive or behavioural performance measures because these can include sedation and/or fatigue, which themselves directly impair performance and undermine internal study validity. Dietary, or amino acid, challenges, including tryptophan depletion and tyrosine/phenylalanine depletion, were developed, in part, to elucidate the neurobiology of mood disorders and the mechanisms of therapeutic response to antidepressant pharmacotherapy. While acute drug challenges have been widely used to study craving and other addiction-related processes, amino acid challenges have rarely been applied outside the study of major psychopathology (see Booij *et al.* 2003). Amino acid challenges interfere with neurotransmitter function broadly because they reduce the availability of the relevant amino acid precursor presynaptically and, therefore, influence neurotransmission across all receptor types. Behavioural effects, when there are going to be present, reliably occur within 5 h. These well-validated techniques have the potential to tell us much about neurobehavioural mechanisms that underlie the core cognitive processes that drive nicotine addiction.

Drug challenge paradigms have been used to examine the role of dopamine in substance-related disorders and have been applied to nicotine dependence only within the last 10 years. Dopamine agonists reduce smoking while dopamine antagonists increase smoking (Dawe et al. 1995; Caskey et al. 1999, 2002; Jarvik et al. 2000). Caskey et al. (2002) found that the dopamine agonist, bromocriptine, decreased self-reported craving and smoking behaviour, as indicated by alterations in smoking topography, whereas the dopamine antagonist, haloperidol, increased cravings and smoking behaviour. Selegiline, which enhances dopaminergic function via inhibition of monoamine oxidase, has been found to enhance short-term abstinence (Houtsmuller et al. 2002; George et al. 2003) and to decrease self-reported craving in smoking treatment (Houtsmuller et al. 2002). Neurobiological challenges such as tyrosine/phenylalanine depletion that antagonize neurotranmitter function broadly, as opposed to targeting a specific receptor(s), have been used to examine the role of catecholamines (dopamine/norepinephrine) in craving-related processes in only one published study. In that study, Leyton et al. (2005) examined the effect of acute tyrosine/phenylalanine depletion (ATPD) on cue-induced cocaine craving and cocaine self-administration after ingesting a balanced amino acid mixture, an ATDP mixture or an ATPD mixture plus delayed co-administration of L-dopa. The latter condition was used as a second control manipulation in that restoration of dopaminergic neurotransmission by L-dopa after acute depletion was expected to restore the craving response. Both ATPD conditions decreased cocaine craving but had no influence on duration of cocaine self-administration or the euphoric response to consumption. Leyton and colleagues suggested that L-dopa might have failed to reverse the craving response to ATPD because of its recently recognized ability to antagonize catecholaminergic function acutely. In their earlier study of alcohol drinking, Leyton et al. (2000) found that ATPD decreased intake acutely in healthy women (see Table 8.2). In general, these findings are consistent with those from animal and human laboratory studies involving drug challenges that demonstrate an association between catecholaminergic function and broadly defined craving. The extent to which catecholaminergic dysfunction maintains specific cognitive processes

Table 8.2 Studies of monoamine depletion targeting craving and drug use behaviour

Study	Country	Substance	Challenge	Primary measure	Participants	Results
Satel et al. (1995)	USA	Cocaine	ATD	Cue-induced craving (images)	25 cocaine-dependent male inpatients	ATD attenuated craving but only among patients who completed ATD after placebo
Leyton et al. (2005)	Canada	Cocaine	ATPD	Cue-induced craving (cocaine), euphoria, self-administration	8 non-dependent, non-treatment-seeking cocaine users	ATPD attenuated craving, but had no effect on euphoria or duration of self-administration
Leyton et al. (2000)	Canada	Alcohol	ATD, ATPD	Alcohol self-administration	39 healthy female social drinkers	ATPD, but not ATD, decreased alcohol consumption relative to placebo
Petrakis et al. (2001)	USA	Alcohol	ATD	Cue-induced craving (alcohol)	16 alcohol-dependent patients, in treatment, 1–4 months abstinence	No difference in craving between ATD and placebo conditions
Petrakis et al. (2002)	USA	Alcohol	ATD	Alcohol self-administration	12 alcohol-abusing ordependent subjects, not in treatment	No difference in alcohol consumption between ATD and placebo conditions
Pierucci-Lagha et al. (2004)	USA	Alcohol	ATD	Cue-induced craving (alcohol)	14 alcohol-dependent patients with co-morbid MD; responders to 5-HT drug treatment (8 active drug, 4 placebo)	ATD increased craving response (operationalized as urge to drink), as compared with placebo
Perugini et al. (2003)	Canada	Nicotine	ATD	Acute nicotine withdrawal	18 male smokers, 5 h of abstinence	No effect of ATD on withdrawal or its reversal after acute smoking
Hitsman et al. (2005a)	USA	Nicotine	ATD	Cigarette smoking topography	19 smokers (8 healthy, 11 schizophrenic), minimal abstinence	ATD altered smoking topography for both groups (increased puff duration)
Hitsman et al. (2005b)	USA	Nicotine	ATD	Attentional salience of smoking-related cues (words)	34 smokers (19 healthy, 15 with history of MD, minimal abstinence	ATD heightened the salience of smoking words, but not NA or PA words

ATD = acute tryptophan depletion; ATPD = acute tyrosine/phenylalanine depletion; MD = major depression; 5HT = 5-hydroxytryptamine (serotonin); NA = negative affect, hypothesized to underlie nicotine craving has not been evaluated to date. Future research in this area should examine the extent to which the previously observed effect of ATPD on cravings extends to underlying attentional-associated cognitive mechanisms in smokers.

As with the catecholamines, the role of serotonergic neurotransmission can also be studied effectively and safely in the context of the human laboratory. Only a handful of human experimental studies have examined the role of serotonin in craving and drug use. Most of these have manipulated seroto-nergic neurotransmission acutely using the tryptophan depletion technique. Tryptophan depletion has been a valuable technique for studying serotonergic involvement in major psychopathology, such as major depressive disorder (Booij *et al.* 2003; Hood *et al.* 2005). Only recently has it been used to understand the neurobiology of addictive behaviour. As with the amino acids tyrosine and phenylalanine and their synthesis of brain catecholamines, the synthesis of serotonin is dependent on the availability of its amino acid precursor, L-tryptophan. Tryptophan is an essential amino acid present in food. Since brain serotonin levels depend directly on the amount of the exogenous tryptophan available for its synthesis, amounts of tryptophan may be manipulated through diet to alter the corresponding levels of brain serotonin. Depletion of tryptophan occurs in the brain because the ingested amino acids induce protein synthesis, and as plasma tryptophan is incorpor-ated into protein, the level of tryptophan declines rapidly. Administration of a tryptophan-deficient amino acid mixture reduces plasma tryptophan by over 80 per cent within 5–7 h (Delgado *et al.* 1990), resulting in greater than 90 per cent reduction in availability of serotonin in the brain (Nishizawa *et al.* 1997).

The majority of acute tryptophan depletion (ATD) studies in the area of addiction have focused on self-reported craving and consumption among alcohol-dependent patients. Petrakis and colleagues did not find that ATD influenced alcohol cue-induced craving among either abstinent alcoholics (1–3 months of sobriety; Petrakis *et al.* 2001) or non-treatment-seeking alcoholics (Petrakis *et al.* 2002). The latter study also observed no influence on the number of drinks consumed or subjective intoxication. In contrast, Pierucci *et al.* (2004) found that ATD, as compared with placebo challenge, increased alcohol cue-induced craving (operationalized as urge to drink) and depressive symptoms among alcoholics with co-morbid major depression (Pierucci *et al.* 2004). Moreover, *5-HTTLPR* genotype was a significant moderator of depressed mood following ATD, such that subjects homozygous for the L allele showed a greater change in depressed mood than those heterozygous or homozygous for the S allele. It was unclear in the report, however, whether genotype moderated the effect of ATD on craving. The genetic influence of the *5-HTTLPR* polymorphism on the depressive response to ATD is consistent with an earlier study by Moreno *et al.* (2002) of non-medicated patients in clinical remission from major depression.

The Piercucci *et al.* (2004) study raises the important question of the degree to which drug cue-provoked craving is mediated by acute change in negative mood. To date, this important issue has not been directly tested by dietary challenge studies. One hypothesized addiction-related vulnerability mechanism that has been looked at, though, is impulsive behaviour. LeMarquand *et al.* (1999) found that males with a family history of alcoholism exhibited greater behavioural disinhibition following ATD than after placebo. In a second study, ATD impaired performance on a behavioural inhibition task among males with a family history of alcoholism, but in a separate analysis had no overall effect on mood. Surprisingly, it improved inhibition among those with no family history of alcoholism (Crean *et al.* 2002). Taken together, it appears that behavioural disinhibition among individuals who are genetically predisposed to alcoholism is associated with serotonergic neurotransmission. There has been at least one ATD study of cocaine dependence. Consistent with the Leyton *et al.* (2005) ATPD study of cocaine craving, Satel *et al.* (1995) found that ATD decreased cue-induced craving in cocaine-dependent inpatients. One limitation of these results, however, is that the craving response to ATD was found only within the subgroup of subjects who underwent ATD after having completed the placebo challenge first (Satel *et al.* 1995).

Even fewer studies have focused on the problem of nicotine dependence, which is the focus of this chapter. To our knowledge, Spring and colleagues were the first to apply the ATD technique to examine serotonergic involvement in appetitive and affective processes among smokers (Spring *et al.* 1997, 2006). In this initial study, euthymic smokers and non-smokers with and without a history of major depression (MD) underwent ATD and taste-matched placebo challenges. Smoking was standardized across and within test conditions to prevent extraneous affective variability due to nicotine withdrawal. Five hours after consuming the mixture (around the time of maximal tryptophan depletion), subjects underwent a negative mood induction procedure (Litt *et al.* 1990). ATD, as compared with placebo challenge, increased depressive symptoms among subjects with a history of MD, but not among those without MD. The depressive response was greatest among smokers. It is plausible that chronic smoking worsened affective vulnerability through further compromising the serotonergic dysfunction underlying depression (Balfour and Ridely 2000). However, it is also possible that serotonin is directly involved in smoking behaviour independently of its role in mood.

In a second study, Hitsman *et al.* (2005*a*) examined the influence of ATD on smoking behaviour among schizophrenic and non-psychiatric smokers. ATD, but not the taste-matched placebo challenge, influenced smoking topography in both groups (increased puff duration). No effect of ATD on depressive symptoms was observed in a separate analysis of mood, suggesting (albeit indirectly) that the effect on smoking behaviour was not triggered by negative mood. ATD may have influenced smoking behaviour in this study by increasing the urge to smoke.

We evaluated this hypothesis using data collected as a part of Spring and colleagues' initial study. Specifically, we tested whether tryptophan depletion also influenced the attentional salience of smoking-related word cues (Hitsman *et al.* 2005*b*). Attentional salience was operationalized as the tendency of smoking-related word cues to be distracting and to interfere with speed of colour naming on a modified Stroop task (Gross *et al.* 1993). Smokers with ($n = 15$) and without ($n = 19$) a history of MD completed the Stroop 5 h after consumption of the ATD and placebo mixtures. Subjects named as quickly and accurately as possible the colours of words appearing on four different cue cards: smoking-related (e.g. puff, ashtray), negative affect (e.g. low, failure), positive affect (e.g. glad, cheerful) and furniture words (e.g. chair, dresser, rocker). An interference score was calculated for each word cue type that equalled the response time to the motivationally salient word lists minus the response time for the neutral word list. Scores for each cue type are shown in Fig. 8.4 as a function of amino acid condition and history of MD.

As hypothesized, interference was significantly greater following the ATD challenge than after the placebo challenge for the smoking-related cues, but not for negative affect cues or the positive affect cues, although the latter two effects approached significance. These results were unchanged when controlling for change in dysphoric mood from baseline to 5 h during each condition. As compared with smokers without a vulnerability to MD, smokers with a history of MD had nearly twice the delay in response times for smoking-related word cues following ATD (5.8 $\pm$ 1.1 versus 2.8 $\pm$ 1.4), but this difference failed to reach statistical significance. With respect to negative affect word cues, both groups tended to show greater interference times during ATD compared with placebo. The greater interference observed among smokers with a history of MD during both conditions, as compared with those without a MD history, was probably associated with their heightened response to the

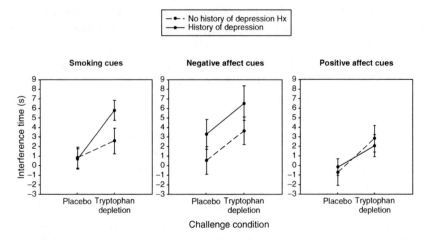

Fig. 8.4 Interference time (in seconds) for smoking-related, negative affect and positive affect word cues among smokers as a function of lifetime history of major depression and amino acid challenge condition. Attentional salience of smoking-related and negative and positive affect word cues, operationalized as the difference between response time to each of the motivationally salient word lists and that to a neutral word list (furniture), for smokers with and without a history of depression after placebo and tryptophan depletion.

negative mood induction procedure completed prior to the Stroop task. The difference in interference between ATD and placebo for the positive affect word cues was even smaller than that observed for the negative affect cues.

The ATD studies reviewed in this section contribute to the basic science and clinical literature linking serotonergic neurotransmission in nicotine dependence, and provide direct evidence indicating that serotonin plays an important role in regulating craving-related processes in smokers. The specific mechanisms underlying these processes are unknown. Nicotine withdrawal probably can be ruled out as a factor, as ATD has not been found to exacerbate withdrawal among smokers (Perugini *et al.* 2003), despite evidence indicating that drug withdrawal is mediated by serotonergic mechanisms (Watkins *et al.* 2000).

One possible mechanism for the association between reduced serotonin availability and heightened attentional bias to smoking cues is that ATD altered reward and incentive motivation processes (Berridge and Robinson 2003). In rats, enhancement of serotonergic function decreases sensitivity to rewarding brain stimulation (Lee and Kornetsky 1998) and reverses the reward deficit associated with acute nicotine withdrawal (Harrison *et al.* 2001). Evidence of serotonergic modulation of reward and incentive motivation processes in humans has been found in an experiment by Rogers *et al.* (2003). In that study, ATD altered behavioural decision making by reducing subjects' ability to discriminate between different magnitudes of reward on a gambling task.

A second possible mechanism, perhaps related to the first, is that ATD interfered with brain inhibitory control systems. There is substantial evidence from basic science and preclinical studies indicating that the serotonin system modulates impulsivity and, more specifically, inhibitory control over cognition and behaviour (Olausson *et al.* 2002). Smokers may have been distracted by smoking-related cues because ATD impaired their ability to suppress conditioned responses.

In sum, the research reviewed in this chapter represents only the initial steps to understand the core neurobehavioural mechanisms underlying drug craving and the mediating attentional and affective processes. Neuroimaging, molecular genetic techniques, and safe and effective neurobiological challenges, such as tyrosine/phenylalanine and tryptophan depletion, have the potential to shed light on this complex problem. Experimental paradigms that combine these challenges with neuroimaging techniques, molecular genetic methods and cognitive performance measures to identify endophenotypes represent the future of human laboratory research on nicotine craving and addiction.

Acknowledgements

This chapter was supported in part by a Mentored Clinical Scientist Development Award (K08 DA017145) to B.H. Any correspondence concerning this chapter should be addressed to Dr Marcus Munafò, Department of Experimental Psychology, University of Bristol, Bristol, BS8 1TN.

References

Balfour, D.J. (2004). The neurobiology of tobacco dependence: a preclinical perspective on the role of the dopamine projections to the nucleus accumbens. *Nicotine and Tobacco Research*, 6, 899–912.

Balfour, D.J.K. and Ridley, D.L. (2000). The effects of nicotine on neural pathways implicated in depression: a factor in nicotine addiction? *Pharmacology, Biochemistry and Behaviour*, 66, 79–85.

Balfour, D.J.K., Wright, A.E., Benwell, M.E.M. and Birrell, C.E. (2000). The putative role of extra-synaptic mesolimbic dopamine in the neurobiology of nicotine dependence. *Behavioural Brain Research*, 113, 73–83.

Berridge, K.C. and Robinson, T.E. (2003). Parsing reward. *Trends in Neuroscience*, 26, 507–513.

Blendy, J.A., Strasser, A., Walters, C.L., Perkins, K.A., Patterson, F., Berkowitz, R. and Lerman, C. (2005). Reduced nicotine reward in obesity: cross-comparison in human and mouse. *Psychopharmacology (Berlin)*, 180, 306–315.

Booji, L., Van Der Does, A.J., and Riedel, W.J. (2003) Monoamine depletion in psychiatric and healthy populations: Review. *Molecular Psychiatry*, 8, 951–973.

Bowen, D.J., Spring, B. and Fox, E. (1991). Tryptophan and high carbohydrate diets as adjuncts to smoking cessation therapy. *Journal of Behavioural Medicine*, 41, 97–110.

Bradley, B.P., Mogg, K., Wright, T. and Field, M. (2003). Attentional bias in drug dependence: vigilance for cigarette-related cues in smokers. *Psychology of Addictive Behaviors*, 17, 66–72.

Bradley, B., Field, M., Mogg, K. and De Houwer, J. (2004). Attentional and evaluative biases for smoking cues in nicotine dependence: component processes of biases in visual orienting. *Behavioral Pharmacology*, 15, 29–36.

Breiter, H.C. and Rosen, B.R. (1999). Functional magnetic resonance imaging of brain reward circuitry in the human. *Annals of the New York Academy of Sciences*, 877, 523–547.

Brody, A.L., Mandelkern, M.A., London, E.D., Childress, A.R., Lee, G.S., Bota, R.G., Ho, M.L., Saxena, S., Baxter, L.R. Jr, Madsen, D. and Jarvik, M.E. (2002). Brain metabolic changes during cigarette craving. *Archives of General Psychiatry*, 59, 1162–1172.

Brody, A.L., Mandelkern, M.A., Lee, G., Smith, E., Sadeghi, M., Saxena, S., Jarvik, M.E. and London, E.D. (2004). Attenuation of cue-induced cigarette craving and anterior cingulate cortex activation in bupropion-treated smokers: a preliminary study. *Psychiatry Research*, 130, 269–281.

Caskey, N.H., Jarvik, M.E. and Wirshing, W.C. (1999). The effects of dopaminergic D2 stimulation and blockade on smoking behaviour. *Experimental and Clinical Psychopharmacology*, 7, 72–78.

Caskey, N.H., Jarvik, M.E., Wirshing, W.C., Madsen, D.C., Iwamoto-Schaap, P.N., Eisenberger, N.I., Huerta, L., Terrace, S.M. and Olmstead, R.E. (2002). Modulating tobacco smoking rates by dopaminergic stimulation and blockade. *Nicotine and Tobacco Research*, 4, 259–266.

Ciccocioppo, R. (1999). The role of serotonin in craving: from basic research to human studies. *Alcohol and Alcoholism*, 34, 244–253.

Cooper, J.R., Bloom, F.E. and Roth, R.H. (2003). *The biochemical basis of neuropharmacology*. Oxford: Oxford University Press.

Covey, L.S., Glassman, A.H., Stetner, F., Rivelli, S. and Stage, K. (2002). A randomized trial of sertraline as a cessation aid for smokers with a history of major depression. *American Journal of Psychiatry*, 159, 1731–1737.

Crean, J., Richards, J.B. and de Wit, H. (2002). Effect of tryptophan depletion on impulsive behaviour in men with and without a family history of alcohol dependence. *Behavioural Brain Research*, 136, 349–357.

David, S.P., Munafò, M.R., Johansen-Berg, H., Smith, S.M., Rogers, R.D., Matthews, P.M. and Walton, R.T. (2005). Ventral striatum/nucleus accumbens activation to smoking-related pictorial cues in smokers and nonsmokers: a functional magnetic resonance imaging study. *Biological Psychiatry*, 58, 488–494.

Dawe, S., Gerada, C., Russell, M.A.H. and Gray, J.A. (1995). Nicotine intake in smokers increases following a single dose of haloperidol. *Psychopharmacology (Berlin)*, 117, 110–115.

Delgado, P.L., Charney, D.S., Price, L.H., Aghajanian, G.K., Landis, H. and Heninger, G.R. (1990). Serotonin function and the mechanism of antidepressant action. *Archives of General Psychiatry*, 47, 411–418.

Domino, E.F., Minoshima, S., Guthrie, S.K., Ohl, L., Ni, L., Koeppe, R.A., Cross, D.J. and Zubieta, J. (2000a). Effects of nicotine on regional cerebral glucose metabolism in awake resting tobacco smokers. *Neuroscience*, **101**, 277–282.

Domino, E.F., Minoshima, S., Guthrie, S., Ohl, L., Ni, L., Koeppe, R.A. and Zubieta, J.K. (2000b). Nicotine effects on regional cerebral blood flow in awake, resting tobacco smokers. *Synapse*, **38**, 313–321.

Due, D.L., Huettel, S.A., Hall, W.G. and Rubin, D.C. (2002). Activation in mesolimbic and visuospatial neural circuits elicited by smoking cues: evidence from functional magnetic resonance imaging. *American Journal of Psychiatry*, **159**, 954–960.

Ehrman, R.N., Robbins, S.J., Bromwell, M.A., Lankford, M.E., Monterosso, J.R. and O'Brien, C.P. (2002). Comparing attentional bias to smoking cues in current smokers, former smokers and non-smokers using a dot-probe task. *Drug and Alcohol Dependence*, **67**, 185–191.

Field, M., Mogg, K. and Bradley, B.P. (2004). Eye movements to smoking-related cues: effects of nicotine deprivation. *Psychopharmacology (Berlin)*, **173**, 116–123.

Field, M., Mogg, K. and Bradley, B.P. (2005). Alcohol increases cognitive biases for smoking cues in smokers. *Psychopharmacology (Berlin)*, **180**, 63–72.

George, T.P., Verrico, C.D. and Roth, R.H. (1998). Effects of repeated nicotine pretreatment on mesoprefrontal dopaminergic and behavioural responses to acute footshock stress. *Brain Research*, **801**, 36–49.

George, T.P., Verrico, C.D., Picciotto, M.R. and Roth, R.H. (2000). Nicotinic modulation of mesoprefrontal dopamine systems: pharmacologic and neuroanatomic characterization. *Journal of Pharmacology and Experimental Therapeutics*, **295**, 58–66.

George, T.P., Vessicchio, J.C., Termine, A., Jatlow, P.I., Kosten, T.R. and O'Malley, S.S. (2003). A preliminary placebo-controlled trial of selegiline hydrochloride for smoking cessation. *Biological Psychiatry*, **53**, 136–143.

Gilbert, D.G. and Gilbert, B.O. (1995). Personality, psychopathology and nicotine response as mediators of the genetics of smoking. *Behavior Genetics*, **25**, 133–147.

Gross, T.M., Jarvik, M.E. and Rosenblatt, M.R. (1993). Nicotine abstinence produces content-specific Stroop interference. *Psychopharmacology (Berlin)*, **110**, 333–336.

Harrison, A.A., Liem, Y.T.B. and Markou, A. (2001). Fluoxetine combined with a serotonin-1a receptor antagonist reversed reward deficits observed during nicotine and amphetamine withdrawal in rats. *Neuropsychopharmacology*, **25**, 55–71.

Heath, A.C., Madden, P.A., Slutske, W.S. and Martin, N.G. (1995). Personality and the inheritance of smoking behavior: a genetic perspective. *Behavior Genetics*, **25**, 103–117.

Heinz, A., Siessmeier, T., Wrase, J., Hermann, D., Klein, S., Grusser, S.M., Flor, H., Braus, D.F., Buchholz, H.G., Grunder, G., Schreckenberger, M., Smolka, M.N., Rosch, F., Mann, K. and Bartenstein, P. (2004). Correlation between dopamine D(2) receptors in the ventral striatum and central processing of alcohol cues and craving. *American Journal of Psychiatry*, **161**, 1783–1789.

Hitsman, B., Pingitore, R., Spring, B., Mahableshwarkar, A., Mizes, J.S., Segraves, K.A., Kristeller, J.L. and Xu, W. (1999). Antidepressant pharmacotherapy helps some cigarette smokers more than others. *Journal of Consulting and Clinical Psychology*, **67**, 547–554.

Hitsman, B., Spring, B., Pingitore, R., McChargue, D.E. and Doran, N. (2005a). Serotonergic involvement in the attentional salience of cigarette cues. Paper presented at the 11th

Annual Meeting of the Society for Research on Nicotine and Tobacco, Prague, Czech Republic, 2005.

Hitsman, B., Spring, B., Wolf, W., Pingitore, R., Crayton, J.W. and Hedeker, D. (2005*b*). Effects of acute tryptophan depletion on negative symptoms and smoking topography in nicotine dependent schizophrenics and non-psychiatric controls. *Neuropsychopharmacology*, 30, 640–648.

Hogarth, L.C., Mogg, K., Bradley, B.P., Duka, T. and Dickinson, A. (2003). Attentional orienting towards smoking-related stimuli. *Behavioral Pharmacology*, 14, 153–160.

Hood, S.D., Bell, C.J. and Nutt, D.J. (2005). Acute tryptophan depletion. Part I: rationale and methodology. *Australian and New Zealand Journal of Psychiatry*, 39, 558–564.

Houtsmuller, E.J., Thornton, J.A. and Stitzer, M.L. (2002). Effects of selegiline (l-deprenyl) during smoking and short-term abstinence. *Psychopharmacology (Berlin)*, 163, 213–220.

Hutchison, K.E., Niaura, R. and Swift, R. (2000). The effects of smoking high nicotine cigarettes on prepulse inhibition, startle latency and subjective responses. *Psychopharmacology (Berlin)*, 150, 244–252.

Jarvik, M.E., Caskey, N.H., Wirshing, W.C., Madsen, D.C., Iwamoto-Schaap, P.N., Elins, J.L., Eisenberger, N.I. and Olmstead, R.E. (2000). Bromocriptine reduces cigarette smoking. *Addiction*, 95, 1173–1183.

Johnsen, B.H., Thayer, J.F., Laberg, J.C. and Asbjornsen, A.E. (1997). Attentional bias in active smokers, abstinent smokers and nonsmokers. *Addictive Behaviors*, 22, 813–817.

Killen, J.D., Fortmann, S.P., Schatzberg, A.F., Hayward, C., Sussman, L., Rothman, M., Strausberg, L. and Varady, A. (2001). Nicotine patch and paroxetine for smoking cessation. *Journal of Consulting and Clinical Psychology*, 68, 883–889.

Koob, G.F. and Le Moal, M. (2001). Drug addiction, dysregulation of reward and allostasis. *Neuropsychopharmacology*, 24, 97–129.

Kubota, K., Yamaguchi, T., Fujiwara, T. and Matsuzawa, T. (1987). Effects of smoking on regional cerebral blood flow in cerebral vascular disease patients and normal subjects. *Tohoku Journal of Experimental Medicine*, 151, 261–268.

Lachman, H.M. (1996). Human catechol-O-methyltransferase pharmacogenetics: description of a functional polymorphism and its potential application to neuropsychiatric disorders. *Pharmacogenetics*, 6, 243–250.

Lee, K. and Kornetsky, C. (1998). Acute and chronic fluoxetine treatment decreases the sensitivity of rats to rewarding brain stimulation. *Pharmacology, Biochemistry and Behaviour*, 60, 539–544.

LeMarquand, D.G., Benkelfat, C., Pihl, R.O., Palmour, R.M. and Young, S.N. (1999). Behavioural disinhibition induced by tryptophan depletion in non-alcoholic young men with multigenerational family histories of paternal alcoholism. *American Journal of Psychiatry*, 156, 1771–1779.

Leyton, M., Young, S.N., Blier, P., Baker, G.B., Pihl, R.O. and Benkelfat, C. (2000). Acute tyrosine depletion and alcohol ingestion in healthy women. *Alcoholism: Clinical and Experimental Research*, 24, 459–64.

Leyton, M., Casey, K.F., Delaney, J.S., Kolivakis, T. and Benkelfat, C. (2005). Cocaine craving, euphoria and self-administration: a preliminary study of the effect of catecholamines precursor depletion. *Behavioural Neuroscience*, 119, 1619–1627.

Litt, M.D., Cooney, N.L., Kadden, R.M. and Gaupp, L. (1990). Reactivity to alcohol cues and induced moods in alcoholics. *Addictive Behaviours*, 15, 137–146.

McClernon, F.J., Hiott, F.B., Huettel, S.A. and Rose, J.E. (2005). Abstinence-induced changes in self-report craving correlate with event-related fMRI responses to smoking cues. *Neuropsychopharmacology*, 30, 1940–1947.

Mogg, K. and Bradley, B.P. (2002). Selective processing of smoking-related cues in smokers: manipulation of deprivation level and comparison of three measures of processing bias. *Journal of Psychopharmacology*, 16, 385–392.

Mogg, K., Bradley, B.P., Field, M. and D.Houwer, J. (2003). Eye movements to smoking-related pictures in smokers: relationship between attentional biases and implicit and explicit measures of stimulus valence. *Addiction*, 98, 825–836.

Mogg, K., Field, M. and Bradley, B.P. (2005). Attentional and approach biases for smoking cues in smokers: an investigation of competing theoretical views of addiction. *Psychopharmacology (Berlin)*, 180, 333–341.

Moreno, F.A., Rowe, D.C., Kaiser, B., Chase, D., Michaels, T., Gelernter, J. and Delgado, P.L. (2002). Association between a serotonin transporter promoter region polymorphism and mood response during tryptophan depletion. *Molecular Psychiatry*, 7, 213–216.

Munafò, M.R. and Flint, J. (2004). Meta-analysis of genetic association studies. *Trends in Genetics*, 20, 439–444.

Munafò, M., Mogg, K., Roberts, S., Bradley, B.P. and Murphy, M. (2003a). Selective processing of smoking-related cues in current smokers, ex-smokers and never-smokers on the modified Stroop task. *Journal of Psychopharmacology*, 17, 310–316.

Munafò, M.R., Clark, T.G., Moore, L.R., Payne, E., Walton, R. and Flint, J. (2003b). Genetic polymorphisms and personality in healthy adults: a systematic review and meta-analysis. *Molecular Psychiatry*, 8, 471–484.

Munafò, M., Clark, T., Johnstone, E., Murphy, M. and Walton, R. (2004). The genetic basis for smoking behavior: a systematic review and meta-analysis. *Nicotine and Tobacco Research*, 6, 583–597.

Munafò, M., Roberts, K., Johnstone, E., Walton, R. and Yudkin, P. (2005). Association of serotonin transporter gene polymorphism with nicotine dependence: no evidence for an interaction with trait neuroticism. *Personality and Individual Differences*, 38, 843–850.

Munafò, M.R., Johnstone, E.C., and Mackintosh, G. (2005) Association of serotonin transporter genotype with selective processing of smoking-related stimuli in current smokers and ex-smokers. *Nicotine and Tobacco Research*, 7, 773–778.

Nestler, E.J. (2000). Genes and addiction. *Nature Genetics*, 26, 277–281.

Nishizawa, S., Benkelfat, C., Young, S.N., Leyton, M., Mzengeza, S., De Montigny, C., Blier, P. and Diksic, M. (1997). Differences between males and females in rates of serotonin synthesis in human brain. *Proceedings of the National Academy of Sciences of the USA*, 94, 5308–5313.

Olausson, P., Engel, J.A. and Soderpalm, B. (2002). Involvement of serotonin in nicotine dependence: processes relevant to positive and negative regulation of drug intake. *Pharmacology, Biochemistry and Behaviour*, 71, 757–771.

Perugini, M., Mahoney, C., Ilivitsky, V., Young, S.N. and Knott, V. (2003). Effects of tryptophan depletion on acute smoking abstinence symptoms and the acute smoking response. *Pharmacology, Biochemistry and Behaviour*, 74, 513–522.

Petrakis, I.L., Trevisan, L., Boutros, N.N., Limoncelli, D., Cooney, N.L. and Krystal, J.H. (2001). Effect of tryptophan depletion on alcohol cue-induced craving in abstinent alcoholic patients. *Alcoholism: Clinical and Experimental Research*, 25, 1151–1155.

Petrakis, I.L., Buonopane, A., O'Malley, S., Cermik, O., Trevisan, L., Boutros, N.N., Limoncelli, D. and Krystal, J.H. (2002). The effect of tryptophan depletion on alcohol self-administration in non-treatment-seeking alcoholic individuals. *Alcoholism: Clinical and Experimental Research*, 26, 969–975.

Pierucci-Lagha, A., Feinn, R., Modesto-Lowe, V., Swift, R., Nellissery, M., Covault, J. and Kranzler, H.R. (2004). Effects of rapid tryptophan depletion on mood and urge to drink in patients with co-morbid major depression and alcohol dependence. *Psychopharmacology (Berlin)*, 171, 340–348.

Pomerleau, O.F. (1995). Individual differences in sensitivity to nicotine: implications for genetic research on nicotine dependence. *Behavior Genetics*, 25, 161–177.

Pontieri, F.E., Tanda, G., Orzi, F. and DiChiara, G. (1996). Effects of nicotine on the nucleus accumbens and similarity to those of addictive drugs. *Nature*, 382, 255–257.

Redden, D.T., Shields, P.G., Epstein, L., Wileyto, E.P., Zakharkin, S.O., Allison, D.B. and Lerman, C. (2005). Catechol-O-methyl-transferase functional polymorphism and nicotine dependence: an evaluation of nonreplicated results. *Cancer Epidemiology, Biomarkers and Prevention*, 14, 1384–1389.

Robinson, T.E. and Berridge, K.C. (1993). The neural basis of drug craving: an incentive-sensitization theory of addiction. *Brain Research Reviews*, 18, 247–291.

Robinson, T.E. and Berridge, K.C. (2000). The psychology and neurobiology of addiction: an incentive-sensitization view. *Addiction*, 95 (Suppl. 2), S91–117.

Rogers, R.D., Blackshaw, A.J., Middleton, H.C., Matthews, K., Hawtin, K., Crowley, C., Hopwood, A., Wallace, C., Deakin, J.F., Sahakian, B.J. and Robbins, T.W. (1999). Tryptophan depletion impairs stimulus–reward learning while methylphenidate disrupts attentional control in healthy young adults: implications for the monoaminergic basis of impulsive behaviour. *Psychopharmacology (Berlin)*, 146, 482–491.

Rogers, R.D., Tunbridge, E.M., Bhagwager, Z., Drevets, W.C., Sahakian, B.J.and Carter, C.S. (2003). Tryptophan depletion alters the decision-making of healthy volunteers through altered processing of reward cues. *Neuropsychopharmacology*, 28, 153–162.

Rose, J. E., Behm, F.M., Westman, E.C., Mathew, R.J., London, E.D., Hawk, T.C., Turkington, T.G. and Coleman, R.E. (2003). PET studies of the influences of nicotine on neural systems in cigarette smokers. *American Journal of Psychiatry*, 160, 323–333.

Salokangas, R.K.R., Vilkman, H., Ilonen, T., Taiminen, T., Bergman, J., Haaparanta, M., Solin, O., Alanen, A., Syvalahti, E. and Hietala, J. (2000). High levels of dopamine activity in the basal ganglia of cigarette smokers. *American Journal of Psychiatry*, 157, 632–634.

Satel, S.L., Krystal, J.H., Delgado, P.L., Kosten, T.R. and Charney, D.S. (1995). Tryptophan depletion and attenuation of cue-induced craving from cocaine. *American Journal of Psychiatry*, 152, 778–783.

Sayette, M.A., Shiffman, S., Tiffany, S.T., Niaura, R.S., Martin, C.S. and Shadel, W.G. (2000). The measurement of drug craving. *Addiction*, 95 (Suppl. 2), S189–S210.

Sayette, M.A., Martin, C.S., Wertz, J.M., Shiffman, S. and Perrott, M.A. (2001). A multidimensional analysis of cue-elicited craving in heavy smokers and tobacco chippers. *Addiction*, 96, 1419–1432.

Spring, B., Wurtman, J., Gleason, R., Wurtman, R. and Kessler, K. (1991). Weight gain and withdrawal symptoms after smoking cessation: a preventive intervention using d-fenfluramine. *Health Psychology*, 10, 216–223.

Spring, B., Pingitore, R., Kessler, K., Hitsman, B., Gunnarsdottir, D., Corsica, J. and Pergadia, M. (1997). Parallel depressive response to serotonergic challenge and nicotine withdrawal. *Annals of Behavioural Medicine*, 19, S036.

Spring, B., Hitsman, B., Pingitore, R., McChargue, D.E., Gunnarsdottir, D., Corsica, J., Pergadia, M., Doran, N., Kessler, K., Crayton, J.W., Baruah, S. and Hedeker, D. (2006). Effect of tryptophan depletion on smokers with and without a history of major depression. *Biological Psychiatry*.

Stein, E.A., Pankiewicz, J., Harsch, H.H., Cho, J.K., Fuller, S.A., Hoffmann, R.G., Hawkins, M., Rao, S.M., Bandettini, P.A. and Bloom, A.S. (1998). Nicotine-induced limbic cortical activation in the human brain: a functional MRI study. *American Journal of Psychiatry*, 155, 1009–1015.

Stewart, J., de Wit, H. and Eikelboom, R. (1984). Role of unconditioned and conditioned drug effects in the self-administration of opiates and stimulants. *Psychological Review*, 91, 251–268.

Stitzel, J.A., Lu, Y., Jimenez, M., Tritto, T. and Collins, A.C. (2000). Genetic and pharmacological strategies identify a behavioral function of neuronal nicotinic receptors. *Behavioural Brain Research*, 113, 57–64.

Terborg, C., Birkner, T., Schack, B. and Witte, O.W. (2002). Acute effects of cigarette smoking on cerebral oxygenation and hemodynamics: a combined study with near-infrared spectroscopy and transcranial Doppler sonography. *Journal of the Neurological Science*, 205, 71–75.

Tiffany, S.T. (1990). A cognitive model of drug urges and drug-use behavior: role of automatic and nonautomatic processes. *Psychological Review*, 97, 147–168.

Tiffany, S.T., Cox, L.S. and Elash, C.A. (2000). Effects of transdermal nicotine patches on abstinence-induced and cue-elicited craving in cigarette smokers. *Journal of Consulting and Clinical Psychology*, 68, 233–240.

Trimmel, M. and Wittberger, S. (2004). Effects of transdermally administered nicotine on aspects of attention, task load and mood in women and men. *Pharmacology Biochemistry and Behavior*, 78, 639–645.

Veenstra-VanderWeele, J., Anderson, G.M. and Cook, E.H., Jr (2000). Pharmacogenetics and the serotonin system: initial studies and future directions. *European Journal of Pharmacology*, 410, 165–181.

Volkow, N.D., Fowler, J.S. and Wang, G.J. (2003). The addicted human brain: insights from imaging studies. *Journal of Clinical Investigation*, 111, 1444–1451.

Waters, A.J. and Feyerabend, C. (2000). Determinants and effects of attentional bias in smokers. *Psychology of Addictive Behaviors*, 14, 111–120.

Waters, A.J., Shiffman, S., Bradley, B.P. and Mogg, K. (2003a). Attentional shifts to smoking cues in smokers. *Addiction*, 98, 1409–1417.

Waters, A.J., Shiffman, S., Sayette, M.A., Paty, J.A., Gwaltney, C.J. and Balabanis, M.H. (2003b). Attentional bias predicts outcome in smoking cessation. *Health Psychology*, 22, 378–387.

Waters, A.J., Shiffman, S., Sayette, M.A., Paty, J.A., Gwaltney, C.J. and Balabanis, M.H. (2004). Cue-provoked craving and nicotine replacement therapy in smoking cessation. *Journal of Consulting and Clinical Psychology*, 72, 1136–1143.

Watkins, S.S., Koob, G.F. and Markou, A. (2000). Neural mechanisms underlying nicotine addiction: acute positive reinforcement and withdrawal. *Nicotine and Tobacco Research*, 2, 19–37.

Wertz, J.M. and Sayette, M.A. (2001). Effects of smoking opportunity on attentional bias in smokers. *Psychology of Addictive Behaviors*, 15, 268–271.

Yamamoto, Y., Nishiyama, Y., Monden, T., Satoh, K. and Ohkawa, M. (2003). A study of the acute effect of smoking on cerebral blood flow using 99mTc-ECD SPET. *European Journal of Nuclear Medicine and Molecular Imaging*, 30, 612–614.

Yoshida, K., Hamajima, N., Kozaki, K., Saito, H., Maeno, K., Sugiura, T., Ookuma, K. and Takahashii, T. (2001). Association between the dopamine D2 receptor A2/A2 genotype and smoking behavior in the Japanese. *Cancer Epidemiology Biomarkers and Prevention*, 10, 403–405.

Clinical relevance of implicit cognition in addiction

Andrew J. Waters and Adam M. Leventhal

9.1 Introduction

Recent research has used paradigms derived from experimental cognitive psychology to assess implicit cognitive processes in the addictions. Much of this research has focused on how these measures help us achieve theoretical insights into the psychological mechanisms underlying addictive behaviours. Here, we focus more on the clinical relevance of the implicit measures. In particular, if these cognitive measures are to prove useful in the clinical domain, they should display the following characteristics: (1) they should be associated with measures of use or dependence; (2) they should be prospectively associated with clinically relevant outcomes, and show incremental prediction over self-report measures; and (3) they should be modifiable by drug, behavioural or cognitive interventions. In addition, the measure will be particularly useful if it can be shown that the effects of treatments on clinical outcome are mediated by effects on implicit cognitive processes. We review the evidence relating to each of these issues, and consider how further work might proceed within the clinical domain.

9.1.1 Cognition and addictive behaviour

Much research has examined the social, economic and environmental variables that are associated with addictive behaviour. Thus, a great deal is known about the risk factors for drug-use initiation (e.g. Hawkins *et al.* 1992), and drug-use cessation (e.g. Jarvis 1997). It is generally assumed that these variables do not directly cause the drug-use behaviour (initiation or cessation). Rather, these variables can be conceptualized as 'distal' predictors of drug use. The effects of the distal variables on drug use are presumably mediated in part by more 'proximal' psychological variables (Stacy 1997; Baron and Kenny 1986; Hine *et al.* 2002). Thus, the distal variables may cause changes in cognitive and affective processes, which, in turn, cause changes in behaviour. If the changes in cognition are well understood and easy to measure, these cognitions might provide an ideal locus for a clinical intervention. In fact, from a clinical perspective, the cognitive processes may potentially be easier to change than the distal variables, which may reflect relatively immutable features of the environment or individual.

9.1.2 Automatic and controlled cognitive processes

As noted elsewhere in this volume, a fundamental distinction has been drawn in cognitive psychology between controlled and automatic psychological processes (e.g. Schneider and Shiffrin 1977). Controlled processes are typically slow, serial, effortful and driven by a conscious appraisal of events. These types of processes may be captured reasonably well by self-report measures. In contrast, automatic processes are fast, parallel, effortless and may not engage conscious awareness. As noted by De Houwer (2006), several of the properties of automatic processes are also properties that are attributed to implicit processes. For example, both implicit and automatic processes are considered: (1) uncontrollable (participants cannot control the outcome of the process); and (2) unconscious (in the sense that participants are unaware of the fact that an effect was due to a particular process, or in the sense that they do not have conscious access to the process that produced the effect). Thus, for the purposes of the rest of this chapter, implicit and automatic processes are considered co-terminous. The same is true for explicit and controlled processes.

9.1.3 Implicit cognition and addiction

Over the past few years, there have been two important developments in the cognition and addiction literature. First, research has suggested that self-report measures of cognitive and affective processes are unlikely to explain all the variance in addictive behaviour. For example, with regard to affective processes, it has generally been assumed that consciously experienced negative affect, as assessed by self-report, has a large role to play in relapses to smoking. Although a sudden rise of negative affect is an important risk factor for some relapse episodes (Shiffman and Waters 2004), many lapses occur when participants report being in a neutral or positive mood (Shiffman et al. 1996). Thus, not all relapses to smoking may be explained by a sudden rise in selfreported negative affect.

Secondly, there has been growing interest in automatic processes in addiction. As noted by DiChiara (2002), 'both cognitive, conscious (explicit/declarative), as well as associative, unconscious (implicit/procedural) mechanisms contribute to purposeful behaviour' (p. 76). Indeed, recent articles have stressed the importance of automatic processes (Tiffany 1990; Robinson and Berridge 1993; Baker *et al.* 2004). These two developments have stimulated interest in alternative measures that may index automatic cognitive, affective and motivational processes.

9.1.4 Implicit cognition in a clinical setting

In a clinical setting, tasks that assess automatic psychological processes (implicit tasks) may be particularly useful for two reasons. First, patients may under-report substance abuse symptoms (e.g. Stein and Graham 1999), and exaggerate treatment progress to appease therapists (e.g. Chisholm *et al.* 1997). Several authors have noted that—due to the stigmatized character of substance use—the validity of explicit measures may be undermined by self-presentational bias (Swanson *et al.* 2001; Huijding *et al.* 2005). For example, users may in fact hold positive attitudes toward the substances they abuse, but are unwilling to reveal them. It is noteworthy that implicit and explicit attitudes are generally not strongly correlated in the addictions literature (e.g. Swanson *et al.* 2001; Wiers *et al.* 2002), suggesting that implicit tasks may provide information that is not captured by self-report. This information may be useful in order to monitor treatment progress.

Secondly, implicit tasks attempt to assess automatic psychological processes that cannot be assessed with self-report measures. As noted earlier, given the recent theoretical interest in implicit processes, this is particularly relevant (e.g. Tiffany 1990; Robinson and Berridge 1993; Stacy 1997; Baker *et al.* 2004).

9.1.5 Interventions targeting implicit and explicit processes

Many interventions directly target explicit processes. For example, in relapse prevention, patients are taught to identify situations that may precipitate relapse. They are also instructed to rehearse coping techniques, such as consciously reflecting on the negative consequences of substance use and the positive effects of abstinence (Marlatt 1979). After relapse prevention strategies are learned, patients are expected intentionally to search long-term memory and retrieve this information when in relapse-provoking environments. In sum, many explicit processes may be engaged to prevent a relapse from occurring.

Thus, in many interventions, such as relapse prevention, talk therapy, counselling, contingency management and psycho-education, the patient is aware that he or she is receiving an intervention. In addition, the patient is also generally aware of how the intervention is meant to work. He or she is encouraged to engage in conscious information processing to create new cognitions (e.g. expectancies) or action plans (e.g. coping strategies) that are intended to assist in behaviour change.

Recently, interventions have been proposed that directly target implicit processes (e.g. MacLeod *et al.* 2002; see Fig. 9.1). As discussed later, interventions such as attentional re-training directly target automatic attentional processes. In these types of intervention, the patient may or may not be aware that he or she is receiving an intervention. They will have little or no understanding of how the intervention is meant to work. New attentional responses or cognitive associations are formed through processes that may occur outside the patient's awareness. We will discuss these interventions in a later section.

9.1.6 Potential clinical significance for research into implicit processes

It is likely that both implicit and explicit processes underlie addictive behaviours (Fig. 9.1). The study of implicit cognition and addiction is important for the following reasons. First, study of implicit cognition may suggest methods to identify individuals at high risk for relapse who might not be so identified using self-report measures. For example, a smoker may exhibit an elevated attentional bias, but may not score highly on standard self-report measures of dependence or withdrawal. Thus, it is plausible that the only clue to the high risk of relapse may be found in the implicit measures.

Secondly, knowledge about implicit processes may facilitate the development of more effective interventions. For example, interventions that are able to modify both explicit and implicit processes may do better than those that are only able to change explicit processes. Moreover, interventions that directly target implicit processes may be useful.

Thirdly, implicit measures may assist in the tailoring of interventions to individuals. For example, one individual may benefit more from an intervention that targets implicit processes, whereas a second individual may benefit more from an intervention that targets explicit processes. In particular, future research could examine whether the effects of interventions are moderated by implicit and explicit measures, assessed at baseline.

9.1.7 Scope of chapter

Many articles have focused on the clinical relevance of explicit cognition in the addictions (e.g. Crawford 1987; Jones *et al.* 2001). Here, our central focus concerns the clinical relevance of implicit cognition (Fig. 9.1). First, we consider the literature that has examined the associations between use and dependence and implicit measures (link 1 in Fig. 9.1). These studies provide initial support that implicit measures are associated with use or dependence. We then evaluate the studies that have examined prospective relationships between implicit measures and clinical outcomes (link 2 in Fig. 9.1). These studies provide more direct support that implicit measures may be useful in a clinical setting. Thirdly, we evaluate the effects of interventions (pharmacological and non-pharmacological) on implicit and explicit cognition (link 3 in Fig. 9.1). We consider whether the effects of the interventions may be mediated by their effects on cognition. Consistent with our clinical focus, our emphasis is on implicit processes in drug-use cessation, rather than in drug-use initiation. Nonetheless, we briefly consider the relevance of implicit cognition in drug-use initiation in a later section.

9.2 Associations between drug dependence and implicit measures

Many studies have examined whether there is a cross-sectional association between measures of drug use/dependence and implicit measures. This literature is briefly described below. Indeed, research has investigated a

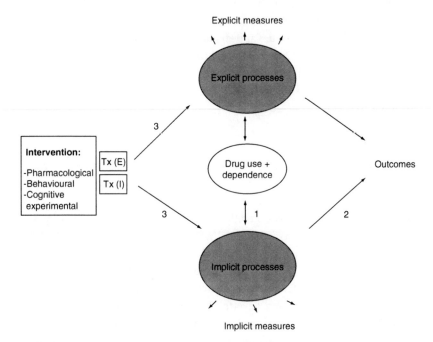

Fig. 9.1 Schematic representation of links between explicit/implicit processes and measures relevant to drug dependence.

number of implicit cognitive processes, including: (1) automatic attention capture (attentional bias); (2) automatic evaluations (implicit attitudes); and (3) automatic retrieval (or 'accessibility') of cognitions, such as expectations. As noted elsewhere in this volume, a number of tasks have been used to assess attentional bias, including the modified Stroop task and the visual dot-probe task. A number of tasks have also been used to assess implicit attitudes, including the implicit association task, the stimulus–response compatibility test and the priming task. Accessibility of outcome expectancies has been assessed using the expectancy accessibility task, and memory association tests.

9.2.1 **Alcohol**

Jones and Bruce (2006) conducted a detailed review of the literature examining the relationships between attentional bias to alcohol cues and alcohol use/dependence. These associations were examined across a range of attentional bias tasks. They concluded that there is 'strong evidence' that excessive (e.g. treatment-seeking) drinkers exhibit greater attentional bias than social drinkers. They also argued that there is 'emerging evidence' that heavier social drinkers exhibit greater attentional bias than lighter social drinkers. In sum, attentional bias to alcohol cues appears to be robustly associated with drinking behaviour. Wiers *et al.* (2006) reviewed the literature examining implicit and explicit alcohol associations. They concluded that the measures derived from the arousal Implicit Association Test (IAT) are associated with heaviness of drinking (see Wiers *et al.* 2002).

9.2.2 **Smoking**

Waters and Sayette (2006) reviewed the literature examining the relationships between attentional bias to smoking cues and smoking behaviour. They concluded that there is good evidence that smokers exhibit significant attentional bias, and this bias is generally greater than the bias exhibited by non-smokers. Studies that have examined associations between attentional bias and measures of use/dependence in samples of smokers have, however, yielded more mixed findings. Thus, within smokers, it is not yet clear whether measures of use/dependence are associated with attentional bias in a straightforward manner. Waters and Sayette (2006) note that studies have reported associations between heaviness of smoking and IAT effects (Sherman *et al.* 2003), and heaviness of smoking and expectancy accessibility (Palfai 2002). However, this research is still in its infancy.

9.2.3 **Other addictions**

Ames *et al.* (2006) have reviewed the literature on implicit cognition and other drugs of abuse. They reported a number of studies documenting associations between drug use and attentional bias in heroin addiction (e.g. Franken *et al.* 2000; Lubman *et al.* 2000), and marijuana use (e.g. Jones *et al.* 2003). Ames *et al.* documented one study that has reported that marijunana non-users have significantly more negative implicit attitudes on an IAT than marijuana users (Field *et al.* 2004). They also summarized a number of studies that have reported associations between measures of memory associations and drug use across a wide variety of populations.

9.2.4 **Summary**

There is strong evidence that drug users exhibit significantly larger attentional biases, and more positive (or arousal-related) IAT effects, than non-users. Within each drug use population, the evidence for associations between use/dependence and implicit measures is emerging, and will become clearer in due course.

9.3 **Relationships with clinical outcomes**

Of course, the cross-sectional associations noted above do not unequivocally demonstrate that implicit processes play a causal role in addictive behaviour. The cognitive biases may be a consequence, or correlate, of addictive behaviour, rather than a causal factor. Prospective studies that demonstrate an association between implicit measures and subsequent clinical outcome would provide additional support for a causal role. They would also provide more direct evidence for the clinical relevance of implicit cognition. These studies are reviewed below.

9.3.1 **Alcohol**

Cox *et al.* (2002) assessed attentional bias in a group of alcohol abusers ($n = 14$) attending a treatment programme. A control group ($n = 16$), drawn from the staff of the treatment unit, was also assessed. A modified Stroop task was administered on two occasions: on entry to the treatment programme, and at discharge, about 4 weeks later. The control group was also assessed twice over a similar period. The Stroop task contained alcohol words ('individualized' to reflect the beverage preferences of each participant) and words reflecting participants' current concerns (also individualized; see Fadardi *et al.* 2006). The alcohol abusers were followed up 3 months after discharge.

There were two main results. First, individuals who had relapsed at follow-up ('relapsers', $n = 9$) demonstrated an increase in attentional bias towards alcohol stimuli between the first and second assessment. In contrast, non-relapsers at follow-up ($n = 5$) tended to show a decrease in attentional bias. The same was true for the control group. Thus, an increase in attentional bias to alcohol-related stimuli is associated with subsequent relapse. Secondly, alcohol abusers who dropped out of the treatment programme had significantly higher interference times for the concern-related category (at the first test session) than those who completed the treatment programme, and controls. Thus, Stroop effects for concern-related material at baseline predicted treatment retention.

9.3.2 Smoking

Waters and colleagues (2003a) administered a smoking Stroop task to participants enrolled in a research smoking cessation clinic ($n = 158$). Participants completed the smoking Stroop task on the first day of a quit attempt. Attentional bias predicted whether participants lapsed in the first week of a quit attempt, and also predicted time to first lapse using survival analysis. Moreover, attentional bias continued to predict cessation outcomes when controlling for self-reported urge. However, a limitation of the study was that significant relationships with cessation outcomes were only observed for a Stroop measure that utilized the first subset of smoking words. The same sample also performed a visual probe task roughly 2 weeks before the quit attempt. However, attentional bias assessed with this task did not predict cessation outcomes (Waters et al. 2003b).

9.3.3 Other drugs

Marissen and colleagues administered a heroin Stroop task to participants attending a treatment centre (Marissen et al. 2005). Participants ($n = 110$) were randomly assigned to receive nine sessions of cue exposure therapy (CET) or control therapy (see Marissen et al. 2006). They completed a heroin Stroop task both before and after completing therapy. Attentional bias assessed before treatment predicted relapse status at a 3-month follow-up. Attentional bias continued to predict relapse when controlling for self-reported craving at the time of the test.

9.3.4 Summary

Relatively few studies have examined associations between implicit measures and subsequent treatment outcome. Thus, it is too soon to draw any firm conclusions. Three studies, however, have reported associations between attentional bias and treatment outcome. Further research is required to determine the robustness of the associations with treatment outcomes.

9.4 **Effects of treatment on implicit measures**

Data from prospective studies are consistent with the hypothesis that implicit biases causally influence drug use. However, they do not prove it. It remains possible that a third variable, such as a risk-enhancing personality trait (e.g. impulsivity), independently influences both implicit processes and drug-taking behaviour, resulting in their prospective relationship. More robust support for the clinical relevance of implicit biases would be derived from studies that experimentally manipulate implicit processes and demonstrate effects on drug-use behaviour. The first step in this process would be to demonstrate that treatments do indeed impact drug-related implicit cognition. The second step would be to show that the effects of such interventions on drug use are mediated by changes in implicit cognition.

Of course, a given intervention may influence both implicit and explicit processes. Here we consider the effects of three types of intervention: pharmacological, behavioural and cognitive experimental. We also briefly consider the impact of messages, and other real-world interventions, on implicit measures. Given that much of this research is in its infancy, some of the following discussion is speculative.

9.4.1 **Pharmacological interventions**

Pharmacological interventions could be expected to target implicit processes. For example, Franken (2003) has argued that attentional bias is mediated by mesolimbic dopamine. Thus, drugs that influence mesolimbic dopamine transmission might be expected to impact on automatic attention capture.

Franken and colleagues examined the effects of a dopamine antagonist on attentional bias to heroin cues (Franken *et al.* 2004). Using a double-blind, randomized crossover design, 17 detoxified heroin-dependent patients received a single oral dose of haloperidol (2 mg), and placebo. Patients performed a heroin Stroop task. In addition, self-reported craving was assessed. Patients exhibited a smaller heroin Stroop effect in the haloperidol condition than in the placebo condition. However, no effect of haloperidol was found on self-reported craving. These findings provide evidence that attentional bias in heroin-dependent humans is mediated by dopaminergic mechanisms. The data also suggest that attentional bias, as assessed by the Stroop task, is modifiable by a pharmacological intervention.

Hitsman and colleagues examined the effects of a serotoninergic challenge on attentional bias to smoking cues (Hitsman *et al.* 2005). Using a double-blind, crossover design, participants drank a tryptophan-depleting drink and taste-matched placebo drink, 1 week apart. Five hours after consumption of each drink, subjects performed a modified Stroop task that assessed attentional bias to smoking words and negative affect words. The tryptophan-depleting drink significantly increased attentional bias to smoking and negative affect words. The results suggest that serotoninergic neurotransmission may play a role in attentional biases. These data provide further evidence that attentional bias is modifiable by a pharmacological intervention.

In a quasi-experimental study, Zack et al. (1999) reported that anxious 'problem drinkers', who were unmedicated, exhibited significant priming between negative affect words and alcohol targets on a semantic priming task. In contrast, anxious participants, who were medicated on benzodiazepines, did not show this priming effect. Although participants were not randomly assigned to medication condition, these data suggest that automatic evaluations (assessed by a semantic priming task) may be modifiable by benzodiazepines. Further research is required to substantiate this finding.

9.4.2 Behavioural interventions—expectancy challenge

Expectancy challenges are interventions that attempt to change expectations about drug use outcomes (Darkes and Goldman 1993, 1998; Dunn et al. 2000; Cruz and Dunn 2003; Musher-Eizenman and Kulick 2003). In the alcohol expectancy challenge procedure, a group of drinkers are invited to participate in meetings involving social interactions (e.g. playing board games). During these sessions, participants drink an alcohol-flavoured beverage. They are informed that the drinks of some participants will contain alcohol, whereas the drinks of other participants will not contain alcohol (a placebo). After drinking the beverage and engaging in the social interaction, participants have to decide which individuals (including themselves) actually consumed alcohol.

Typically, participants identify some individuals as intoxicated who, in fact, did not receive alcohol. In addition, they identify other individuals as sober, who, in fact, did receive alcohol. The intent of the challenge is to provide drinkers with a direct experience of the 'true' pharmacological effects of alcohol, uncontaminated by effects caused by outcome expectancies. Through group discussion, and a didactic lecture on expectancy theory, drinkers learn that many of the desired behavioural effects are due more to expectations than pharmacology (for a review of the protocol, see Darkes and Goldman 1993). Expectancy challenges are effective in weakening positive alcohol outcome expectancies, and strengthening negative alcohol outcome expectancies, as assessed by standard self-report measures (Dunn et al. 2000).

It is of interest to determine whether expectancy challenges influence implicit expectancies, as well as explicit expectancies. This was examined by

Wiers and colleagues. They randomly assigned heavy drinkers ($n = 92$) to an expectancy challenge procedure or a control procedure in the same laboratory (a sham alcohol experiment) (discussed in Wiers *et al.* 2004). Implicit and explicit assessments were taken 1 week before and 1 week after the intervention. Participants in the expectancy challenge condition exhibited a significant reduction in both explicit and implicit arousal associations (on an arousal-IAT). In contrast, explicit and implicit arousal remained unchanged in the control condition. The changes in implicit and explicit expectancies were uncorrelated.

Traditionally, cognitive–behavioural therapies such as expectancy challenge and relapse prevention are thought to work by changing the content of memory. These techniques engage explicit processes in order to help patients form new abstinence-promoting beliefs (e.g. 'Substance abuse has ruined my relationships with family') and counteract old relapse-precipitating beliefs (e.g. 'Using right now will make my problems go away'). Consequently, patients can actively (and explicitly) reflect on abstinence-promoting beliefs when they feel at risk of relapse. Given the emphasis on explicit processes, one might therefore wonder why they might also target implicit processes (as suggested above). In addition to changing memory content, it is possible that these therapies might also change the structure of memory. For example, while reviewing a recent lapse, a therapeutic discussion of abstinence-promoting beliefs might create a new mental association between use-eliciting and abstinence-promoting representations. With sufficient practice, abstinence-promoting beliefs might become automatically activated in use-eliciting situations. These changes might be reflected in implicit measures.

9.4.3 Behavioural interventions—cue exposure therapy (CET)

CET attempts to extinguish craving and other responses to drug-related cues, and thereby reduce the motivation to use drugs (Drummond and Glautier 1994; Havermans and Jansen 2003). This treatment involves frequent presentations of drug-related cues to elicit craving, while drug use is prohibited. Unreinforced cue exposure was originally believed to cause extinction through the elimination or 'unlearning' of previously learned associations between cues and reward. However, more recent theories argue that CET, and other extinction treatments, generate new associations between drug-related stimuli and the absence of rewarding outcomes (non-reward). These new drug–non-reward associations are stored in memory along with the older drug–reward associations (Wiers *et al.* 2004). Thus, drug cues can acquire an ambiguous meaning, being associated with both the presence and absence of reward.

It is thought that the meaning of the extinguished stimulus is disambiguated by the context. Within the treatment context, the drug–non-reward association is likely to be retrieved from memory, whereas outside this context the drug–reward association may be retrieved (Wiers *et al.* 2004). In fact, a review of CET efficacy demonstrated that current CETs do not improve treatment outcomes in the addictions (Conklin and Tiffany 2002). Indeed, one study reported that CET may even be harmful in the treatment of heroin addiction (Marissen *et al.* 2006).

One study has examined the effect of CET on attentional bias (Marissen *et al.* 2005). Participants ($n = 110$) were randomly assigned to receive nine sessions of CET or control therapy (see Marissen *et al.* 2006). They completed a heroin Stroop task both before and after completing therapy. Attentional bias was reduced in both groups after therapy, independent of therapy condition. Treatment condition did not have a significant effect.

It is likely that drug–non-reward associations are accessible in the treatment context, while drug–reward associations remain accessible in other contexts, such as relapse-provoking environments (Havermans and Jansen 2003). Therefore, expectancy accessibility tasks (e.g. Palfai and Wood 2001) may be useful to examine the relative accessibility of drug–reward and drug–non-reward associations in individuals being treated by CET. Presumably, for CET to be successful, accessibility of drug–non-reward associations will need to remain high across many contexts. Thus, implicit assessments may provide a marker for the effectiveness of CET.

9.4.4 Behavioural interventions—implementation intention therapy

Implementation intentions are planning strategies that prospectively link specific cues to actions. They take the form: 'When I encounter x at time y, I will do z' (Gollwitzer 1999). A number of studies have shown that goals are more likely to be realized when participants form specific plans about when, where and how the goals are to be pursued (e.g. Sheeran and Orbell 1999). Implementation intention treatments (IITs) have recently been applied to addictive disorders (Murgraff *et al.* 1996; Higgins and Conner 2003). Individuals are required to determine when and where they may encounter use-provoking situations, and plan what actions to take to avoid drug use in these situations (Prestwitch *et al.* 2005). Higgins and Conner (2003) tested the efficacy of an IIT designed to help children refuse a cigarette when offered. None of the 51 children treated with IIT tried smoking during the subsequent 8 weeks, whereas three of the 53 children (6 per cent) tried smoking in the control condition. Murgraff *et al.* (1996) tested the efficacy of an IIT designed to help undergraduate drinkers refuse an alcoholic drink (e.g. they were to say: 'No thanks, I have to get up early tomorrow'). This IIT caused a significant reduction in drinking behaviour in comparison with a control group. Thus, IITs may be an effective treatment for addiction.

To the best of our knowledge, no studies in the addictions have investigated the effects of IITs on explicit and implicit processes. However, Aarts et al. (1999) demonstrated that the effects of IITs on a mundane (non-addiction) behaviour may be mediated by changes in implicit cognition. Participants were asked to collect a coupon on their journey from the laboratory to the cafeteria. Some participants were exposed to an IIT designed to facilitate execution of the coupon collection, and others were assigned to a no intervention control condition. Before embarking on the journey, participants completed a lexical decision task that indexed the accessibility of cognitions relevant to the implementation intention. Some of the words were related to where the coupon was to be collected (e.g. 'corridor'). The average reaction time (RT) for the goal-relevant words tapped their accessibility. The results showed that the IIT participants were more effective in goal pursuit (collecting the coupon) than the control group participants. Moreover, after controlling for RT on the lexical decision task, the effect of the IIT on goal completion became non-significant. This suggests that the beneficial effect of the IIT on goal completion was mediated by the heightened accessibility of specific contextual cues. Thus, this study provides 'proof of concept' that effects of IITs may be mediated by implicit cognition. It would be useful to examine the effects of IITs on implicit cognition in the addictions.

9.4.5 Behavioural intervention—motivational counselling

Motivational counselling (MC; Cox and Klinger 2004) is a technique for assessing and changing non-drug motivations (i.e. 'current concerns') in order to treat addictive behaviours. According to the motivational theory of alcoholism (Cox and Klinger 1988; Cox et al. 2002), if alcohol abusers can be automatically distracted by compelling, positive concerns in other areas of their lives, they might have better treatment outcomes. In contrast, if abusers are distracted by negative concerns, with little hope of resolving, this may be problematic because they will be unmotivated to reduce their drinking. Thus, MC may change alcohol use through its effects on non-drug-related motivations.

As noted earlier, Cox et al. (2002) found that attentional bias toward concern-related stimuli at baseline was associated with subsequent premature treatment discharge. The authors hypothesized that the negative nature of patients' current concerns contributed to a low motivation to quit drinking. Although MC was not the primary intervention in this study, its findings suggest that measuring attentional biases for current concerns may be useful for identifying individuals who are at risk of relapse or dropping out of treatment. It will be interesting to determine whether MC can significantly impact on attentional bias.

9.4.6 Behavioural interventions—mindfulness-based therapy

Mindfulness can be described as a meta-cognitive state of 'awareness that emerges through paying attention on purpose, in the present moment, and non-judgmentally to the unfolding of experience moment by moment' (Kabat-Zinn 2003, p. 145). Mindfulness-based treatments have been shown to be effective in reducing a variety of disorders, including anxiety, depression and ratings of chronic pain (Baer 2003). Recent research indicates that mindfulness-based treatments may be useful for treating substance use disorders (e.g. Marlatt *et al.* 2004).

The psychological mechanisms underlying the efficacy of mindfulness are not well understood. However, some authors suggest that mindfulness may attenuate the influence of automatic processes (Bishop *et al.* 2004). We are currently examining the effects of mindfulness-based therapy (MBT) on implicit measures in smokers attempting to quit. MBT fosters the ability to recognize that cognitions are 'just thoughts' and emotions are 'just feelings', and that these phenomena are not necessarily an accurate reflection of self or reality. Thus, MBT does not attempt to change the content of cognitions. Rather it attempts to change how those cognitions are perceived and interpreted by the individual. Individuals attempt to achieve a 'de-centred perspective' on mental events. We are using the IAT to examine the strength of mental associations between smoking concepts and me/not me concepts. We hypothesize that MBT will reduce the strength of associations between the concepts of smoking and me, and increase the relative strength of associations between not smoking and me (relative to a control group).

9.4.7 Behavioural interventions—cognitive experimental interventions

There is recent interest in interventions that directly target implicit processes through the use of cognitive re-training (e.g. Mathews and MacLeod 2002). The basic idea is that cognitive experimental paradigms are modified so that they can actually change implicit processes (rather than simply measuring them). Participants perform the 'modified' versions of the cognitive experimental paradigm. Researchers then examine whether the intervention has had the desired effect on the implicit process, by assessing performance on the 'standard' version of the cognitive experimental task. Researchers may also assess whether the cognitive experimental intervention influences post-intervention self-report and behavioural measures.

Interest in this area has been fuelled by the work of MacLeod and colleagues. They have used a modified visual probe task to train participants automatically to attend towards, or away from, threat-related stimuli ('attentional re-training' AR). In the AR procedure used by MacLeod *et al.*, for some participants the position of the probe always replaced negative words (attend-negative group), and for other participants the position of the probe always replaced neutral words (attend-neutral group). Both groups received an extensive number of training trials (576).

There were two important results. First, AR influenced attentional bias to negative stimuli, as assessed by the standard visual probe procedure. Participants in the attend-negative condition tended to be faster on trials in which the probe replaced negative words, whereas participants in the attend-neutral condition tended to be faster on trials in which the probe replaced neutral words. Thus, the AR did significantly influence attentional bias, even to new material (negative and neutral words that were not included in training trials). Secondly, participants assigned to the attend-negative condition reported significantly greater anxiety and depression on a subsequent stressful task (an anagram stress task) compared with those in the attend-neutral condition. Because attentional bias has been directly manipulated as an independent variable in this study, the result indicates that attentional bias may have causal effects on emotional vulnerability.

The results presented by MacLeod *et al.* (2002) have sparked interest in whether cognitive experimental procedures such as AR can reduce maladaptive attentional bias, or other implicit biases, in other areas of psychopathology. For example, two studies have demonstrated that a dot-probe-based AR reduced bias toward threat-related stimuli and led to symptom reduction in patients with generalized anxiety disorder (Vasey *et al.* 2002; Amir *et al.* 2004). Moreover, researchers have started to apply AR techniques to patients with addictive disorders.

For example, Field and Eastwood (2005) randomly assigned heavy social drinkers to an attend-alcohol AR condition (probe always replaced alcohol stimuli) or an attend-neutral AR condition (probe always replaced neutral stimuli). After AR, the magnitude of attentional biases, assessed on a standard visual probe task, was significantly increased in the attend-alcohol group, but was significantly reduced in the avoid-alcohol group. Moreover, self-reported alcohol craving increased after AR in the attend-alcohol condition, compared with a pre-training baseline. However, self-reported craving did not decrease in the avoid-alcohol group. Finally, after AR, participants were given the opportunity to drink some beer. The attend-alcohol group consumed significantly more beer than the avoid-alcohol group.

Consistent with the results of MacLeod *et al.* (2002), these results demonstrate that attentional bias for alcohol-related cues can be experimentally manipulated. They also demonstrate that attentional bias can increase the motivation to drink alcohol, as assessed by self-reported craving and a behavioural measure of alcohol consumption. This suggests that attentional bias may directly cause self-reported craving. Further research should investigate AR in the addictions.

Evaluative conditioning refers to the process by which changes in affective reactions to stimuli are acquired though simple co-occurrences of a conditioned stimulus (CS) and a valenced unconditioned stimulus (US) (De Houwer *et al.* 2001). In contrast to some forms of Pavlovian conditioning, evaluative conditioning (EC) is thought to be largely independent of contingency awareness (i.e. individuals do not need to be conscious of the CS–US association) (De Houwer 2001). Although a large number of stimuli can function as the US (e.g. electric shock, the presentation of food, exposure to a gustatory stimulus), the use of computerized tasks in psychological research has facilitated picture–picture or word–word EC procedures. In these formats, the CS and the US are either both pictures or both words.

De Houwer *et al.* (2000) used a picture–picture EC procedure. On each trial, a CS was presented for 1 s, followed by a blank screen for 2 s, followed by a US for 1 s. Although participants were not aware of the pairings during the presentation sequences, the study demonstrated that EC with positive and negative US led to increases and decreases, respectively, in self-reported ratings of previously neutral CS.

Recent studies have performed EC under subliminal presentation conditions (e.g. Lovibond and Shanks 2002; Diksterhuis 2004). In these studies, a CS is presented very briefly, followed by a meaningless mask. The mask ensures that the stimulus cannot be consciously perceived. Dijksterhuis (2004) paired the word 'I' with positive trait terms during a subliminal EC procedure. This brief intervention enhanced implicit self-esteem as assessed by three different measures (compared with a control condition, in which 'I' was paired with XXXX). The EC procedure also made participants less sensitive to negative social feedback. In sum, just as AR appears to be a useful tool for modifying automatic attentional processes, subliminal and supraliminal EC appear to be useful tools for modifying automatic evaluative processes.

Studies have not yet examined whether EC influences addiction-related associations. However, EC procedures that attempt to engender automatic negative evaluations of drug-related stimuli may be useful in a clinical context. EC is distinct from classical conditioning in that it is more resistant to extinction, does not require conscious awareness of the relationship between US and CS, and is less affected by context (which is a problem for treatments such as CET, see above) (De Houwer *et al.* 2001). Therefore, EC may be a potentially useful tool for addiction treatment.

Subliminal cue exposure (SCE) is a procedure in which repeated exposure to valenced cues under subliminal presentation conditions results in the affective habituation to these cues (Dijkstehuis and Smith 2002). In SCE, emotionally valenced masked cues are briefly presented on a computer screen, and the participant is not aware of the exposure. Dijksterhuis and Smith (2002) used a SCE paradigm in which the participants' task was to identify whether random letter strings (i.e. 'noahlief') began with vowels or consonants. Unknown to participants, letter strings were followed by masked positive or negative words. SCE caused these words to become less extreme than words that had not been presented, as assessed by implicit and explicit evaluation measures.

The authors suggested that SCE may be relevant for exposure treatment for phobias. They noted that the early stages of exposure treatment can be highly distressing for some patients. SCE may be a method to expose patients to relevant stimuli while at the same time preventing unpleasant emotional reactions. The same procedures may be useful in the treatment of addictive disorders. In traditional CET, there have been concerns that direct exposure to drug-related cues may provoke such severe craving that patients experience discomfort or may even relapse (Ehrman *et al.* 1998). Indeed, Ehrman *et al.* (1998) recommended that clinicians use active strategies (i.e. 'talk down' sessions) to reduce craving after CET. SCE may be useful as a primer treatment used prior to explicit cue exposure, in order to reduce the severity of craving responses. In addition to reducing the adverse reactions to explicit CET, SCE may function as a treatment in its own right. However, the effect of SCE on subsequent cue reactivity or substance use is yet to be examined.

9.4.8 Other interventions

In the wider implicit cognition literature, studies have examined the effects of brief messages or 'real-world' interventions on implicit and explicit measures (e.g. Rudman *et al.* 2001; Maio *et al.* 2002). In the addictions, it is of interest to determine whether brief messages can be effective in changing individuals' implicit and explicit attitudes to their drug use. However, we are only aware of one study that has addressed this issue (Rudman and Heppen 2005; see also

Rudman 2004). In this study, one group of participants were assigned to read a newspaper article that reports a tragic story about the consequences of smoking. Another group were assigned to read a 'control' article that discussed smoking in a non-emotional manner. Measures of implicit and explicit attitudes were administered after the intervention. Participants who read the emotional message exhibited a significantly more negative implicit attitude (IAT effects become more negative) compared with the control participants. However, they did not exhibit significantly different self-reported attitudes to smoking. This study shows that the effects of brief messages may be apparent in implicit measures, but not explicit measures. Of course, it would also be of interest to determine whether 'real-world' interventions, such as smoking prevention programmes, impact on explicit and implicit attitudes. However, we are not aware of any studies that have addressed this issue.

9.5 Discussion

In general, there is good evidence that implicit measures are associated with drug use/dependence (link 1). There is some initial evidence for prospective associations between implicit measures and clinical outcomes (link 2). However, this research is in its infancy, and much further research is required to determine the robustness of these associations. There is some evidence that implicit measures are modifiable by interventions (link 3). This is true for pharmacological (e.g. Franken *et al.* 2004), behavioural (e.g. Wiers *et al.* 2004) and cognitive experimental (e.g. Field and Eastwood 2005) interventions. However, we are not aware of any study that has demonstrated that the effect of the intervention on an addiction-related outcome is mediated by its effects on implicit cognition. Thus, the causal role of implicit processes is not yet proven.

As noted above, there is much interest in the use of cognitive experimental interventions that target implicit processes, and there will doubtless be reports of a number of studies appearing soon. We now discuss some other issues that may become important in the next few years (see Fig. 9.2). Because this research is in its infancy, much of this discussion is speculative.

9.5.1 Clinical issues 1: effects of multiple interventions

If implicit and explicit processes are both reliably associated with clinical outcomes, then a combination of an intervention that affects an explicit process with one that affects an implicit process may do better than a single intervention. For example, the advantages of combining pharmacotherapy with psychotherapy (e.g. Carroll 1997) may derive, in part, from the fact that these therapies target implicit and explicit processes, respectively.

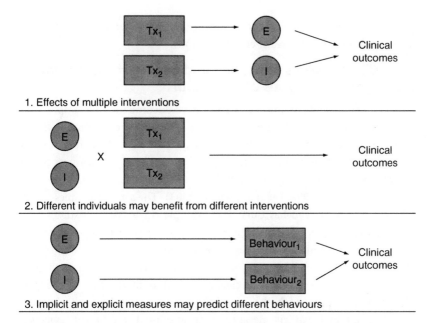

Fig. 9.2 Schematic representation of inter-relationships between explicit/implicit measures, treatment variables, and clinical outcomes.

An advantage of examining the effects of pre-existing interventions on implicit and explicit processes is that the results might suggest ways in which pre-existing interventions can be usefully combined. Thus, future research can examine whether the beneficial effect of a combination treatment on clinical outcome is mediated by its dual effects on implicit and explicit processes.

9.5.2. Clinical issues 2: different individuals may benefit from different interventions

Future research can determine whether individuals can be matched to different interventions on the basis of performance on a battery of implicit and explicit tasks (administered at baseline). For example, an individual with an elevated attentional bias might benefit most from an AR intervention, or a medication that reduces attentional bias. Alternatively, an individual with highly positive implicit attitudes to drug use might benefit most from interventions that stimulate an automatic aversive response, such as evaluative conditioning. On the other hand, some individuals may benefit most from interventions that target explicit processes. Thus, future research can examine whether the effects of treatment are moderated by implicit measures (i.e. test implicit measure by treatment interaction) or by explicit measures (i.e. test explicit measure by treatment interaction).

9.5.3 Clinical issues 3: implicit and explicit measures may predict different types of behaviour

Studies in the social cognition literature have documented that implicit and explicit measures predict different types of behaviour. For example, implicit measures predict automatic behaviours (e.g. blinking and interpersonal eye contact) better than explicit measures. In contrast, explicit measures predict controlled behaviours (e.g. verbal evaluations) better than implicit measures (e.g. Dovidio *et al.* 2002; Fazio and Olson 2003). The same principles may be relevant to addictive behaviours. For example, in a cue-exposure setting, implicit measures may predict automatic behaviours such as facial gestures when exposed to a lit cigarette (Sayette *et al.* 2003), whereas explicit measures may predict self-reported evaluations of the cigarette. If participants are given the opportunity to smoke the cigarette, then both implicit and explicit measures might predict whether or not the cigarette is smoked. If it is smoked, then implicit and explicit measures may also predict different aspects of smoking behaviour. Implicit measures may predict the time taken to light up, and perhaps the regularity of smoking (as assessed by the consistency of inter-puff intervals), whereas explicit measures may predict whether the cigarette is stubbed out early or is smoked to the end.

Accordingly, it is possible that interventions that target explicit processes primarily influence non-automatic (controlled) substance-use behaviours, whereas those that target implicit processes primarily influence automatic behaviours. The relationships between different behaviour types and clinical outcomes will become clearer in due course. Relatedly, recently developed multi-dimensional measures of tobacco dependence contain separate subscales for assessing automatic and non-automatic drug-use behaviour (Piper *et al.* 2004; Shiffman *et al.* 2004). When administered at baseline, these dependence subscales could be useful to evaluate whether certain individuals (e.g. those with high ratings on automaticity scales) might benefit more from an intervention that targets implicit processes (i.e. test dependence type by treatment interactions).

9.5.4. Measurement issues 1: implicit measures may capture both implicit and explicit processes

In common with all psychological assessments, implicit tasks have limitations. First, although implicit tasks purport to assess implicit processes, implicit measures may be contaminated with controlled 'strategic' processes. Under certain conditions, performance on the modified Stroop task may be influenced by strategic processes (Mathews and Sebastian 1993), and this may obscure what the task is trying to measure (automatic attention capture). Recently, it has been suggested that performance on the IAT may also sometimes involve strategic processes (Rothermund and Wentura 2004), and these processes may complicate the interpretation of data from the IAT. Similarly, responses on expectancy accessibility tasks and memory association tasks may also be influenced by controlled processes. For example, a participant may deliberately withhold or delay a response on these tasks. In sum, implicit tasks may not cleanly assess implicit processes. Future research can ascertain the extent to which implicit measures reflect explicit as well as implicit processes.

9.5.5 Measurement issues 2: little is known on psychometric properties of implicit measures

The application of psychometric principles is routine in many areas of clinical psychology (e.g. clinical rating scales). Indeed, the psychometric properties of explicit cognitive assessments have been carefully examined (e.g. Copeland et al. 1995; Cox et al. 2001). However, the psychometric properties of the implicit cognitive assessments have been less rigorously scrutinized in the addiction literature (though see Greenwald et al. 2003 for an analysis of IAT effects in other domains). Thus, the internal and test–retest reliabilities of many of the implicit measures are not known (see Kindt et al. 1996).

Furthermore, a few studies have examined the correlations between implicit measures that purport to assess a similar construct. These correlations have generally been non-significant in the addiction literature (Mogg and Bradley 2002; Sherman et al. 2003). This is consistent with what is known in the implicit attitude, implicit memory and implicit self-esteem literatures (Bosson et al. 2000; Buchner and Wippich 2000; Fazio and Olson 2003). The lack of association may indicate that the different implicit tasks measure subtly different constructs. The lack of association may also, however, reflect the measurement error associated with the implicit measures. Cunningham et al. (2001) used latent variable analysis and reported that association between implicit attitude measures improved substantially when separating measurement error from estimates of stability. This latent variable approach might be useful in examining the associations between implicit measures in the addictions.

9.5.6 Measurement issues 3: effects of repeated measurements

In a clinical setting, it is obviously desirable to administer the implicit assessments at multiple time points within a clinical trial. For example, researchers are interested in examining the effects of treatment on implicit measures over time. However, there is reason to believe that there will be effects of repeated measurements on at least some of the implicit measures. For example, in one of our studies, smokers ($n = 203$) completed a smoking-Stroop task at two sessions, conducted a few days apart. Smoking-Stroop effects were much larger ($P < 0.0001$) at the first session (74 ms) than at the second session (21 ms) (see also Munafò et al. 2003). In another ongoing study of implicit attitudes in smokers ($n = 54$), we have noted that IAT effects are significantly more negative ($P < 0.001$) on the first administration (IAT effect $= -459$ ms) than the subsequent three test sessions (IAT effects $= -216$, -277 and -164 ms, respectively).

In a sense, the effect of repeated assessments is not surprising. The effects of practice on the classic Stroop task were noted in the original article (Stroop 1935; see also Melara and Mounts 1993), and practice effects have been observed on other selective attention tasks (e.g. Harris and Pashler 2004). Nonetheless, these effects underscore an important difference between psychological and biological measurements. For example, the measurement of blood nicotine levels, assessed from a blood sample, is not affected if blood nicotine levels have been measured on previous occasions. In contrast, the assessment of implicit cognitive processes may be affected if there is a history of previous assessment. Participants may perform the tasks in subtly different ways on repeated occasions, or else the stimuli may not have the same impact if they are presented repeatedly. It would also be useful to examine whether the time between two assessments influences the size of repeated-measures effects.

To counter these effects, treatment studies should be careful to include control groups to account for the effects of repeated assessments. Even if control groups are employed, the repeated-measures effect may make it more difficult to detect the effects of treatment. For example, if a repeated-measure effect reduces an addiction-Stroop effect dramatically in an untreated group, it may be difficult to demonstrate further reduction in the treated group. Thus, there may be research settings in which it is best to administer the implicit assessments on just a single occasion. For those laboratory studies that employ repeated assessments, it would also be useful to examine whether the associations between implicit measures and key external variables (such as drug use/dependence) is robust across repeated assessments (i.e. implicit measure by assessment–number interaction). Finally, the psychological processes underlying the repeated measurement effects may be interesting in their own right.

9.5.7 Implicit cognition and drug-use initiation

Finally, although the focus of this chapter has been on drug-use cessation, implicit assessments may be useful in the study of drug-use initiation. For example, in tobacco research, many studies have examined adolescents' explicit attitudes to smoking and their subsequent risk of smoking initiation. In the susceptibility to smoking measure, adolescents are asked to respond to the following questions on a 4-point scale, ranging from 'definitely not' to definitely yes': Do you think that you will smoke a cigarette soon?; Do you think you will smoke a cigarette in the next year?; If one of your best friends were to offer you a cigarette, would you smoke it? A respondent who endorses 'definitely not' to all items is classified as 'not susceptible'; all other respondents are classified as 'susceptible'.

This simple measure of future intentions and expectations about future behaviour is a surprisingly robust predictor of future smoking. For example, in a multi-ethnic study of 5624 high school students in Houston and Austin, Texas, 19.4 per cent of the adolescents classified as susceptible at baseline became ever smokers at the 12-month follow-up, whereas only 1.6 per cent of their not susceptible counterparts did (Prokhorov et al. 2002; see also Choi et al. 1997). This finding clearly attests to the utility of self-report measures in predicting smoking initiation. Nonetheless, not all susceptible individuals progress to smoking. It is possible that implicit measures provide additional information that can help to identify individuals at high risk for smoking initiation (see Stacy 1997). Prevention interventions can then be directed at those individuals who are most at risk, as identified by a combination of explicit and implicit measures.

9.5.8 Conclusion

Research on the clinical relevance of implicit cognition in the addictions is in its infancy. Further research is required to determine whether the implicit measures have sufficient reliability and validity to be useful within the clinical domain. For example, it will be important to determine whether the various implicit measures are associated with treatment outcomes. We anticipate that, over the next few years, there will a large number of studies published that will address these very questions.

Acknowledgements

Any correspondence concerning this chapter should be addressed to Dr Andrew Waters, Department of Behavioral Science, University of Texas, Houston TX 77030, USA.

References

Aarts, H., Dijksterhuis, A. and Midden, C. (1999). To plan or not to plan: goal achievement of interrupting the performance of mundane behaviors. *European Journal of Social Psychology*, 29, 971–979.

Ames, S.L., Franken, I.H.A. and Coronges, K. (2006). Implicit cognition and drugs of abuse. In R.W. Wiers and A.W. Stacy (ed.), *Handbook of implicit cognition and addiction.* Thousand Oaks, CA: Sage, pp. 363–378.

Amir, N.B., Beard, C., Klumpp, H., and Elias, J. (2004). Training of attentional bias in social phobia. *International Journal of Psychology,* **39**, 12 Suppl. S.

Baer, R.A. (2003). Mindfulness training as a clinical intervention: A conceptual and empirical review. *Clinical Psychology: Science and Practice,* 10, 125–143.

Baker, T.B., Piper, M.E., McCarthy, D.E., Majeskie, M.R., and Fiore, M.C. (2004). Addiction Motivation Reformulated: An Affective Processing Model of Negative Reinforcement. *Psychological Review,* 111, 33–51.

Baron, R.M. and Kenny, D.A. (1986). The moderator–mediator variable distinction in social psychological research: conceptual, strategic and statistical considerations. *Journal of Personality and Social Psychology,* 51, 1173–1182.

Bishop, S.R., Lau, M., Shapiro, S., Carlson, L., Anderson, N.D., Carmody, J., Segal, Z.V., Abbey, S., Speca, M., Velting, D. and Devins, G. (2004). Mindfulness: a proposed operational definition. *Clinical Psychology: Science and Practice,* 11, 230–241.

Bosson, J.K., Swann, W.B. and Pennebaker, J.W. (2000). Stalking the perfect measure of implicit self-esteem: the blind men and the elephant revisited. *Journal of Personality and Social Psychology,* **79**, 631–643.

Buchner, A. and Wippich, W. (2000). On the reliability of implicit and explicit memory measures. *Cognitive Psychology,* **40**, 227–259.

Carroll, K.C. (1997). Integrating psychotherapy and pharmacotherapy to improve drug abuse outcomes. *Addictive Behaviors,* **22**, 233–245.

Chisholm, S.M., Crowther, J.H. and Ben-Porath, Y.S. (1997). Selected MMPI-2 scales' ability to predict premature termination and outcome from psychotherapy. *Journal of Personality Assessment,* **69**, 127–144.

Choi, W.S., Pierce, J.P., Gilpin, E.A., Farkas, A.J.and Berry, C.C. (1997). Which adolescent experimenters progress to established smoking in the United States. *American Journal of Preventive Medicine,* 13, 385–391.

Conklin, C.A. and Tiffany, S.T. (2002). Applying extinction research and theory to cue-exposure addiction treatments. *Addiction,* 97, 155–167.

Copeland, A.L., Brandon, T.H. and Quinn, E.P. (1995). The Smoking Consequences Questionnaire-Adult: measurement of smoking outcome expectancies of experienced smokers. *Psychological Assessment,* 7, 484–494.

Cox, W.M. and Klinger, E. (2004). A motivational model of alcohol use: determinants of use and change. In W.M. Cox and E. Klinger (ed.), *Handbook of motivational counseling: concepts, approaches and assessment.* London: John Wiley and Sons Ltd, pp. 121–138.

Cox, L.S., Tiffany, S.T. and Christen, A.G. (2001). Evaluation of the brief questionnaire of smoking urges (QSU-brief) in laboratory and clinical settings. *Nicotine and Tobacco Research,* 3, 7–16.

Cox, W.M., Hogan, L.M., Kristian, M.R. and Race, J.H. (2002). Alcohol attentional bias as a predictor of alcohol abusers' treatment outcome. *Drug and Alcohol Dependence,* 68, 237–243.

Crawford, A. (1987). Attitudes about alcohol: a general review. *Drug and Alcohol Dependence,* 19, 279–311.

Cruz, I.Y. and Dunn, M.E. (2003). Lowering risk for early alcohol use by challenging alcohol expectancies in elementary school children. *Journal of Consulting and Clinical Psychology*, 7, 493–503.

Cunningham, W.A., Preacher, K.L. and Banaji, M.R. (2001). Implicit attitude measures: consistency, stability and convergent validity. *Psychological Science*, 12, 163–170.

Darkes, J. and Goldman, M.S. (1993). Expectancy challenge and drinking reduction: experimental evidence for a mediational process. *Journal of Consulting and Clinical Psychology*, 61, 344–353.

Darkes, J. and Goldman, M.S. (1998). Expectancy challenge and drinking reduction: process and structure in the alcohol expectancy network. *Experimental and Clinical Psychopharmacology*, 6, 64–76.

De Houwer, J. (2006). What are implicit measures and why are we using them? In R. W. Wiers and A.W. Stacy (ed.), *Handbook of implicit cognition and addiction*. Thousand Oaks, CA: Sage, pp. 275–286.

De Houwer, J., Baeyens, F., Vansteenwegen, D. and Eelen, P. (2000). Evaluative conditioning in the picture–picture paradigm with random assignment of conditioned stimuli to unconditioned stimuli. *Journal of Experimental Psychology: Animal Behavior Processes*, 26, 237–242.

De Houwer, J., Thomas, S. and Baeyens, F. (2001). Association learning of likes and dislikes: a review of 25 years of research on human evaluative conditioning. *Psychological Bulletin*, 127, 853–869.

DiChiara, G. (2002). Nucleus accumbens shell and core dopamine: differential role in behavior and addiction. *Behavioural Brain Research*, 137, 75–114.

Dijksterhuis, A. (2004). I like myself but I don't know why: enhancing implicit self-esteem by subliminal evaluative conditioning. *Journal of Personality and Social Psychology*, 86, 345–355.

Dijksterhuis, A. and Smith, P.K. (2002). Affective habituation: subliminal exposure to extreme stimuli decreases their extremity. *Emotion*, 2, 203–214.

Dovidio, J.F., Kawakami, K. and Gaertner, S.L. (2002). Implicit and explicit prejudice and interracial interaction. *Journal of Personality and Social Psychology*, 82, 62–68.

Drummond, D.C. and Glautier, S. (1994). A controlled trial of cue exposure treatment in alcohol dependence. *Journal of Consulting and Clinical Psychology*, 62, 809–817.

Dunn, M.E., Lau, H.C. and Cruz, I.Y. (2000). Changes in activation of alcohol expectancies in memory in relation to changes in alcohol use after participation in an expectancy challenge program. *Experimental and Clinical Psychopharmacology*, 8, 566–575.

Ehrman, R.N., Robbins, S.J., Childress, A.R., Goehl, L., Hole, A.V. and O'Brien, C.P. (1998). Laboratory exposure to cocaine cues does not increase cocaine use by outpatient subjects. *Journal of Substance Abuse Treatment*, 15, 431–435.

Fadardi, J.S., Cox, W.M. and Klinger, E. (2006). Individualized versus general measures of addiction-related implicit cognitions. In R.W. Wiers and A.W. Stacy (ed.), *Handbook of implicit cognition and addiction*. Thousand Oaks, CA: Sage.

Fazio, R.H., Jackson, J.R., Dunton, B.C. and Williams, C.J. (1995). Variability in automatic activation as an unobstrusive measure of racial attitudes: a bona fide pipeline? *Journal of Personality and Social Psychology*, 69, 1013–1027.

Fazio, R.H. and Olson, M.A. (2003). Implicit measures in social cognition research: their meaning and use. *Annual Review of Psychology*, 43, 297–327.

Field, M. and Eastwood, B. (2005) Experimental manipulation of induced attentional bias increases the motivation to drink alcohol. *Psychopharmacology (Berlin)*, 183, 350–357.

Field, M., Mogg, K. and Bradley, B.P. (2004). Cognitive bias and drug craving in recreational cannabis users. *Drug and Alcohol Dependence*, 74, 105–111.

Franken, I.H.A. (2003). Drug craving and addiction: integrating psychological and neuro-psychopharmacological approaches. *Progress in Neuro-Psychopharmacology and Biological Psychiatry*, 27, 563–579.

Franken, I.H.A., Kroon, L.Y., Wiers, R.W. and Jansen, A. (2000). Selective cognitive processing of drug cues in heroin dependence. *Journal of Psychopharmacology*, 145, 395–400.

Franken, I.H.A., Hendriks, V.M., Stam, C.J. and van den Brink, W. (2004). A role for dopamine in the processing of drug cues in heroin dependent patients. *European Neuropsychopharmacology*, 14, 503–508.

Gollwitzer, P.M. (1999). Implementation intentions: Strong effects of simple plans. *American Psychologist*, 54(7), 493–503.

Greenwald, A.G., Nosek, B.A. and Banaji, M.R. (2003). Understanding and using the Implicit Association Test: I. An improved scoring algorithm. *Journal of Personality and Social Psychology*, 85, 197–216.

Harris, C.R. and Pashler, H.E. (2004). Attention and the processing of emotional words and names. *Psychological Science*, 15, 171–178.

Havermans, R.C. and Jansen, A.T.M. (2003). Increasing the efficacy of cue exposure treatment in preventing relapse of addictive behavior. *Addictive Behaviors*, 28, 989–994.

Hawkins, J.D., Catalano, R.F. and Miller, J.Y. (1992). Risk and protective factors for alcohol and other drug problems in adolescence and early adulthood: implications for substance abuse prevention. *Psychological Bulletin*, 112, 64–105.

Higgins, A. and Conner, M. (2003). Understanding adolescent smoking: the role of the theory of planned behaviour and implementation intentions. *Psychology, Health and Medicine*, 8, 173–186.

Hine, D.W., McKenzie-Richer, A., Lewko, J., Tilleczek, K. and Perreault, L. (2002). A comparison of the mediational properties of four adolescent smoking expectancy measures. *Journal of Addictive Behaviors*, 16, 187–195.

Hitsman, B., Spring, B., Pingitore, R., McChargue, D.E. and Doran, N. (2005). Serotonergic involvement in the attentional salience of cigarette cues. Paper presented at the 11th Annual Meeting of the Society for Nicotine and Tobacco, Prague, Czech Republic, 2005.

Huijding, J., de Jong, P.J., Wiers, R.W. and Verkooijen, K. (2005). Implicit and explicit attitudes toward smoking in a smoking and a nonsmoking setting. *Addictive Behaviors*, 30, 949–961.

Jarvis, M.J. (1997). Patterns and predictors of smoking cessation in the general population. In Bolliger, C.T., Fagerström, K.O. (ed.), *The tobacco epidemic*. Basel: Karger.

Jones, B.T. and Bruce, G. (2006). Methods, measures and findings of attentional bias in substance use, abuse and dependence. In R.W. Wiers and A.W. Stacy (ed.), *Handbook of implicit cognition and addiction*. Thousand Oaks, CA: Sage.

Jones, B.T., Corbin, W. and Fromme, K. (2001). A review of expectancy theory and alcohol consumption. *Addiction*, **96**, 57–72.

Jones, B.T., Jones, B.C., Smith, H. and Copley, N. (2003). A flicker paradigm for inducing change blindness reveals alcohol and cannabis information processing biases in social users. *Addiction*, **98**, 235–244.

Kabat-Zinn, J. (2003). Mindfulness-based interventions in context: Past, present, and future. *Clinical Psychology: Science and Practice*, **10**, 144–156.

Kindt, M., Bierman, D. and Brosschot, J.F. (1996). Stroop versus Stroop: comparison of a card format and a single-trial format of the standard color-word Stroop task and the emotional Stroop task. *Personality and Individual Differences*, **21**, 653–661.

Lovibond, P.F. and Shanks, D.R. (2002). The role of awareness in Pavlovian conditioning: empirical evidence and theoretical implications. *Journal of Experimental Psychology: Animal Behavior Processes*, **28**, 3–26.

Lubman, D.I., Peters, L.A., Mogg, K., Bradley, B.P. and Deakin, J.F.W. (2000). Attentional bias for drug cues in opiate dependence. *Psychological Medicine*, **30**, 169–175.

MacLeod, C., Rutherford, E., Campbell, L., Ebsworthy, C. and Holker, L. (2002). Selective attention and emotional vulnerability: assessing the causal basis of their association through the experimental manipulation of attentional bias. *Journal of Abnormal Psychology*, **111**, 107–123.

Marissen, M.A.E., Franken, I.H.A., Waters, A.J., Blanken, P., van den Brink, W. and Hendriks, V.M. (2005). Attentional bias predicts relapse in heroin dependence following treatment. Manuscript under review.

Marissen, M.A.E., Franken, I.H.A., Blanken, P., van den Brink, W. and Hendriks, V. M. (2006). Cue exposure therapy for the treatment of opiate addiction might be harmful: results from a randomised controlled trial. *Drug and Alcohol Dependence*, in press.

Maio, G., Watt, S.E. and Hewstone, M. (2002). Effects of anti-racism messages on explicit and implicit intergroup attitudes: the moderating role of attitudinal ambivalence. Paper presented at the European Association of Experimental Social Psychology General Meeting, San Sebastian, Spain, 2002.

Marlatt, G.A. (1979). A cognitive–behavioral model of the relapse process. *NIDA Research Monograph*, **25**, 191–200.

Marlatt, G.A., Witkiewitz, K., Dillworth, T.M., Bowen, S., Parks, G.A., MacPherson, L.M., Lonczak, H.S., Larimer, H.E., Simpson, T., Blume, R.W. and Crutcher, R. (2004). Vipassana meditation as a treatment for alcohol and drug use disorders. In S.C. Hayes, M.M. Linehan and V.M. and Follette (ed.), *Mindfulness and acceptance: expanding the cognitive–behavioral tradition*. New York, NY: Guildford Press, pp. 261–287.

Mathews, A.M., and Sebastian, S. (1993). Suppression of emotional Stroop effects by fear-arousal. *Cognition & Emotion*, **7**, 517–530.

Mathews, A., and MacLeod, C. (2002). Induced processing biases have causal effects on anxiety. *Cognition & Emotion*, **16**, 331–354.

Melara, R.D. and Mounts, J.R. (1993). Selective attention to Stroop dimensions: effects of baseline discriminability, response mode and practice. *Memory and Cognition*, **21**, 627–645.

Mogg, K. and Bradley, B.P. (2002). Selective processing of smoking-related cues in smokers: manipulation of deprivation level and comparison of three measures of attentional bias. *Journal of Psychopharmacology*, **16**, 385–392.

Munafò, M., Mogg, K., Roberts, S., Bradley, B.P. and Murphy, M. (2003). Selective processing of smoking-related cues in current smokers, ex-smokers and never-smokers on the modified Stroop task. *Journal of Psychopharmacology*, **17**, 310–316.

Murgraff, V., White, D. and Phillips, K. (1996). Moderating binge drinking: it is possible to change behaviour if you plan it in advance. *Alcohol and Alcoholism*, **31**, 577–582.

Musher-Eizenman, D.R. and Kulick, A.D. (2003). An alcohol expectancy-challenge prevention program for at risk college women. *Psychology of Addictive Behaviors*, **17**, 163–166.

Palfai, T.P. (2002). Positive outcome expectancies and smoking behavior: the role of expectancy accessibility. *Cognitive Research and Therapy*, **26**, 317–333.

Palfai, T. and Wood, M.D. (2001). Positive alcohol expectancies and drinking behavior: the influence of expectancy strength and memory accessibility. *Psychology of Addictive Behaviors*, **15**, 60–67.

Piper, M.E., Piasecki, T.M., Federman, E.B., Bolt, D.M., Smith, S.S., Fiore, M.C. and Baker, T.B. (2004). A multiple motives approach to tobacco dependence: the Wisconsin Inventory of Smoking Dependence Motives (WISDM-68). *Journal of Consulting and Clinical Psychology*, **72**, 139–154.

Prestwich, A., Conner, M., and Lawton, R.J. (2005). Implementation Intentions: Can They Be Used to Prevent and Treat Addiction? In R.W. Wiers and A.W. Stacy (ed.), *Handbook of implicit cognition and addiction*. Thousand Oaks, CA: Sage, pp. 455–469.

Prokhorov, A.V., de Moor, C.A., Hudmon, K.S., Hu, S., Kelder, S.H. and Gritz, E.R. (2002). Predicting initiation of smoking in adolescents: evidence for integrating the stages of change and susceptibility to smoking constructs. *Addictive Behaviors*, **27**, 697–712.

Robinson, T.E. and Berridge, K.C. (1993). The neural basis of craving: an incentive-sensitization theory of addiction. *Brain Research Review*, **18**, 247–291.

Rothermund, K. and Wentura, D. (2004). Underlying processes in the Implicit Association Test: dissociating salience from associations. *Journal of Experimental Psychology: General*, **133**, 139–165.

Rudman, L.A. (2004). Sources of implicit attitudes. *Current Directions in Psychological Science*, **13**, 79–82.

Rudman, L.A., Ashmore, R.D. and Gary, M.L. (2001). Automatic biases: the malleability of implicit prejudice and stereotypes. *Journal of Personality and Social Psychology*, **81**, 856–868.

Sayette, M.A., Wertz, J.M., Martin, C.S., Cohn, J.F., Perrott, M.A. and Hobel, J. (2003). Effects of smoking opportunity on cue-elicited urge: a facial coding analysis. *Experimental and Clinical Psychopharmacology*, **11**, 218–227.

Schneider, W. and Shiffrin, R.M. (1977). Controlled and automatic human information processing: In. Detection, search and attention. *Psychological Review*, **84**, 1–66.

Sheeran, P., and Orbell, S. (1999). Augmenting the theory of planned behavior: Roles for anticipated regret and descriptive norms. *Journal of Applied Social Psychology*, **29**, 2107–2142.

Sherman, S.J., Rose, J.S. and Koch, K. (2003). Implicit and explicit attitudes toward cigarette smoking: the effects of context and motivation. *Journal of Social and Clinical Psychology*, **22**, 13–39.

Shiffman, S., Paty, J.A., Gnys, M., Kassel, J.A. and Hickcox, M. (1996). First lapses to smoking: within-subjects analysis of real-time reports. *Journal of Consulting and Clinical Psychology*, **64**, 366–379.

Shiffman, S. and Waters, A.J. (2004). Negative affect and smoking lapses: a prospective analysis. *Journal of Consulting and Clinical Psychology*, **7**, 192–201.

Shiffman, S., Waters A.J. and Hickcox, M. (2004). The Nicotine Dependence Syndrome Scale: a multi-dimensional measure of nicotine dependence. *Nicotine and Tobacco Research*, **6**, 327–348.

Stacy, A.W. (1997). Memory activation and expectancy as prospective predictors of alcohol and marijuana use. *Journal of Abnormal Psychology*, **106**, 61–73.

Stein, L.A.R. and Graham, J.R. (1999). Detecting fake-good MMPI-A profiles in a correctional facility. *Psychological Assessment*, **11**, 386–395.

Stroop, J.R. (1935). Studies of interference in serial verbal reactions. *Journal of Experimental Psychology*, **18**, 643–661.

Swanson, J.E., Rudman, L.A. and Greenwald, A.G. (2001). Using the implicit association test to investigate attitude–behaviour consistency for stigmatized behaviour. *Cognition and Emotion*, **15**, 207–230.

Tiffany, S.T. (1990). A cognitive model of drug urges and drug-use behavior: role of automatic and nonautomatic processes. *Psychological Review*, **97**, 147–168.

Vacey, M.W., Hazen, R., and Schmidt, N.B. (2002). Attentional retraining for chronic worry and generalized anxiety disorder (GAD). Paper presented at the 36th annual conference of the American Association of Behavior Therapy.

Waters, A.J. and Sayette, M.A. (2006). Implicit cognition and tobacco addiction. In R.W. Wiers and A.W. Stacy (ed.), *Handbook of implicit cognition and addiction*. Thousand Oaks, CA: Sage, pp. 309–338.

Waters, A.J., Shiffman, S., Sayette, M.A., Paty, J.A., Gwaltney, C.J. and Balabanis, M.H. (2003a). Attentional bias predicts outcome in smoking cessation. *Health Psychology*, **22**, 378–387.

Waters, A.J., Shiffman, S., Bradley, B.P. and Mogg, K. (2003b). Attentional shifts to smoking cues in smokers. *Addiction*, **98**, 1409–1417.

Wiers R.W., Van Woerden, N., Smulders, F.T. Y. and De Jong, P.J. (2002). Implicit and explicit alcohol-related cognitions in heavy and light drinkers. *Journal of Abnormal Psychology*, **111**, 648–658.

Wiers, R.W., Wood, M.D., Darkes, J., Corbin, W.R., Jones, B.T. and Sher, K.J. (2003). Changing expectancies: cognitive mechanisms and context effects. *Alcohol: Clinical and Experimental Research*, **27**, 186–197.

Wiers, R.W., de Jong, P.J., Havermans, R. and Jelicic, M. (2004). How to change implicit drug use-related cognitions in prevention: a transdisciplinary integration of findings from experimental psychopathology, social cognition, memory and experimental learning psychology. *Substance Use and Misuse*, **39**, 1625–1684.

Wiers, R.W., Houben, K., Smulders, F.T.Y., Conrod, P.J. and Jones, B.T. (2006). To drink or not to drink: the role of automatic and controlled cognitive processes in the etiology of alcohol-related problems. In R.W. Wiers and A.W. Stacy (ed.), *Handbook of implicit cognition and addiction*. Thousand Oaks, CA: Sage, pp. 369–362.

Zack, M., Toneatto, T. and MacLeod, C.M. (1999). Implicit activation of alcohol concepts by negative affective cues distinguishes between problem drinkers with high and low psychiatric distress. *Journal of Abnormal Psychology,* **108**, 518–53.

Appetite lost and found: cognitive psychology in the addiction clinic

Frank Ryan

> Two principles of human nature reign;
> Self-love, to urge, and reason to restrain;
> Nor this a good, nor this a bad we call,
> Each works its end, to move or govern all.
>
> Alexander Pope (1733) *An essay on man* (Butt 1963).

Summary

Cognitive neuroscience characterizes addiction as a disorder of appetite resulting in the development of compulsive patterns of drug acquisition and consumption. The cognitive processes that govern the resulting sensitivity to incentive-related cues are relatively automatized. The outcomes of this processing are thus unavailable to introspection and resistive to modification. Further, because repeated drug intoxication is associated with deficits in executive control of cognitive functions such as attentional switching, the capacity of addicted individuals to overcome habitual responding is compromised. Evidence-based therapeutic interventions, including cognitive–behaviour therapy (CBT), must accommodate findings revealed in this monograph in order to maintain empirical status and the potential to deliver enhanced outcomes. There are three broad implications for therapeutic practice. First, CBT needs to acknowledge fully the autonomy of addiction in the face of efforts to change. This sets the context for a more resilient therapeutic alliance in the face of inevitable setbacks. Secondly, the primary therapeutic focus needs to be on efforts to enhance impulse control in advance of, or at least parallel with, attempts to improve affect regulation or interpersonal functioning. Thirdly, therapy should aim to modulate the impact of cognitive biases either directly through re-training or rehearsal, or more generally through mindfulness or other meta-cognitive approaches. Conversely, changes in implicit cognitive processes such as selective attention should, hypothetically, index therapeutic gain. Evidence of a reciprocal link between memory span and attentional processing suggests that interventions that infiltrate working memory (WM) are more likely to disrupt the cognitive processing of potential drug cues which can lead to relapse. There is a lack of prospective studies that evaluate the trajectory and predictive utility of cognitive processes in therapeutic contexts. This gap between experimental cognitive psychology and clinical research and practice needs to be filled.

10.1 Introduction

I encountered a former client in the waiting room of a Treatment Centre for Drug Dependence. This 32-year-old woman with a history of heroin addiction had remained drug free for 4 years following detoxification. She told me that during this period she had enrolled in a University degree programme and graduated with first class honours. However, she had since resumed heroin use and she was now once again attending the Treatment Centre. I offered her reassurance, re-referral and encouragement to continue with the treatment programme. The question addressed in this chapter is whether, in the light of recent findings in the cognitive sciences, there are grounds for re-conceptualizing current treatment models and, ultimately, the help available to people like this woman whose lives are disrupted by addiction. This applies not just to opiate dependence, but across the broad spectrum of addictive behaviour addressed here. I propose that the answer is a cautious, but nonetheless optimistic, 'yes'. The caution is due firstly to the fact that the emergent findings in cognition and addiction are largely cross-sectional in nature. Accordingly, the trajectory and predictive power of these primarily implicit cognitive processes in therapeutic contexts remain a matter for conjecture. Secondly, assuming clinical significance can be assigned to these hitherto neglected cognitive processes, the remedial mechanism is unclear. The promise of gaining additional leverage over the intransigence of addiction is, combined with the findings revealed in this monograph, sufficient justification to confront these challenges.

10.2 Cognitive behaviour therapy for addiction in transition

Given the subject of this monograph, CBT is understandably under scrutiny. This is because CBT accounts for the acquisition, maintenance and regulation of dysfunctional behaviour by invoking mechanisms of learning, a cognitive process. Pioneers of CBT were quick to adapt therapeutic procedures to modify addictive behaviour. Notable applications have been Marlatt and Gordon (1985) and Beck *et al.* (1993). These approaches were developed to focus on the patterns of thought and action that appeared to characterize addiction and co-occurring problems. The therapeutic objective was to supplant them with more adaptive coping mechanisms. Essentially, addiction was adduced from persistent attempts to cope with maladaptive core beliefs such as 'I am helpless' or 'I am unlovable' (Beck *et al.* 1993, p. 52). However, drug use, and subsequently dependence or misuse, are ultimately unsuccessful compensatory strategies in this regard, and the client is propelled into treatment. The prevalence of emotional distress among those seeking help for addictive behaviour problems seemed to confirm this: change the negative beliefs, modulate the mood, and the motivation for drug use will diminish or disappear. However, this is *ex consequentia* reasoning: because those seeking help for addiction show emotional dysregulation linked to negative beliefs, causality cannot be inferred in retrospect. Moreover, it is possible to challenge robustly the conceptual and empirical basis of social learning theory and its clinical applications in the treatment of addictive disorders (see Chapter 5).

In this chapter, I argue that, while affect and addiction are inextricably linked, in therapeutic contexts it is helpful to address them as separate components, with priority assigned to the latter. This is a heuristic stance which clinical judgement will sometimes over-ride. Nonetheless, there is a need to acknowledge the autonomy and potency of addiction, regardless of co-existing, or pre-existing, problems in other domains. Table 10.1 outlines the changes in emphasis that can be derived from recent findings in cognition and addiction. Essentially, the traditional model, derived from social learning accounts, emphasizes controlled cognitive processes, whereas the emergent model ascribes a fulcral role to automatic or uncontrolled cognition. Thus far, this is a relatively straightforward business, although inevitably an oversimplified dichotomy results. The real challenge is to derive effective therapeutic strategies that can alter the elusive cognitive processes that are nonetheless deemed crucial to the persistence of addiction and the almost inevitable reversals that occur following therapeutic gain.

10.3 **Neuroscience perspectives**

Insights from neuroscience research point to the involvement of both subcortical and cortical processing in addiction. Grant *et al.* (1996), for example, conducted neural imaging research with cocaine-addicted partici-pants. They found evidence of more global cognitive processing in areas such as dorsolateral and medial orbitofrontal cortex following exposure to drug-related stimuli. This contrasted with the activation of components of the dopaminergic system consistent with more direct pharmacological effects of actual drug intake. Consistent with this, Volkow *et al.* (2002) proposed that

Table 10.1 Contrasting emphases between CBT and cognitive neuroscience models of addiction

Factor	Current CBT model (influenced by social learning theory)	Emerging CBT model (influenced by cognitive neuroscience)
Causation	Multiple causation due to diverse learning opportunities or vulnerability experiences	Relatively specific causal mechanisms via neural reward pathways, but shaped by diverse environments
Individual differences	Non-specific role for individual differences; emphasis on environmental influences	Specific role for individual differences in susceptibility to addiction. Tendencies towards impulsivity or novelty seeking, and avoidance of dysphoria acting to increase incentive value of drug effects.
Appraisal process	Secondary appraisal: flexible, controlled process; focal attention	Primary appraisal: inflexible, automatic process evaluates drug cue as positive. Pre-attentive, but allocates processing priority
Role of affect	Addiction is compensatory. Dysphoria precedes drug use which is conditioned by negative reinforcement	Addiction develops because drugs are primary reinforcers. Dysphoria increases incentive value of drug and is consequential to drug use
Motivation	Drug users are motivated because they *like* the effects of drugs. This construct can be indexed by self-report.	Addicted drug users 'want' drugs whether they like it or not. This construct is not *directly* amenable to self-report, but is indexed by behaviour, arousal and implicit cognitions.
Attention	Non-specific role for attentional processes	Incentive salience implies hypervigilance: salient stimuli recruit focal attention and preferential access to working memory
Neuro-cognitive deficits	No formal role assigned to neurocognitive deficits	Neurocognitive deficits from drug effects contribute to the persistence of addiction
Sequence of craving	Craving precedes drug seeking	Drug seeking can precede craving or occur in its absence

This highlights the contrasting assumptions in the respective models. It perhaps exaggerates the dichotomy, but there are fundamental differences nonetheless. In particular, the existing CBT model assigns drug use an instrumental rather than a primary role. Drug use is a response to something else, such as traumatic exposure, maladaptive schemas or interpersonal difficulties. This diverts the therapist, and the client, away from the addictive behaviour. Further, the current CBT model does not specify a role for implicit cognition. The existing therapeutic approaches do deliver gain, but this does not validate some of the assumptions, or justify the omissions.

the highly recurrent aspect of drug administration subverts normal self-regulation processes via altered dopaminergic functions in mesolimbic and mesocortical circuits involved in memory and decision making. The marked fluctuations in dopamine availability following intoxication are seen as a necessary but, ultimately, insufficient, condition of the emergence of addiction. As Volkow and her colleagues note, a role for dispositional and environmental variables must also be delineated. This is because, fortunately, only a minority of individuals exposed to drug-based reinforcement develop the compulsive habits of addiction. Individual differences in susceptibility to addiction will be addressed in a subsequent section.

Neuroimaging findings implying deficits in executive control among addicted populations are highly relevant from a therapeutic perspective, because they signal vulnerability to relapse. Recent findings have provided insight into the cognitive processes underlying this susceptibility of executive functions such as WM to overload and the deleterious effect this can have on the control of attention. Hester and Garavan (2005), working with a non-clinical sample, showed that increasing the WM load by adding a secondary distracter task reduced control over attentional engagement and disengagement. Deficits in executive control, indexed by difficulty in task switching and inhibition, were particularly pronounced when stimulus items coincided with those in WM. These authors speculated that this could echo the experience of addicted people who tend to maintain drug-related thoughts and images in WM. May *et al.* (2004) emphasized the key role of mental images in facilitating the recycling of appetitive impulses in their Elaborated Intrusion theory of craving. Both of these accounts and related empirical findings suggest that altering the contents of WM in the context of addiction would lessen the distracting or intrusive effect of drug cues and the ensuing reaction. Thus, a self-perpetuating cycle of active rehearsal in WM and increased attentional salience of drug cues could increase the frequency and duration of episodes of cue reactivity.

These findings help bridge the gap between the rather reflexive cognitive processes that engender cues with positive valence and the subsequent persistence and ingenuity with which drug-addicted individuals pursue their favourite drugs. They also sound a cautionary note for inferences derived from research with non-human species that lack the capacity to encode drug cues in abstract codes or represent them in higher levels of consciousness (see, for example, Teasdale and Barnard 1993; Damasio 1999). Applied to a client attempting restraint from drug use, the implication is that they have to battle on two fronts. First, 'bottom-up' stimulus-driven processes trigger cue reactivity. Secondly, the inhibitory influence of 'top-down' processes governed by higher cortical systems is compromised. This could be because competing non-drug rewards have relatively diminished appetitive motivational properties and do not capture attention quickly enough to ensure a more equitable share of scarce WM capacity. Lack of representation in WM span further reduces the likelihood of non-drug cues capturing attention and leaves the door open for drug cues.

10.4 **Treatment outcome research**

If empirical support for CBT for addiction had been more consistent, there would be less need for the re-conceptualization outlined here. Thus, Carroll (1996) concluded after reviewing 24 randomized controlled trials focusing on smoking, alcohol, marihuana and cocaine addiction that CBT applications such as relapse prevention skills training (RP) was more effective than no treatment but not superior to approaches such as supportive therapy. The robust effect sizes routinely observed in other applications of CBT have been the exception rather than the rule when individuals seek help for addiction. In a review of 26 comparative treatment studies involving 9504 participants, Irwin *et al.* (1999) found, on average, a small but reliable effect size ($r = 0.14$) for reducing substance misuse. In contrast, the reliably observed effect size ($r = 0.48$) for improved psychosocial functioning was notably larger. CBT thus appears to be more useful, compared wih other approaches, in addressing issues such as poor affect regulation and interpersonal coping skills rather than the compromised impulse control that defines addictive behaviour. Clearly, this has been a major contribution to treating addicted clinic populations, but leaves room for improvement in addressing the key issue of addictive behaviour.

The conclusions from controlled trials comparing CBT with other approaches are also equivocal (Newman 2004). Thus, in one controlled study (Crits-Christoph *et al.* 1999), both cognitive therapy and brief psychodynamic therapy led to significantly poorer outcomes compared with traditional Twelve Step counselling. This was despite the fact that the Twelve Step clients attended significantly fewer sessions. In another well-designed study, Litt *et al.* (2003) found that good coping skills were predictive of good outcome in their cohort of 128 alcohol-dependent individuals. However, specific coping skills training as part of a CBT programme did not appear necessary for the acquisition of these clearly relevant skills, as participants allocated to a condition not involving explicit coping skills training showed a similar learning curve. A recent UK trial (UKATT 2005) compared social behaviour and network therapy, which utilizes problem drinkers' social networks to initiate and to maintain change, and motivational enhancement therapy, a manualized version of motivational interviewing. Again, both groups (total $n = 742$) showed substantial, but equivalent, decreases of 44 per cent in alcohol consumption when followed up after 12 months.

Clearly, therapeutic intervention delivers gain, but the mechanisms of such change are obscure. The fact that interventions based on apparently mutually exclusive concepts such as acknowledging 'powerlessness' in Twelve Step Facilitation and supporting self-efficacy in Social Learning Theory can generate similar outcomes (Project Match 1997) is an enigma. Non-specific factors such as therapeutic engagement or motivation could account for these uniform findings. Litt *et al.* (2003), for example, speculated that therapeutic programmes simply provide an opportunity for individuals to capitalize on their motivation to change. Rather like an undiscerning theatregoer, the action on stage is apparently less important than the commitment shown in buying a ticket and the experience of sitting quietly for a couple of hours.

Implicit cognitive processing could, however, be another candidate as a non-specific factor. The assumption is that, regardless of the therapeutic discourse proceeding at a conscious level, implicit cognitive processes might show equivalent plasticity in a given group of clients despite diverse clinical protocols. Thus, whether a client attends a Twelve Step programme, a motivational enhancement session or engages in cognitive therapy, changes in cognitive processes can occur at multiple levels. Existing accounts, for instance those derived from social learning theory, tend to oversimplify this, in particular overlooking the role of processes such as attention and implicit memory. It is argued here that this cognitive processing could be an authentic index of therapeutic change, although its latency makes it easy to overlook. This gives rise to two implications. First, measuring implicit cognitive processes is deemed clinically relevant. Secondly, interventions that appear to exert more influence on this level of cognition might be superior to those that exert less. These hypothetical claims need to be tested by conducting further research, in clinical settings. If proved valid, the subsequent challenge will be to devise therapeutic protocols that acknowledge the potentially decisive influence of these cognitive processes.

10.5 Time for *CHANGE*: re-conceptualizing cognitive–behaviour therapy for addiction

Because implicit cognition is an elusive target for therapeutic procedures, the contemporary therapist has little option but to use tried and tested techniques for changing cognition, behaviour and affect. The implications for therapeutic intervention stemming from recent findings on cognition and addiction are so far a matter for speculation, albeit relying on a growing empirical base (see Chapter 9). A further constraint stems from the fact that addressing implicit cognitive processes can be repetitive and the procedures lack face validity. These procedures could, of course, be embedded in more elaborate and engaging therapeutic manoeuvres. Nonetheless, there are at present only indicative findings justifying the importing of experimental cognitive and behavioural techniques into the addiction clinic. Accordingly, a pragmatic approach to psychological intervention is proposed here. The focus of therapy should be on coping with the inevitable intrusions triggered by cognitive processes that, by default, continue to detect and select drug cues. This involves first preparing the client for this eventuality and encouraging them to view these 'drug alerts' as a symptom of the disorder rather than a sign of weakness or lack of will power. Secondly, devising strategies to maintain the goal of restraint in WM seems justified. Hypothetically, influencing the contents of WM will impact on attentional switching from a drug-seeking focus and thus free scarce cognitive resources necessary for the deployment of coping skills. This broad cognitive behavioural framework, part blueprint, part prototype, is known as CHANGE. This is an acronym for Change Habits and Negative Generation of Emotion. The aim is to address impulse control and affect regulation sequentially, but as part of an integrated treatment plan.

The components of this model will, no doubt, be familiar. In one sense, the cognitive processing perspective further legitimizes the conventional therapeutic focus on relatively specific person–drug–situation interactions. This is because, ultimately, overcoming drug addiction is about consistently making the 'right' decisions and having the skills to enact them. The theoretical assumptions underlying existing CBT approaches and the CHANGE model informed by cognitive neuroscience research are nonetheless distinct and occasionally contradictory (see Table 10.1). Emerging data on cognitive processes in addiction focus on cognitive processes that are antecedents to these decision points. The evidential basis for this will now be briefly reviewed under headings corresponding to positive and negative reinforcement accounts of addiction. Next, the role of individual differences and neurocognitive deficiencies in therapeutic contexts will be addressed. The implications for therapeutic engagement, assessment and intervention will then be outlined in more detail.

10.6 Enhanced reward detection

Cognitive appraisal of cues predicting threat or reward occurs rapidly and pre-consciously (see Whalen 1998; LeDoux 1999). The evolutionary significance of such crude but effective 'reward radar' in the environments where access to food and water is limited is clear. Schultz (2002, p. 257), however, pointed out that drugs *directly* target neural reward-processing centres, compared with natural rewards such as food or water. This results in elevation of dopamine levels in areas such as the nucleus accumbens which is significantly greater than that elicited by conventional rewards. Crucially, this dopamine release consolidates associative learning. It appears, therefore, that exposure to the reinforcing effects of drugs can activate elementary learning mechanisms from the outset. With repetition, and possibly dependence, exposure to stimuli predictive of drug availability often precipitates relapse (Rohsenow *et al.* 1992; Carter and Tiffany 1999). Enhanced detection (i.e. attentional bias) of these cues is thus a relatively direct implication of the motivational properties that they acquire via associative learning.

Until recently, the role of processes such as selective attention in cue exposure has been largely overlooked. As indicated above, invoking a role for biased cognitive processing is entirely consistent with appetitive motivational accounts of addiction. Robinson and Berridge (1993, 2003) have proposed that the transition from controlled drug use to compulsive drug use is due to the induction of hypersensitivity in neural reward mechanisms. This is differentiated, both phenomenologically and in terms of the neural substrate, from the hedonic or 'liked' effects of a drug. The attribution of incentive salience invests cues with powerful motivational properties because they become associated with the reinforcing effects of dopamine release, specifically in the mesolimbic dopamine pathway. This activation ultimately recruits more global cognitive processing but more immediately it imposes its own imperative on attentional deployment. In this regard, hypersensitivity implies hypervigilance. Moreover, because sensitization is subserved by morphological and neurochemical changes in neural reward circuitry, it is an enduring vulnerability factor for relapse.

This assumed persistence of incentive salience indicates that appetitive cues can trigger drug seeking long after the motivational effects of withdrawal dysphoria have receded. Incentive salience is also assumed to outlast the hedonic or 'liked' aspects of drug use. Accordingly, a person can end up wanting drugs more but liking them less.

In summary, concepts such as incentive salience have direct implications for psychological processes in addiction because they infer a fulcral role for associative learning and attentional processing. Further, they provide insight into the dissociation between the expressed intentions of the client, which index like or dislike, and subsequent drug use, which reflects wanting or compulsion. This is a key reference point for constructing an evidence-based protocol for psychological intervention with addictive behaviour.

10.7 **Negative reinforcement**

In recent years, the pendulum has swung towards appetitive motivational accounts of addiction that emphasize the role of positive reinforcement (Stewart *et al.* 1984; Robinson and Berridge 1993, 2003). Baker *et al.* (2004) have attempted to reverse this and they have recently reformulated the negative reinforcement model. They propose that escape from or avoidance of negative affect is the primary motive for addictive drug use. Thus, negative affect, withdrawal severity and stress all prove predictive of relapse. These theorists have pointed out that affectively valenced information is prioritized pre-consciously, thus ensuring that drug users have a steady supply of mainly interoceptive cues which motivate drug seeking.

Clearly, both positive and negative affect are intimately involved with drug motivation: drugs are used, at least initially, because they change affect or arousal. It is beyond the scope of the present chapter to address the issues raised in any depth. However, Baker *et al.* (2004, p. 47) acknowledge that at high levels of negative affect drugs have 'great incentive value'. Regardless, therefore, of the contradictions between negative and positive reinforcement accounts, both emphasize the capacity of somewhat diverse cues to motivate drug-acquisitive behaviour after acute withdrawal symptoms have diminished. Both accounts also emphasize the efficient and preferential way in which these cues can be processed, regardless of whether they are exteroceptive or interoceptive. As indicated by the chapter title, my view is that addictive behaviour is essentially an abnormal appetite, albeit one that is accentuated in the context of adverse mood. The appetitive essence of addiction has not, however, been allocated the prominence it deserves in theoretical accounts of therapeutic intervention and clinical applications.

10.8 **Individual differences**

Despite the involvement of powerful conditioning mechanisms, most individuals who are exposed to drugs do not become addicted (Anthony *et al.* 1994). Existing accounts invoke a role for 'low frustration tolerance' or 'sensitivity to unpleasant feelings' (Beck *et al.* 1993, p. 39) as vulnerability factors for addiction. It is assumed that these emerge from environmental events experienced as invalidating or abusive. Encoded as core beliefs (e.g. 'I am vulnerable'), the scene is set for avoidance or safety-seeking behaviour such as drug use. Recent findings indicate that individual variation in susceptibility to addiction appears partially due to genetic factors. Early reports of allelic association of the dopamine D2 receptor gene and alcoholism (Blum *et al.* 1990) have generally been supported by subsequent findings and meta-analytic studies (see Young *et al.* 2004). Converging findings that polymorphisms of the D2 receptor gene are associated with low D2 receptor density in areas such as the striatum and related structures (Noble *et al.* 1991) point to a common neurobiological substrate that might make individuals more sensitive to alcohol- or other drug-induced reinforcement. The inference is that those deemed susceptible by virtue of their allelic status might experience stronger impulses to repeat drug-seeking and drug-taking behaviours. This could be reflected in cognitive processes such as attention and motivation. Accordingly, Young *et al.* (2004) found that problem drinkers positive for the dopamine D2 A1 allele exhibited lower self-efficacy regarding their drink refusal skills than a cohort without the polymorphism.

Tendencies towards novelty-seeking or anti-social behaviour are often observed in cohorts of substance misusers. This suggests that there might be a common genetic susceptibility to novelty seeking or dysphoria that also invests psychoactive drugs with more incentive value. Hiroi and Agatsuma (2005) concluded that postulating pre-existing differences, presumably partial expressions of genetic variation, had heuristic value in accounting for individual variation in addiction liability. It appears, therefore, that activation of neural reward mechanisms is a prerequisite rather than a determinant of addiction. This is because individuals vary in their need for stimulation or avoidance of emotional distress. These sensitivities suggest that there are at least two potential pathways to addictive disorders, driven by positive and negative reinforcement. This complicates matters for addiction theorists, not least because these susceptibilities are not mutually exclusive. It matters little, at least initially, to those who find the effects of drugs hedonic, remedial or both. What does eventually count is that the acquisitive behaviour is repeated, and becomes compulsive and inextricably linked to affective regulation. Parallel with this, the cognitive processes regulating this behaviour become increasingly automatized and selective. This processing is intimately involved in the regulation and persistence of addictive behaviour. It is therefore important to construct a theoretical framework that models this processing in advance of evaluating the therapeutic implications.

10.9. Neurocognitive deficits

Neuropsychological impairment is commonly observed in the context of chronic substance misuse. Despite a burgeoning literature, the extent to which observed neurocognitive deficits either pre-date, or result from chronic exposure to drug effects is not established (Rogers and Robins 2001). These reviewers concluded that, with the exception of chronic alcohol excess, neuropsychological deficits in attentional and mnemonic functions appear to be rather subtle with regard to drug preference. A number of studies (e.g. Grant *et al.* 2000) found that chronic drug users manifest intact performance on standard clinical neuropsychological tests such as the Wisconsin Card Sorting Test, but impaired decision making when required to estimate short- and long-term gains on a gambling task: about two-thirds of the substance-dependent group showed deficits comparable with those shown by patients with ventromedial prefrontal cortex lesions. The assumption is that this reflects a failure to acquire implicit knowledge, indexed by anticipatory 'somatic markers', about the emergent contingencies between choices made on the gambling task and outcome. However, other learning mechanisms such as reversal learning or inhibitory failure cannot be discounted (see Dunn *et al.* 2006). Further, intact WM is also important for the development of effective strategies on the gambling task. Regardless of the underlying mechanism, the possibility exists that relatively subtle deficits in reasoning could undermine a therapeutic procedure that tasks the client with acquiring relatively advanced self-regulation skills in pursuit of both short- and long-term goals.

Further, impaired executive control among substance misusers has been linked to characteristic explanatory styles. Garcia *et al.* (2005) reported that performance on cognitive flexibility and inhibition tasks was positively associated with a more adaptive cognitive style and inversely related to some components of a 'learned helplessness' attributional style (Abramson *et al.* 1978). This suggests a theoretical mechanism through which cognitive deficiencies, whether inherent or acquired, can directly alter the probability of effortful coping being enacted.

10.10 A theoretical framework

I have proposed elsewhere (Ryan 2002) that cognitive processing influences cue reactivity by enhanced detection and biased appraisal. The detection of drug cues is automatically facilitated regardless of the conscious goal of the restrained drug user (e.g. 'I want to avoid triggers for using drugs'). Due to the relentless nature of this hypervigilance, drug cues tend to engage attention even if they are unpredictable and occur over a lengthy period. Once detected, the ensuing evaluative appraisal assigns the cue a positive valence by default. This primary appraisal can elicit components of cue reactivity such as physiological arousal, behaviour and cognitions (e.g. expectancies) in advance of, or parallel with, the recruitment of focal attention. These components of cue reactivity are subjected to more elaborate appraisal mediated by attentional and inferential biases. It was argued that this selective processing contributed to the weak relationship between physiological arousal and subjective reports of craving noted by Carter and Tiffany (1999) when they conducted a meta-analysis of 41 cue reactivity studies. They found modest effect sizes for physiological activation such as skin conductance and significantly larger effect sizes for subjective reports of craving. The experience of craving thus emerges as a rather indirect and cognitively synthesized representation of essential components of cue reactivity. Further, the dissociation across components of cue reactivity was assumed to increase in proportion to the chronicity of addiction and the concomitant strength of cognitive biases. The implications for the addicted individual aiming to abstain were 2-fold. First, he or she was liable to detect a wide range of appetitive cues regardless of conscious intentions or preferences. Even if these stimuli were ambiguous, inferential biases guided appraisal to evaluate them as predicting reward. Secondly, once cue reactivity was initiated, it too was subjected to highly prepared cognitive processing.

This account, emphasizing the role of attentional processes and primary appraisal mechanisms, provides at best only a partial insight into cognitive processing in addiction. In particular, the role of WM was not delineated, although it was assumed that access to this scarce resource was the prize awarded to selected drug cues. WM capacity appears, however, to be crucial in governing attentional switching, inhibitory control and goal maintenance in the face of distraction. Kane and Engle (2003) characterized this aspect of WM as an 'executive-attention' mechanism that maintains stimulus representations, action plans and goals. These researchers found that individuals with high WM capabilities evidenced less interference on versions of a Stroop task (Stroop 1935) with many congruent trials (e.g. high proportions of the word *RED* printed in red ink) as well as better performance on the standard colour-conflict version. Thus, good WM functioning predicted good attentional control.

10.11 Applying new findings to psychological interventions

The emerging theme from the foregoing is that both implicit and explicit cognitive and motivational processes are influential in the governance of addictive behaviour. However, the former have been neglected, mainly because they are neither reportable nor easily measurable (see Table 10.1). Processes such as attentional deployment are latent factors, invisible to both therapist and client. An attempt is made here to redress the balance by focusing on the role of implicit cognitive processes. Superficially, ascribing a key role to implicit, and hence largely unreportable, cognitive processes challenges much of what addiction therapists and their clients do in therapeutic settings. For instance, if there is a discrepancy between what a client knows they would like (to remain drug free) and what they want (more drugs), eliciting self-efficacious statements might be less relevant as a predictor of outcome. In this regard, even though the client knows what she likes, she does not in fact know what she wants. This is because the 'wanting', in the addictive sense, is not directly available to introspection.

The question addressed in the following section, and indicated in the opening paragraph, is what 'added value' could be adduced from applying new findings in cognitive psychology. Beginning with an exploration of the implications of recent cognitive findings for the therapeutic alliance, the section will go on to explore possible avenues of therapeutic intervention suggested by recent findings in cognition and addiction. It is firstly assumed that involuntary processing of addictive cues will potentially elude remedial strategies, but intervention can address the consequences of this intrusive cognitive activity in a variety of ways. Secondly, the approach accentuates the existing emphasis on ready access to and deployment of coping skills in situations where drugs are available. This is influenced by findings of relatively subtle cognitive impairment in areas such as decision making and evaluating the likely utility of planned action. More generally, it is proposed that recent findings in cognitive processes in addiction provide a more constructive but less potentially judgemental framework for forming and maintaining a robust therapeutic alliance.

10.12 The role of cognitive processes in therapeutic engagement

Recent findings relating to the role of cognition in addiction offer insight as to why apparently genuine commitment to change offered in the clinic does not reliably generalize to real life situations: these undertakings are based on conscious, deliberate cognitive processing, whereas susceptibility to cue reactivity and hence relapse is supported by automatized processes that are not directly represented in conscious awareness. There is a risk that both client and therapist are relegated to the role of onlookers as automaticity takes its course. The concept of 'loss of control' has a long history in addiction and resonates in 'Twelve Step' approaches. Cognitive science aims to delineate the specific mechanisms of volition or its absence. Consider the following brief transcript from a therapy session with 'JJ' a 55-year-old man who had completed a detoxification from alcohol a week earlier:

I've left everybody down again. I'm such an idiot. There must have been about twenty or so different staff-nurses, doctors, therapists- involved in getting me sorted. What did I do? After three days I bumped into a friend on the King's Road and he invited me for a drink. I wasn't even thinking about a drink. I wasn't craving for a drink. I had two drinks but then I went home. I drank more alcohol yesterday. I feel I've left myself and everybody else down, after all the help I got. I'm a real idiot.

JJ, 55-year-old male

Existing approaches (e.g. Marlatt and Gordon 1985) offer very cogent advice to therapists on how to approach this eventuality. Imagine, however, that this scenario was repeated several more times, not just with one client but with many. Even the most resilient therapist can experience frustration at the divergence between the expressed intentions of the client ('I am determined not to use this week-end') and their subsequent report ('I used this week-end'). Addiction therapists have attributed this inconsistency to factors ranging from 'denial' to deficiencies in coping skills, commitment or motivation on the client's part. This attribution is often aided and abetted, if not pre-empted, by the client. This misattribution can signal disengagement from treatment and decrease the self-efficacy of the therapist in the longer term.

Inevitably, assertions such as this are oversimplifications. Nonetheless, when treatment fails or suffers setbacks, attributional judgements result. These often highlight the apparent discrepancies between verbal statements on behalf of the client and his or her subsequent behaviour. If, for instance, a client has agreed to avoid triggers associated with drug use but encounters them anyway, a negative attribution could be made (by the client, the therapist or both) along the lines that the client was 'setting themselves up', or undermining the therapeutic process. If, on the other hand, it is acknowledged that the client *is predisposed* to detect and select such cues due to rapid and involuntary cognitive processes, a less pejorative attribution can be applied. This attribution is more about responsibility than culpability. It serves to focus the minds of both parties on the task of over-riding patterns of thought and behaviour that have become 'second nature'. This, of course, does not in itself prevent a reoccurrence of proscribed appetitive behaviour. Nonetheless, it offers a more constructive framework within which to accommodate setbacks on the road to recovery.

10.13 Assessment of cognitive processes

Directly measuring implicit cognitive processes is difficult, and the techniques are not generally available, or practicable, in the clinic. Regarding attentional processes, for instance, the experimentally derived measurement methods such as the dot-probe or modified Stroop paradigm are necessarily repetitive and therefore potentially tedious. This is merely the establishment of a baseline response: if a procedure designed to *reverse* such putative biases were devised, it is likely that much higher levels of client responding would be required. In this regard, it should be recalled that sensation seeking is a factor implicated in the development of drug dependence. An obvious implication of this is that therapy needs to be as engaging as possible.

First, there are as yet only sporadic reports of the extent to which cognitive biases might serve to predict outcome. Secondly, across the spectrum of research, including that in this text, ecological validity has generally been sacrificed on the altar of internal validity. This is important, because, at the risk of stating the obvious, drug taking takes place in the natural habitat of the drug user. This is an entirely different perceptual world compared with the clinic or psychology laboratory. Surprises are, for instance, difficult to factor into a controlled experiment. This is important because a hypervigilant stance could make 'surprises' distinctly more likely (see Ryan 2002). Combined with findings that relatively unpredictable rewards can trigger increased dopaminergic neuronal activation (Schultz 1998), this suggests a potential vulnerability factor for relapse. Although this research was not with human participants, it is often a chance encounter that triggers resumed drug taking. Thus, a client who had managed over a year's abstinence from opiates reported that he found, by chance, a small amount of heroin hidden under his carpet. He immediately smoked this and subsequently relapsed into daily intravenous use of heroin. This scenario highlights the challenge faced by those who try to overcome their addictive behaviour: one cannot rehearse coping skills for all eventualities.

Accordingly, it seems plausible that individuals attempting restraint will be assailed by cues that elicit reactivity. The fact that this cue reactivity is initiated by partially involuntary cognitive processes could serve to undermine even those most committed to abstinence. The relentless nature of this preoccupation could lead to passivity mediated by appraisals such as 'I can never escape from my craving'; or 'Everywhere I go I see alcohol'. In turn, this can reduce self-efficacy and the generation of effortful coping. There is certainly adequate evidence, including much of the foregoing in this volume, to justify advising clients embarking on relapse prevention programmes that they are susceptible to detect preferentially the very cues that might signal a 'high risk situation'. This supplements rather than discounts earlier accounts (see Marlatt and Gordon 1985) suggesting that high-risk encounters happened either by chance or perhaps through self-deceptive 'seemingly irrelevant decisions' that led the individual to environments where cues were more prevalent. Cues can trigger drug seeking whether they are detected by accident or intent. By informing clients of this, they can re-interpret this and practise disengaging attention from cue reactivity.

10.14 Assessment and formulation

The following statement is that of a 41-year-old man with a history of cocaine addiction who relapsed after 8 months abstinence. He had been unexpectedly offered cocaine after a meal with friends at a restaurant and accepted. His subsequent reaction in the session was as follows:

What is it about me, just when I get things right for a change, I start using again and end up relapsing. I seem to push the 'self-destruct button'. It must be that, deep down, I just want to be a failure.

A psychotherapist trained in any modality would find this statement relevant and dedicate resources to addressing it. However, there is an attributional error possible for both parties. Understandably, the distraught client is seeking to account for his behaviour and the therapist has a contribution to make on this issue. The antecedents of this resumption of cocaine use might be more proximal than distal however, reflecting perhaps the sudden exposure to a cue. This more parsimonious account is also less judgemental. The implication is not that negative self-evaluations such as that uttered by the client should be overlooked. Clearly, these should be carefully acknowledged and addressed in due course. However, this should not allow for either therapist or client to be diverted from the priority, which is to devise coping strategies for dealing with the compulsive nature of drug seeking. If, for example, three or four sessions were used to prioritize tackling negative self-esteem, the result might be an individual with a more positive self-image, who would nonetheless remain vulnerable to resuming drug use. Existing CBT protocols do, of course, address coping with urges and craving. However, in the search for maladaptive core beliefs, relevant though they may be, the centrality of the addictive component is not absolutely prioritized. This could account for findings showing equivalence, or in some cases superiority, of Twelve Step approaches when compared with CBT. What the former possess is a relentless focus on the addictive behaviour, whether it involves compulsive drug use or compulsive gambling.

10.15 Specific therapeutic strategies

10.15.1 Self-help approaches

There is currently sufficient evidence to justify offering advice derived from cognitive research in addiction to clients attempting to quit addictive habits.

This assumes that while it is difficult to alter implicit cognitive processes directly, it is better to be fore-warned, if never entirely fore-armed. A proto-type (Ryan 2001) structures the advice under the heading 'Six Tips'. This was designed to be used as an adjunct to therapeutic work. It attempts to appraise clients about aspects of cognitive processing that might facilitate relapse. The leaflet advises, for example:

The world isn't full of bars and off-licences—it's just that wherever there is one it grabs your attention! This is another trick that your mind can play you by being selective. It's really trying to be helpful as in your drinking days this was what was required. This is a bit like the saying 'your left hand doesn't know what you right hand is doing' and is a telling reminder of how even with loads of willpower we still find ourselves doing precisely what we earlier decided not to. As an exercise, try and imagine your locality (sketch out a map if you like). How frequently do pubs, for example, appear? Now try the same exercise but deliberately concentrate on other significant features: prominent buildings, landmarks or other features.

Other 'tips' refers to memory bias such as forgetting the negative conse-quences of substance misuse or interpreting craving as evidence of lack of commitment to change.

10.15.2 Cognitive processes in therapeutic intervention: managing intrusions

A general theme to emerge from the foregoing is the intrusive nature of the antecedents of addictive impulses and behaviour. This can distract the indi-vidual from the goal of restraint as well as triggering rumination, which in turn primes enhanced detection of appetitive cues. As well as ensuring the prominence of addictive cues, the resulting annexation of WM can lead to suppression or neglect of other goals such as the rehearsal and deployment of coping skills in the face of drug cues. The drug-addicted person can, albeit momentarily, forget (i.e. not retain in WM) that they were aiming for restraint and the coping strategies they have painstakingly learned. Invoking a role for WM thus has two key implications for therapeutic intervention. First, procedures that enhance WM functioning are more likely to deliver lasting therapeutic gain: clients who learn to maintain the abstinent/restraint goal and to regulate attentional switching are more likely to achieve long-term change. Secondly, the neural mechanisms such as the dorsolateral prefrontal cortex essential for intact WM functions are precisely those liable to disruption or impairment when exposed to the repeated neurotoxicity of drugs. Accord-ingly, executive control is compromised when it is most needed. In the parlance of cognitive psychology, those tackling their addiction are in effect engaged in a 'dual-task' paradigm involving the maintenance of inhibition and frequent switches of attention in the face of frequently detected cues.

This provides a novel perspective for conceptualizing and implementing therapeutic strategies. Invoking a role for WM, with its definitive emphasis on rehearsal, suggests one reason why Twelve Step approaches generate outcomes that are as good as, if not superior to those based on clinical psychology such as CBT. The relentless focus on mantras such as 'one drink, one drunk' or 'one day at a time' serves to prioritize the goal of restraint in WM. Another Twelve Step strategy is to attend '90 meetings in 90 days' which serves to keep the goal of restraint prescient in the longer term. Because WM has a limited capacity function, this also reduces the cognitive resources available for processing appetitive cues whether it is one day at a time, 90 days or several years.

10.14.3 Metacognition

There are two possible approaches to addressing implicit cognition therapeutically, one specific and direct which aims to modify; the other more diffuse that aims at helping the individual accept and detach themselves from the inevitable. First, implicit cognitive processes can be made more explicit and thus modifiable by attentional training (Wells 2000) or expectancy challenge (Darkes and Goldman 1998). Secondly, a more global strategy is to encourage mindful acceptance of drug-seeking urges and the negative affect that often precedes them (Breslin *et al.* 2002). Insofar as these approaches emphasize the monitoring and regulation of one's own thought processes, they can be labelled 'meta-cognitive'.

There is, however, a lack of research with populations of addicted individuals that might guide clinical practice. Some cautious generalizations can be made on the basis of findings from anxiety and depression research. Accordingly, while it is argued here that addiction has a distinctive cognitive profile, intrusion and preoccupation are by no means unique features.

Therapeutic protocols effective in managing intrusions in the context of emotional disorder are therefore a good place to start. Evidence supports the efficacy of 'mindfulness-based stress relief' (MBSR) and mindfulness-based cognitive therapy (MBCT) for depression. There are at least two reasons why these findings are relevant to addictive disorders. First, MBCT has been designed as a relapse prevention strategy. It involves working with people with a history of depressive disorders, but who are in remission when introduced to MBCT. Teasdale *et al.* (2000) recruited 145 adults with a history of depression. MBCT appeared to prevent relapse in patients who had experienced three or more episodes of depression: 66 per cent of those who received treatment as usual relapsed in the 60 week study period compared with 37 per cent of those offered MBCT. The selective benefit to those who relapse more frequently suggests that mindfulness might help particularly where selective processing of affective cues follows on with the well-practised lines of 'here I go again; things just don't ever get any better'. Applied to addiction, this raises the possibility that the detachment that is at the core of mindfulness could disrupt the automatic flow of drug-related cognitions that are the key components of cue reactivity. Secondly, negative affect is associated with relapse; therefore, any techniques that help the individual accept or process this should have added value with addictive disorders.

10.14.4 Therapeutic intervention: concluding comments

The implicit nature of the largely implicit cognitive processing highlighted above presents both a challenge and an opportunity to those involved in orchestrating therapeutic change. Here, I have addressed the challenge by outlining some of the implications for therapeutic practice. The opportunity is that those confronted with addiction, whether they are addicted or addiction therapists, can benefit from greater knowledge of the issue they address. Invoking a role for implicit cognitive processes calls for evolution rather than revolution in therapeutic intervention for addiction. Returning to the opening vignette, it is likely that therapy could evolve in a number of ways. First, demonstrable gain in areas such as personal development or affect regulation, while important, is not in itself a 'cure' for addiction. Clearly, changes in these domains are protective factors, but do not indemnify the addicted person against relapse and therefore should not enable complacency to develop. Modifying addictive behaviour must remain the primary, explicitly stated objective. Accordingly, both clients and their carers need to view recovery from addiction as a long-term commitment, with monitoring and follow-up continuing after active psychological and pharmacological components have ceased. A corollary to this is that individual differences, putatively expressed in implicit cognitive processes such as attentional deployment, might confer heightened risk of relapse on subgroups of clients. If measured, these latent factors could provide clinician and client with indications of vulnerability to relapse hitherto invisible. Accordingly, the implication is that a 'one size fits all' approach to relapse prevention needs to be replaced with a more refined risk analysis based on individual differences, whether these are reflected in cognitive processes, cognitive deficits or both. Secondly, while perhaps making a virtue of necessity, much of the foregoing supports conventional psychopharmacological approaches to addiction treatment. However, a more circumspect view of indications of insight and motivation now seems justified: findings revealed in this volume and elsewhere suggest that self-reports can at best be only partial indices of commitment to change addictive behaviour, and at worst misleading. As argued earlier, this dissociation can undermine the therapeutic alliance which is the platform for therapeutic engagement and progress. Thirdly, converging findings implicating implicit cognitive processes highlight points at which the cycle of drug seeking and satiation that defines addiction can be disrupted: I propose that the extent to which evolving therapeutic interventions serve to preoccupy WM will reflect their utility.

10.15 **Conclusion**

The challenge facing those who try to overcome addiction is to outsmart compulsive habits. The empirical evidence presented in this monograph, as well as other findings reviewed in this chapter, suggests that vulnerability to addiction has a distinct and enduring cognitive signature. Once cues, and the drug-induced rewards that they predict, become sought after and overvalued, considerable cognitive resources are captivated by this ultimately futile quest. The approach to therapeutic intervention outlined here implicates cognition in the psychological treatment of addiction in three inter-related ways. First, the incentive properties of drugs engender predictive cues with motivational properties. These cues attract preferential processing manifested in attentional and inferential biases. Secondly, this intrusive processing increases the likelihood of cues gaining access to more elaborative cognitive processes such as WM. This can disrupt other, possibly more adaptive, activities. Thirdly, the relentless nature of this involuntary and selective processing usurps motivation and commitment to change by engendering negative self-evaluations Moreover, subtle cognitive impairment in areas such as forward planning and decision making could contribute to problems in many domains. Addiction can be potentiated by many factors, but affective dysregulation is inextricably linked to drug use genetically, neurophysiologically and psychologically. Accordingly, the prototypical *CHANGE* approach to treatment outlined here addresses impulse control and affect regulation strategies sequentially, but as part of a formulated treatment plan. These assertions have important implications for how therapeutic intervention is conceived and delivered and certainly need to be tested and refined in the clinical arena.

Acknowledgements

Any correspondence concerning this chapter should be addressed to Dr Frank Ryan, Department of Psychology, Central and North West London Mental Health Trust, Hammersmith, London, W6 7DY.

References

Abramson, L., Seligman, M.E.P. and Teasdale, J.D. (1978). Learned helplessness in humans: critique and reformulation. *Journal of Abnormal Psychology*, 87, 49–74.

Anthony, J.C., Warner, L.A. and Kessler, R.C. (1994). Comparative epidemiology of dependence on tobacco, alcohol, controlled substances, and inhalants: basic findings from the national comorbidity survey. *Experimental and Clinical Psychopharmacology*, 2, 244–268.

Baker, T.B., Piper M.E., McCarthy, D.E., Majeskie, M.R. and Fiore, M.C. (2004). Addiction motivation reformulated: an affective processing model of negative reinforcement. *Psychological Review*, 111, 33–51.

Beck, A.T., Wright, F.D., Newman, C.F. and Liese, B.S. (1993). *Cognitive therapy of substance abuse*. New York, NY: Guilford Press.

Breslin, F.C., Zack, M. and McMain, S. (2002). An information-processing analysis of mindfulness: implications for relapse prevention in the treatment of substance abuse. *Clinical Psychology; Science and Practice*, 9, 275–299.

Blum, K., Noble, E.P., Sheridan, P.J., Montgomery, A., Ritchie, T., Jagadeeswaran, P., Nogami, H., Briggs, A.H. and Cohn, J.B. (1990). Allelic association of human dopamine D2 receptor gene in alcoholism. *Journal of the American Medical Association*, 263, 2055–2060.

Butt, J. (ed.) (1963). *The poems of Alexander Pope*. New Haven, CT: Yale University Press.

Carroll, K.M. (1996). Relapse prevention as a psychosocial treatment: a review of controlled trials. *Experimental and Clinical Psychopharmacology*, 4, 46–54.

Carter, B.L. and Tiffany, S.T. (1999). Meta-analysis of cue reactivity in addiction research. *Addiction*, 94, 327–340.

Crits-Christoph, P.Siqueland, L., Blaine, J. Frank, A., Luborsky, L., Onken, L.S., Muenz, L.R., Thase, M.E., Weiss, R.D., Gastfriend, D.R., Woody, G.E., Barber, J.P., Butler, S.F., Daley, D., Salloum, I., Bishop, S., Najavits, L.M., Lis, J., Mercer, D., Griffin, M.L., Moras, K. and Beck, A.T. (1999). Psychosocial treatments for cocaine dependence. National Institute on Drug Abuse collaborative cocaine study. *Archives of General Psychiatry*, 56, 493–502.

Damasio, A. (1999). *The feeling of what happens: body and emotion in the making of consciousness*. Fort Worth, TX: Harcourt College Publishers.

Darkes, J. and Goldman, M.S. (1998). Expectancy challenge and drinking reduction: process and structure in the alcohol expectancy network. *Experimental and Clinical Psychopharmacology*, 6, 64–76.

Dunn, B.D., Dalgleish, T. and Lawrence, A.D. (2006) The somatic marker hypothesis: a critical evaluation. *Neuroscience and Biobehavioural Reviews*, 30, 239–271.

Garcia, A.V., Torrecillas, F.L., de Arcos, F.A. and Garcia, M.P. (2005). Effects of executive impairments on maladaptive explanatory styles in substance abusers: clinical implications. *Archives of Clinical Neuropsychology*, 20, 67–80.

Grant, S., London, E.D., Newlin, D.B., Villemagne, V.L., Liu, X., Contoreggi, C., Phillips, R. L., Kimes, A.S. and Margolin, A. (1996). Activation of memory circuits during cue-elicited cocaine craving. *Proceedings of the National Academy of Sciences of the USA*, 93, 12040–12045.

Grant, S., Contoreggi, C. and. London, E.D. (2000). Drug abusers show impaired performance in a laboratory test of decision-making. *Neuropsychologia*, 38, 1180–1187.

Hester, R. and Garavan, H. (2005). Working memory and executive function: the influence of content and load on the control of attention. *Memory and Cognition*, 33, 221–233.

Hiroi, N. and Agatsuma, S. (2005) Genetic susceptibility to substance dependence. *Molecular Psychiatry*, 10, 336–344.

Irwin, J.E., Bowers, C.A., Dunn, M.E. and Wang, M.C. (1999). Efficacy of relapse prevention: a meta-analytic review. *Journal of Consulting and Clinical Psychology*, 67, 563–570.

Kane, M.J. and Engle, R.W. (2003). Working memory capacity and the control of attention: the contribution of goal neglect, response competition and task set to Stroop interference. *Journal of Experimental Psychology: General*, 13, 47–70.

LeDoux, J. (1999). *The emotional brain*. London: Orion Books.

Litt, M.D., Kadden, R.M. and Cooney, N.L. (2003). Coping skills and treatment outcomes incognitive behavioural interactional group therapy for alcoholism. *Journal of Consulting and Clinical Psychology*, 71, 118–128.

Marlatt, G.A. and Gordon, J.R. (1985). *Relapse prevention: maintenance strategies in the treatment of addictive behaviors*. New York, NY: Guilford Press.

May, J., Andrade, J., Panabokke, N. and Kavanagh, D. (2004). Images of desire: cognitive models of craving. *Memory*, 12, 447–461.

Newman, C. (2004). Substance abuse. In R.L. Leahy (ed.), *Contemporary cognitive therapy*. New York, NY: Guilford Press, pp. 206–227.

Noble, E.P., Blum, K., Ritchie, T., Montgomery, A. and Sheridan P.J. (1991). Allelic association of the D2 dopamine receptor gene with receptor-binding characteristics in alcoholism. *Archives of General Psychiatry*, 48, 648–654.

Project Match Research Group (1997). Matching alcoholism treatments to client heterogeneity: project MATCH post-treatment drinking outcomes. *Journal of Studies on Alcohol*, 58, 7–29.

Rogers, R.D., Everritt, B.J., Baldichino, A., Blackshaw, A.J., Swainson, R., Wynne, K., Baker, N.B., Hunter, J., Carthy, T., Booker, E., London, M., Deakin, J.F., Sahakian, B.J. and Robbins, T.W. (1999). Dissociable deficits in the decision making cognition of chronic amphetamine abusers, opiate abusers, patients with focal damage to prefrontal cortex, and tryptophan-depleted normal volunteers: evidence for monoaminergic mechanisms. *Neuropsychopharmacology*, 20, 322–339.

Robinson, T.E. and Berridge, K.C. (1993). The neural basis of drug craving: an incentive-sensitization theory of addiction. *Brain Research Reviews*, 18, 247–291.

Robinson, T.E. and Berridge, K.C. (2003). Addiction. *Annual Review of Psychology*, 54, 25–53.

Rohsenow, D.J., Monti, P.M., Abrams, D.A., Rubonis, A.V., Niaura, R.S., Sirota, A.D. and Colby, S.M. (1992). Cue-elicited urge to drink and salivation in alcoholics: relationship to individual differences. *Advances in Behaviour Research and Therapy*, 14, 195–210.

Ryan, F. (2001) SIX TIPS: A self-help guide to preventing relapse to addictive behaviour. Poster presentation at the World Congress of Behavioral and Cognitive Therapies, Vancouver, Canada.

Ryan, F. (2002). Detected, selected, and sometimes neglected: cognitive processing of cues in addiction. *Experimental and Clinical Psychopharmacology*, 10, 67–76.

Schultz, W. (1998) Predictive reward signal of dopamine neurons. *Journal of Neurophysiology*, 80, 1–27.

Schultz, W. (2002). Getting formal with dopamine and reward. *Neuron*, 36, 241–263.

Stewart J., de Wit, H. and Eikelboom, R. (1984). The role of unconditioned and conditioned drug effects in the self-administration of opiates and stimulants. *Psychological Review*, 91, 251–268.

Stroop, J.R. (1935) Studies of interference in serial verbal reactions. *Psychological Monographs*, 50, 643–662.

Teasdale, J.D. and Barnard, P.J. (1993). *Affect, cognition, and change: re-modelling depressive thought*. Hillsdale, NJ: Lawrence Erlbaum Associates.

Teasdale, J.D., Segal, Z.V, Williams, J.M., Ridgeway, V.A., Soulsby, J.M. and Lau, M.A. (2000). Prevention of relapse/recurrence in major depression by mindfulness-based cognitive therapy. *Journal of Consulting and Clinical Psychology*, 68, 615–623.

UKATT Research Team (2005). Effectiveness of treatment for alcohol problems: findings of the randomised UK alcohol treatment trial. *British Medical Journal*, 331, 541.

Volkow, N.D., Fowler, J.S., Wang, G.J. and Goldstein, R.Z. (2002). Role of dopamine, the frontal cortex and memory circuits in drug addiction: insights from imaging studies. *Neurobiology of Learning and Memory*, 78, 610–624.

Whalen, P.J. (1998). Fear, vigilance and ambiguity: initial neuroimaging studies of the human amygdala. *Current Directions in Psychological Science*, 7, 177–188.

Wells, A. (2000). *Emotional disorders and metacognition: innovative cognitive therapy*. Chichester, UK: Wiley.

Young, R.M., Lawford, B.R., Feeney, G.F., Ritchie, T. and Noble, E.P. (2004). Alcohol-related expectancies are associated with the D2 dopamine receptor and GABAA receptor β 3 subunit genes. *Psychiatry Research*, 127, 171–183.

Index

acetylcholine 190
acquired preparedness 166
action 108
action schema 75
addiction, defined 2–3
Addiction Attentional Control Training
    Programme (AACTP) 110–11
affective priming 35
Alcohol Expectancy Questionnaire 14,
    201
Alcohol Problems Questionnaire 14
ALDH2 167–8
allostasis 151
Alzheimer's disease 39
amino acid challenge 236–7
amnesia 39–40, 49
amygdala 40, 141, 164–5
anterior cingulate cortex 40, 164, 229
anticipation 151, 152, 153–5
anti-social behaviour 291
anxiety 126, 197
associationism 162
associations 52
association tasks 1
associative memory 36, 41–56
associative network models 45
attention 83–4, 106–7, 108–10
attentional bias 1, 6, 7, 10, 11, 15–20, 34, 41,
    73–99, 109–10, 130, 136, 138, 223–8,
    256–8, 259–60
attentional cueing 81–2
attentional probe task 223–4
attentional re-training/training (see also
    cognitive re-training) 254, 265–6, 299
attitude–behaviour relationship 13
attributional judgements 295, 297
Alcohol Use Disorders Identification Test 14
automaticity of being 141
automatic network theory 135–9
automatic processes 5–6, 7, 12, 32, 33, 74,
    83–4, 107–8, 128–31, 139–40, 252
automatic schema 52
autonomic arousal 138
autonomic cue-reactivity 125–6
avoidance 7
awareness 12, 39

backpropagation 51
basal ganglia 39
behavioural interventions 260–7

benzodiazepines 260
biases 48; *see also* attentional bias, cognitive
    bias
biological factors 103, 164–5
brain
    imaging 228–31, 283, 285
    neural sensitization 40
    permanent/semi-permanent changes 8
brief messages 267–8
Brief Personal Concerns Counseling (PCC-B)
    111–13
bromocriptine 237
Buss–Durkee Hostility Inventory 202

candidate genes 232–3
Cannabis Expectancy Profile 207
categorization 158
CB1 anadamide receptor 192
CHANGE 288
change blindness 15, 34, 82–3
classical conditioning 74, 84–6
cocktail party effect 223
codeine 193
cognition 1, 2, 3–12, 163, 251–2, 294–7
cognitive appraisal 121–2
cognitive architecture 11
cognitive assessment 126–8
cognitive avoidance 91–2
cognitive–behavioural biases 89–91
cognitive–behavioural interventions 123,
    140–1
cognitive–behaviour therapy 281, 282–3, 284,
    286–8, 297
cognitive bias 1, 2, 3, 6–7, 75
cognitive experimental interventions 264–7
cognitive labelling 4–5
cognitive model of drug urges and drug-use
    behaviour 1, 5–7
cognitive–motivational interventions 110–13
cognitive neuroscience 38–41
cognitive performance 14
cognitive processes, measuring 13–14
cognitive processing 106–10, 138–9
cognitive processing model 5
cognitive products 14
cognitive re-training 264
cognitive revolution 42
cognitive social learning theory 120–3
cognitive space 11
cognitive theories 123–5

Lightning Source UK Ltd.
Milton Keynes UK

Lightning Source UK Ltd.
Milton Keynes UK
26 March 2010

151958UK00001B/22/P